Statistical Mechanics

Lecture notes

Statistical Mechanics

Lecture notes

Konstantin K Likharev

IOP Publishing, Bristol, UK

ISBN 978-0-7503-1416-9 (ebook)
ISBN 978-0-7503-1417-6 (print)
ISBN 978-0-7503-1418-3 (mobi)

DOI 10.1088/2053-2563/aaf503

Version: 20190701

IOP Expanding Physics
ISSN 2053-2563 (online)
ISSN 2054-7315 (print)

British Library Cataloguing-in-Publication Data: A catalogue record for this book is available from the British Library.

Published by IOP Publishing, wholly owned by The Institute of Physics, London

IOP Publishing, Temple Circus, Temple Way, Bristol, BS1 6HG, UK

US Office: IOP Publishing, Inc., 190 North Independence Mall West, Suite 601, Philadelphia, PA 19106, USA

Contents

Preface to the EAP Series

Essential Advanced Physics

Essential Advanced Physics (EAP) is a series of lecture notes and problems with solutions, consisting of the following four parts[1]:

- *Part CM*: *Classical Mechanics* (a one-semester course),
- *Part EM*: *Classical Electrodynamics* (two semesters),
- *Part QM*: *Quantum Mechanics* (two semesters), and
- *Part SM*: *Statistical Mechanics* (one semester).

Each part includes two volumes: *Lecture Notes* and *Problems with Solutions*, and an additional file *Test Problems with Solutions.*

Distinguishing features of this series—in brief

- condensed lecture notes (~250 pp per semester)—much shorter than most textbooks
- emphasis on simple explanations of the main notions and phenomena of physics
- a focus on problem solution; extensive sets of problems with detailed model solutions
- additional files with test problems, freely available to qualified university instructors
- extensive cross-referencing between all parts of the series, which share style and notation

Level and precursors

The goal of this series is to bring the reader to a general physics knowledge level necessary for professional work in the field, regardless on whether the work is theoretical or experimental, fundamental or applied. From the formal point of view, this level (augmented by a few special topic courses in a particular field of concentration, and of course by an extensive thesis research experience) satisfies the typical PhD degree requirements. Selected parts of the series may be also valuable for graduate students and researchers of other disciplines, including astronomy, chemistry, mechanical engineering, electrical, computer and electronic engineering, and material science.

The entry level is a notch lower than that expected from a physics graduate from an average US college. In addition to physics, the series assumes the reader's familiarity with basic calculus and vector algebra, to such an extent that the meaning of the formulas listed in appendix A, 'Selected mathematical formulas' (reproduced at the end of each volume), is absolutely clear.

[1] Note that the (very ambiguous) term *mechanics* is used in these titles in its broadest sense. The acronym *EM* stems from another popular name for classical electrodynamics courses: *Electricity and Magnetism.*

Origins and motivation

The series is a by-product of the so-called 'core physics courses' I taught at Stony Brook University from 1991 to 2013. My main effort was to assist the development of students' problem-solving skills, rather than their idle memorization of formulas. (With a certain exaggeration, my lectures were not much more than introductions to problem solution.) The focus on this main objective, under the rigid time restrictions imposed by the SBU curriculum, had some negatives. First, the list of covered theoretical methods had to be limited to those necessary for the solution of the problems I had time to discuss. Second, I had no time to cover some core fields of physics—most painfully general relativity[2] and quantum field theory, beyond a few quantum electrodynamics elements at the end of *Part QM*.

The main motivation for putting my lecture notes and problems on paper, and their distribution to students, was my desperation to find textbooks and problem collections I could use, with a clear conscience, for my purposes. The available graduate textbooks, including the famous *Theoretical Physics* series by Landau and Lifshitz, did not match the minimalistic goal of my courses, mostly because they are far too long, and using them would mean hopping from one topic to another, picking up a chapter here and a section there, at a high risk of losing the necessary background material and logical connections between the course components—and the students' interest with them. In addition, many textbooks lack even brief discussions of several traditional and modern topics that I believe are necessary parts of every professional physicist's education[3].

On the problem side, most available collections are not based on particular textbooks, and the problem solutions in them either do not refer to any background material at all, or refer to the included short sets of formulas, which can hardly be used for systematic learning. Also, the solutions are frequently too short to be useful, and lack discussions of the results' physics.

Style

In an effort to comply with the Occam's Razor principle[4], and beat Malek's law[5], I have made every effort to make the discussion of each topic as clear as the time/space (and my ability :-) permitted, and as simple as the subject allowed. This effort has resulted in rather succinct lecture notes, which may be thoroughly read by a student during the semester. Despite this briefness, the introduction of every new

[2] For an introduction to this subject, I can recommend either a brief review by S Carroll, *Spacetime and Geometry* (2003, New York: Addison-Wesley) or a longer text by A Zee, *Einstein Gravity in a Nutshell* (2013, Princeton University Press).

[3] To list just a few: the statics and dynamics of elastic and fluid continua, the basics of physical kinetics, turbulence and deterministic chaos, the physics of computation, the energy relaxation and dephasing in open quantum systems, the reduced/RWA equations in classical and quantum mechanics, the physics of electrons and holes in semiconductors, optical fiber electrodynamics, macroscopic quantum effects in Bose–Einstein condensates, Bloch oscillations and Landau–Zener tunneling, cavity quantum electrodynamics, and density functional theory (DFT). All these topics are discussed, if only briefly, in my lecture notes.

[4] *Entia non sunt multiplicanda praeter necessitate*—Latin for 'Do not use more entities than necessary'.

[5] 'Any simple idea will be worded in the most complicated way'.

physical notion/effect and of every novel theoretical approach is always accompanied by an application example or two.

The additional exercises/problems listed at the end of each chapter were carefully selected[6], so that their solutions could better illustrate and enhance the lecture material. In formal classes, these problems may be used for homework, while individual learners are strongly encouraged to solve as many of them as practically possible. The few problems that require either longer calculations, or more creative approaches (or both), are marked by asterisks.

In contrast with the lecture notes, the model solutions of the problems (published in a separate volume for each part of the series) are more detailed than in most collections. In some instances they describe several alternative approaches to the problem, and frequently include discussions of the results' physics, thus augmenting the lecture notes. Additional files with sets of shorter problems (also with model solutions) more suitable for tests/exams, are available for qualified university instructors from the publisher, free of charge.

Disclaimer and encouragement

The prospective reader/instructor has to recognize the limited scope of this series (hence the qualifier *Essential* in its title), and in particular the lack of discussion of several techniques used in current theoretical physics research. On the other hand, I believe that the series gives a reasonable introduction to the *hard core* of physics—which many other sciences lack. With this hard core knowledge, today's student will always feel at home in physics, even in the often-unavoidable situations when research topics have to be changed at a career midpoint (when learning from scratch is terribly difficult—believe me :-). In addition, I have made every attempt to reveal the remarkable logic with which the basic notions and ideas of physics subfields merge into a wonderful single construct.

Most students I taught liked using my materials, so I fancy they may be useful to others as well—hence this publication, for which all texts have been carefully reviewed.

[6] Many of the problems are original, but it would be silly to avoid some old good problem ideas, with long-lost authorship, which wander from one textbook/collection to another one without references. The assignments and model solutions of all such problems have been re-worked carefully to fit my lecture material and style.

Preface to *Statistical Mechanics: Lecture Notes*

This graduate-level course of statistical mechanics (including thermodynamics) has more or less traditional structure, with two notable exceptions.

First, because of the recent, still growing interest in nanoscale systems and ultrasensitive physical measurements, relatively large attention is given to fluctuations of various physical variables. Namely, their discussion (in chapter 5) includes not only the traditional topic of the calculation of variances of fluctuations of key variables in thermal equilibrium, but also the characterization of their time dependence (including the correlation function and the spectral density), and also the Smoluchowski and Fokker–Planck equations for dissipative systems. A few parts of this chapter, including the discussion of the most important fluctuation–dissipation theorem (FDT), have unavoidable overlaps with chapter 7 of the quantum-mechanical part of this series, devoted to open quantum systems, but I tried to keep the overlaps to a minimum, for example using a different (and much shorter) derivation of the FDT in this course.

The second, even more significant deviation from the tradition is the brief introduction to physical kinetics in chapter 6, reflecting my belief that such key theoretical tools of this field as the Liouville theorem and the Boltzmann kinetic equation (including its RTA version) have to be in the arsenal of every educated physicist. As another example, it is impossible to carry out, or even contemplate meaningful voltage measurements without understanding the difference between the electrostatic and electrochemical potentials—also discussed in that chapter.

Unfortunately, the additional attention given to these issues, within the strict one-semester confinements of this course, has left less time/space for such traditional topic as the phase transitions—whose barebones discussion is given in chapter 4. Even more regrettably, the coverage of such an important subject as thermodynamics (sometimes taught as a stand-alone discipline) in chapter 1 had to be truncated to the minimum necessary for the discussion of statistical physics in the following chapters.

Let me finish by noting that my (and not only my) teaching experience shows that of all the parts of physics included in this EAP series, the SM part is the hardest one for most students—especially in the aspect of application of the general theory to particular physical systems/models. This is why my sincere advice to the instructors and individual readers of these notes is to focus their efforts on the solution of the problems formulated in the end of each chapter, and at the analysis of their model solutions (published in the accompanying volume) even more than in other parts of the series.

Acknowledgments

I am extremely grateful to my faculty colleagues and other readers of the preliminary (circa 2013) version of this series, who provided feedback on certain sections; here they are listed in alphabetical order[7]: A Abanov, P Allen, D Averin, S Berkovich, P-T de Boer, M Fernandez-Serra, R F Hernandez, A Korotkov, V Semenov, F Sheldon, and X Wang. (Obviously, these kind people are not responsible for any remaining deficiencies.)

A large part of my scientific background and experience, reflected in these materials, came from my education, and then research, in the Department of Physics of Moscow State University from 1960 to 1990. The Department of Physics and Astronomy of Stony Brook University provided a comfortable and friendly environment for my work during the following 25+ years.

Last but not least, I would like to thank my wife Lioudmila for all her love, care, and patience—without these, this writing project would have been impossible.

I know very well that my materials are still far from perfection. In particular, my choice of covered topics (always very subjective) may certainly be questioned. Also, it is almost certain that despite all my efforts, not all typos have been weeded out. This is why all remarks (however candid) and suggestions from readers will be greatly appreciated. All significant contributions will be gratefully acknowledged in future editions.

Konstantin K Likharev
Stony Brook, NY

[7] I am very sorry for not keeping proper records from the beginning of my lectures at Stony Brook, so I cannot list all the numerous students and TAs who have kindly attracted my attention to typos in earlier versions of these notes. Needless to say, I am very grateful to all of them as well.

Notation

Abbreviations	Fonts	Symbols
c.c. complex conjugate	F, $\mathcal{F}$ scalar variables[8]	$\dot{}$ time differentiation operator (d/dt)
h.c. Hermitian conjugate	$\mathbf{F}$, $\boldsymbol{\mathcal{F}}$ vector variables	∇ spatial differentiation vector (*del*)
	$\hat{F}$, $\hat{\mathcal{F}}$ scalar operators	$\approx$ approximately equal to
	$\hat{\mathbf{F}}$, $\hat{\boldsymbol{\mathcal{F}}}$ vector operators	$\sim$ of the same order as
	F matrix	$\propto$ proportional to
	$F_{jj'}$ matrix element	$\equiv$ equal to by definition (or evidently)
		$\cdot$ scalar ('dot-') product
		$\times$ vector ('cross-') product
		$\overline{}$ time averaging
		$\langle\ \rangle$ statistical averaging
		[,] commutator
		{ , } anticommutator

Prime signs

The prime signs (′, ″, etc) are used to distinguish similar variables or indices (such as j and j' in the matrix element above), rather than to denote derivatives.

Parts of the series

Part CM: Classical Mechanics *Part EM: Classical Electrodynamics*
Part QM: Quantum Mechanics *Part SM: Statistical Mechanics*

Appendices

Appendix A: Selected mathematical formulas
Appendix B: Selected physical constants

Formulas

The abbreviation Eq. may mean any displayed formula: either the equality, or inequality, or equation, etc.

[8] The same letter, typeset in different fonts, typically denotes different variables.

IOP Publishing

Statistical Mechanics

Lecture notes

Konstantin K Likharev

Chapter 1

Review of thermodynamics

This chapter starts from a brief discussion of the subject of statistical physics and thermodynamics, and the relation between these two disciplines. Then I proceed to a review of the basic notions and relations of thermodynamics. Most of this material is supposed to be known to the reader from his or her undergraduate studies[1]*, so the discussion is rather brief.*

1.1 Introduction: statistical physics and thermodynamics

Statistical physics (alternatively called 'statistical mechanics') and *thermodynamics* are two different but related approaches to the same goal: an approximate description of the 'internal'[2] properties of very large physical systems, notably those consisting of $N \gg 1$ identical particles—or other components. The traditional example of such a system is a human-scale portion of a gas, with the number N of atoms/molecules of the order of the Avogadro number $N_A \sim 10^{23}$.[3]

The motivation for the statistical approach to such systems is straightforward: even if the laws governing the dynamics of each particle and their interactions were exactly known, and we had infinite computing resources at our disposal, calculating the exact evolution of the system in time would be impossible, at least because it is completely impracticable to measure the exact initial state of each component—in the classical case, the initial position and velocity of each particle. The situation is further exacerbated by the phenomena of chaos and turbulence[4], and the quantum-mechanical uncertainty, which do not allow the exact calculation of final positions

[1] For remedial reading, I can recommend, for example (in the alphabetical order): [1–3].

[2] Here 'internal' is an (admittedly loose) term meaning all the physics unrelated to the motion of the system as a whole. The most important example of the internal dynamics is the thermal motion of atoms and molecules.

[3] For the quantitative definition of this number see, e.g. section 1.4 below.

[4] See, e.g. *Part CM* chapters 8 and 9.

doi:10.1088/2053-2563/aaf503ch1

and velocities of the component particles even if their initial state is known with the best possible precision. As a result, in most situations only statistical predictions about the behavior of such systems may be made, with the *probability theory* becoming a major part of the mathematical tool arsenal.

However, the statistical approach is not as bad as it may look. Indeed, it is almost self-evident that any measurable *macroscopic* variable characterizing a stationary system of $N \gg 1$ particles as a whole (think, e.g. about the stationary pressure P of the gas contained in a fixed volume V) is almost constant in time. Indeed, as we will see below, besides certain exotic exceptions, the relative magnitude of fluctuations—either in time, or among many macroscopically similar systems—of such a variable are of the order of $1/\sqrt{N}$, i.e. for $N \sim N_A$ is extremely small. As a result, the *average values* of appropriate macroscopic variables may characterize the state of the system rather well. The calculation of relations between such average values is the only task of thermodynamics, and the main task of statistical physics. (The analysis of fluctuations is also an important issue, but due to the fluctuation smallness, in most cases it may be based on perturbative approaches—see chapter 5.)

Now let us have a fast look at the typical macroscopic variables the statistical physics and thermodynamics should operate with. Since I have already mentioned pressure P and volume V, let me start with this famous pair of variables. First of all, note that volume is an *extensive variable*, i.e. a variable whose value for a system consisting of several non-interacting (or weakly interacting) parts is the sum of those of its parts. On the other hand, pressure is an example of an *intensive variable* whose value is the same for different parts of a system—if they are in equilibrium. In order to understand why P and V form a natural pair of variables, let us consider the classical playground of thermodynamics, a portion of a gas contained in a cylinder, closed with a movable piston of area A (figure 1.1).

Neglecting the friction between the walls and the piston, and assuming that it is being moved so slowly that the pressure P, at any instant, is virtually the same for all parts of the volume, the elementary work of the external force $\mathscr{F} = PA$, compressing the gas, at a small piston's displacement $dx = -dV/A$, is

$$d\mathscr{W} = \mathscr{F}dx \equiv \left(\frac{\mathscr{F}}{A}\right)(Adx) \equiv -PdV. \tag{1.1}$$

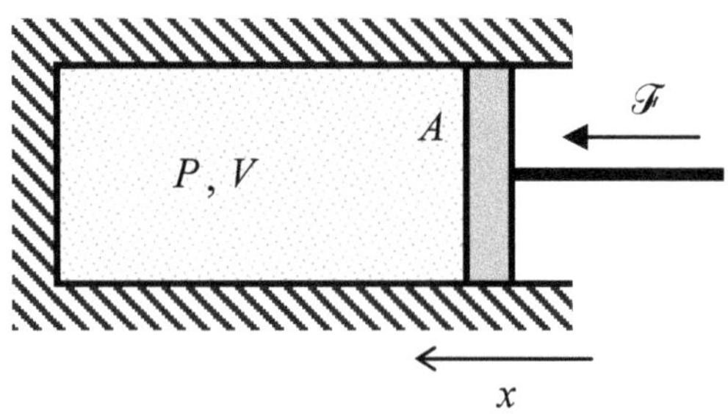

Figure 1.1. Compressing gas.

Of course, the last expression is more general than the model shown in figure 1.1, and does not depend on the particular shape of the system's surface[5]. (Note that in the notation of Eq. (1.1), which will be used through the course, the elementary work done *by* the gas *on* the external system equals $-d\mathscr{W}$.)

From the point of analytical mechanics[6], V and $(-P)$ is just one of many possible *canonical pairs* of generalized coordinates q_j and generalized forces $\mathscr{F}_j$, whose products $d\mathscr{W}_j = \mathscr{F}_j dq_j$ give independent contributions to the total work of the environment on the system under analysis. For example, the reader familiar with the basics of electrostatics knows that if the spatial distribution $\mathscr{E}(\mathbf{r})$ of an external electric field does not depend on the electric polarization $\mathscr{P}(\mathbf{r})$ of the dielectric medium placed into the field, its elementary work on the medium is

$$d\mathscr{W} = \int \mathscr{E}(\mathbf{r}) \cdot d\mathscr{P}(\mathbf{r})\, d^3r \equiv \int \sum_{j=1}^{3} \mathscr{E}_j(\mathbf{r}) d\mathscr{P}_j(\mathbf{r})\, d^3r. \qquad (1.2a)$$

The most important cases when this condition is fulfilled (and hence Eq. (1.2*a*) is valid) are, first, long cylindrical samples in a parallel external field (see, e.g. *Part EM* figure 3.13) and, second, the polarization of a sample (of any shape) due to that of discrete dipoles $\mathscr{p}_k$, whose electric interaction is negligible. In the latter case, Eq. (1.2*a*) may be also rewritten as the sum over the single dipoles, located at points $\mathbf{r}_k$:[7]

$$d\mathscr{W} = \sum_k d\mathscr{W}_k, \qquad \text{with } d\mathscr{W}_k = \mathscr{E}(\mathbf{r}_k) \cdot d\mathscr{p}_k. \qquad (1.2b)$$

Very similarly, and at the similar conditions on the external magnetic field $\mathscr{H}(\mathbf{r})$, its elementary work on a magnetic medium may be also represented in either of two forms[8]:

$$d\mathscr{W} = \mu_0 \int \mathscr{H}(\mathbf{r}) \cdot d\mathscr{M}(\mathbf{r})\, d^3r \equiv \mu_0 \int \sum_{j=1}^{3} \mathscr{H}_j(\mathbf{r}) d\mathscr{M}_j(\mathbf{r})\, d^3r, \qquad (1.3a)$$

$$d\mathscr{W} = \sum_k d\mathscr{W}_k, \qquad \text{with } d\mathscr{W}_k = \mu_0 \mathscr{H}(\mathbf{r}_k) \cdot d\mathscr{m}_k. \qquad (1.3b)$$

where $\mathscr{M}$ and $\mathscr{m}$ are the vectors of, respectively, the medium's magnetization and the moment of a single magnetic dipole. Eqs. (1.2) and (1.3) show that the roles of

[5] In order to prove that, it is sufficient to integrate the scalar product $d\mathscr{W} = d\mathscr{F} \cdot d\mathbf{r}$, with $d\mathscr{F} = -P\mathbf{n}d^2r$, where $d\mathbf{r}$ is the surface displacement vector (see, e.g. *Part CM* section 7.1), and $\mathbf{n}$ is the outer normal, over the surface.

[6] See, e.g. *Part CM* chapters 2 and 10.

[7] Some of my students needed an effort to reconcile the positive signs in Eqs. (1.2) with the negative sign in the well-known relation $dU_k = -\mathscr{E}(\mathbf{r}_k)d\mathscr{p}_k$ for the potential energy of a dipole in an external electric field—see, e.g. *Part EM* Eqs. (3.15). The resolution of this paradox is simple: each term of Eq. (1.2*b*) describes the work $d\mathscr{W}_k$ of the electric field on the *internal* degrees of freedom of the kth dipole, changing its *internal* energy E_k: $dE_k = d\mathscr{W}_k$. This energy change comes from the electric field energy, so that $dU_k = -dE_k$.

[8] Here, as in all my series, I am using the SI units.

generalized coordinates may be played by Cartesian components of the vectors $\mathscr{P}$ (or $\not{p}$) and $\mathscr{M}$ (or $\mathscr{m}$), with the components of the electric and magnetic fields playing the roles of the corresponding generalized forces. This list may be extended to other interactions (such as gravitation, surface tension in fluids, etc). Following tradition, I will use the $\{-P, V\}$ pair in almost all the formulas below, but the reader should remember that they all are valid for any other pair $\{\mathscr{F}_j, q_j\}$.[9]

Again, the specific relations between the variables of each pair listed above are typically affected by the statistics of the components (particles) of the system under analysis, but their definition is not based on statistics. The situation is very different for a very specific pair of variables, *temperature* T and *entropy* S, although these 'sister variables' participate in many formulas of thermodynamics exactly as if they were just one more canonical pair $\{\mathscr{F}_j, q_j\}$. However, the very existence of these two notions is due to statistics. Namely, the temperature T is an intensive variable that characterizes the degree of thermal 'agitation' of system's components. Conversely, the entropy S is an extensive variable that in most cases evades immediate perception by human senses; it is a qualitative measure of *disorder* of the system, i.e. the degree of our ignorance about its exact microscopic state[10].

The reason for the appearance of the $\{T, S\}$ pair of variables in formulas of thermodynamics and statistical mechanics is that the statistical approach to large systems of particles brings some qualitatively new results, most notably the notion of *irreversible* time evolution of collective (*macroscopic*) variables describing the system. On one hand, the irreversibility looks absolutely natural in such phenomena as the diffusion of an ink drop in a glass of water. In the beginning, the ink molecules are located in a certain small part of the system's volume, i.e. to some extent ordered, while at the late stages of diffusion, the position of each molecule in the glass is essentially random. However, as a second thought, the irreversibility is rather surprising[11], taking into account that it takes place even though the laws governing the motion of system's components are *time-reversible*—such as the Newton laws or the basic laws of quantum mechanics. Indeed, if, at a late stage of the diffusion process, we reversed the velocities of all molecules exactly and simultaneously, the ink molecules would again gather (for a moment) into the original spot[12]. The problem is that getting the *information* necessary for the exact velocity reversal is not

---

[9] Note that in systems of discrete particles, most generalized forces, including the fields $\mathscr{E}$ and $\mathscr{H}$, differ from the pressure P in the sense that their work may be explicitly partitioned into single-particle components—see Eqs. (1.2*b*) and (1.3*b*). This fact gives some freedom of choice in carrying out calculations based on thermodynamic potentials—see section 1.4 below.

[10] The notion of entropy was introduced into thermodynamics in the 1850s by R Clausius, on the background of an earlier pioneering work by S Carnot (see section 1.6 below), as a variable related to the 'useable thermal energy' rather than a measure of disorder. In the absence of any clue about the entropy's microscopic origin (which had to wait for decades until the works by L Boltzmann and J Maxwell), this was an amazing intellectual achievement.

[11] Indeed, as recently as in the late 19th century, the very possibility of irreversible macroscopic behavior of microscopically reversible systems was questioned by some serious scientists, notably by J Loschmidt in 1876.

[12] While quantum-mechanical effects, with their intrinsic uncertainty, are quantitatively important in this example, our qualitative discussion does not depend on them. A good similar classical example is the chaotic motion of a ball on a 2D Sinai billiard—see *Part CM* chapter 9 and in particular figure 9.8 and its discussion.

practicable. This example shows a deep connection between the statistical mechanics and the information theory.

A qualitative discussion of the reversibility–irreversibility dilemma requires a strict definition of the basic notion of statistical mechanics (and indeed of the probability theory), the *statistical ensemble*, and I would like to postpone it until the beginning of chapter 2. In particular, in that chapter we will see that the basic law of irreversible behavior is an increase of the entropy S in any closed system. Thus, statistical mechanics, without defying the 'microscopic' laws governing evolution of a system's components, introduces on top of them some new 'macroscopic' laws, intrinsically related to the evolution of *information*, i.e. the degree of our knowledge of the microscopic state of the system.

To conclude this brief discussion of variables, let me mention that as in all fields of physics, a very special role in statistical mechanics is played by the *energy* E. In order to emphasize the commitment to disregard the motion of the system as a whole in this field, the E considered in thermodynamics it is frequently called the *internal energy*, though for brevity, I will skip this adjective in most cases. The simplest example of such E is the sum of kinetic energies of molecules in a dilute gas at their thermal motion, but in general the internal energy also includes not only the individual energies of the system's components, but also their interactions with each other. Besides a few pathological cases of very-long-range interactions (such as the Coulomb interactions in plasma with uncompensated charge density), the interactions may be treated as local; in this case the internal energy is proportional to N, i.e. is an extensive variable. As will be shown below, other extensive variables with the dimension of energy are often very useful as well, including the (Helmholtz) *free energy* F, the *Gibbs energy* G, the *enthalpy* H, and the *grand potential* Ω. (The collective name for such variables is *thermodynamic potentials*.)

Now, we are ready for a brief discussion of the relation between *statistical physics* and *thermodynamics*. While the task of statistical physics is to calculate the macroscopic variables discussed above[13] for this or that particular microscopic model of the system, the main role of thermodynamics is to derive some general relations between the average values of the macroscopic variables (also called *thermodynamic variables*) that do not depend on specific models. Surprisingly, it is possible to accomplish such a feat using a few either evident or very plausible general assumptions (sometimes called the *laws of thermodynamics*), which find their proof in statistical physics[14]. Such general relations allow for a substantial reduction of the amount of calculations we have to do in statistical physics; in most cases it is sufficient to calculate from the statistics just one or two variables, and then use thermodynamic relations to calculate all other properties of interest. Thus the thermodynamics, sometimes snubbed as a phenomenology, deserves every respect

[13] Several other quantities, for example the heat capacity C, may be calculated as partial derivatives of the basic variables discussed above. Also, at certain conditions, the number of particles N in the system may be not fixed and also considered as an (extensive) variable.

[14] Admittedly, some of these proofs are based on other plausible but deeper postulates, for example the central statistical hypothesis (section 2.2), whose final proof is the whole body of experimental data.

not only as a useful theoretical tool, but also as a discipline more general than any particular statistical model. This is why the balance of this chapter is devoted to a brief review of thermodynamics.

1.2 The 2nd law of thermodynamics, entropy, and temperature

Thermodynamics accepts a phenomenological approach to the entropy S, postulating that there is such a unique extensive measure of disorder, and that in a *closed system*[15], it may only grow in time, reaching its constant (maximum) value at equilibrium[16]:

$$dS \geqslant 0. \tag{1.4}$$

This postulate is called the *2nd law of thermodynamics*—arguably its only substantial new law.

Rather surprisingly, this law, together with the additivity of S in composite systems of non-interacting parts (as an extensive variable), is sufficient for a formal definition of temperature, and a derivation of its basic properties that comply with our everyday notion of this key variable. Indeed, let us consider a closed system consisting of two fixed-volume subsystems (figure 1.2) whose internal relaxation is very fast in comparison with the rate of the thermal flow (i.e. the energy and entropy exchange) between the parts. In this case, on the latter time scale, each part is always in some quasi-equilibrium state, which may be described by a unique relation $E(S)$ between its energy and entropy[17].

Neglecting the interaction energy between the parts (which is always possible at $N \gg 1$, and in the absence of long-range interactions), we may use the extensive character of the variables E and S to write the following equalities

$$E = E_1(S_1) + E_2(S_2), \qquad S = S_1 + S_2, \tag{1.5}$$

for the full energy and entropy of the system. Now let us use them to calculate the following derivative:

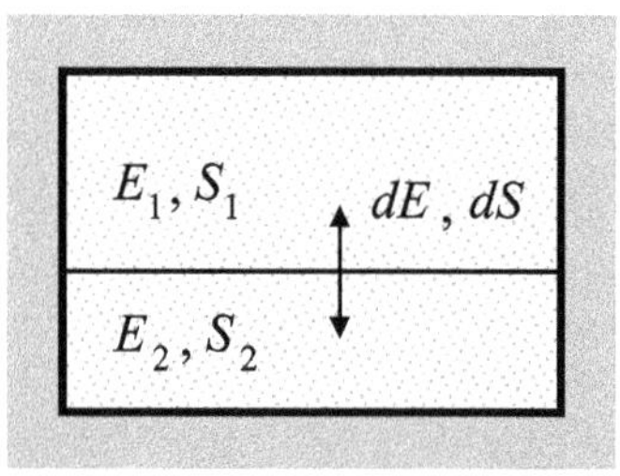

Figure 1.2. A. composite thermodynamic system.

[15] Defined as a system completely isolated from its environment, i.e. the system with its internal energy fixed.
[16] Implicitly, this statement also postulates the existence, in a closed system, of *thermodynamic equilibrium*, an asymptotically reached state in which all macroscopic variables, including entropy, remain constant.
[17] Here we strongly depend on a very important (and possibly the least intuitive) aspect of the 2nd law, namely that the entropy is a *unique* measure of disorder.

$$\frac{dS}{dE_1} = \frac{dS_1}{dE_1} + \frac{dS_2}{dE_1} = \frac{dS_1}{dE_1} + \frac{dS_2}{dE_2}\frac{dE_2}{dE_1} = \frac{dS_1}{dE_1} + \frac{dS_2}{dE_2}\frac{d(E - E_1)}{dE_1}. \tag{1.6}$$

Since the total energy E of the closed system is fixed and hence independent of its redistribution between the subsystems, we have to take $dE/dE_1 = 0$, and Eq. (1.6) yields

$$\frac{dS}{dE_1} = \frac{dS_1}{dE_1} - \frac{dS_2}{dE_2}. \tag{1.7}$$

According to the 2nd law of thermodynamics, when the two parts have reached the thermodynamic equilibrium, the total entropy S reaches its maximum, so that $dS/dE_1 = 0$, and Eq. (1.7) yields

$$\frac{dS_1}{dE_1} = \frac{dS_2}{dE_2}. \tag{1.8}$$

This equality shows that if a thermodynamic system may be partitioned into weakly interacting macroscopic parts, their derivatives dS/dE should be equal in the equilibrium. The reciprocal of such a derivative is called *temperature.* Taking into account that our analysis pertains to the situation (figure 1.2) when both volumes $V_{1,2}$ are fixed, we may write this definition as

$$\left(\frac{\partial E}{\partial S}\right)_V \equiv T, \tag{1.9}$$

the subscript V meaning that volume is kept constant at the differentiation. (Such notation is common and very useful in thermodynamics, with its broad range of variables.)

Note that according to Eq. (1.9), if temperature is measured in energy units (as I will do in this course for the brevity of notation), then S is dimensionless[18]. The transfer to the SI or Gaussian units, i.e. to the temperature T_K measured in kelvins (not 'Kelvins', and not 'degrees Kelvin', please!), is given by the relation $T = k_B T_K$, where the *Boltzmann constant* $k_B \approx 1.38 \times 10^{-23}$ J K^{-1} = 1.38×10^{-16} erg K^{-1}.[19] In those units, the entropy becomes dimensional: $S_K = k_B S$.

The definition of temperature, given by Eq. (1.9), is of course in a sharp contract with the popular notion of T as a measure of the average energy per particle. However, as we will repeatedly see below, is most cases these two notions may be reconciled, with Eq. (1.9) being more general. In particular, let us list some properties of the so-defined T, which are in agreement with our everyday notion of temperature:

[18] Here I have to mention a traditional unit of energy, still used in some fields related to thermodynamics: the *calorie*; in the most common definition (the so-called *thermochemical calorie*) it equals exactly 4.148 J.

[19] For more exact value of this and other constants, see appendix A. Note that both T and T_K define the *absolute* (also called 'thermodynamic') *scale* of temperature, in contrast to such artificial temperature scales as degrees Celsius ('centigrades'), defined as $T_C \equiv T_K + 273.15$, or degrees Fahrenheit: $T_F \equiv (9/5)T_C + 32$.

(i) according to Eq. (1.9), the temperature is an intensive variable (since both E and S are extensive), i.e. in a system of similar particles, is independent of the particle number N;

(ii) temperatures of all parts of a system are equal at equilibrium—see Eq. (1.8);

(iii) in a closed system whose parts are *not* in equilibrium, thermal energy (*heat*) always flows from a warmer part (with higher T) to the colder part.

In order to prove the last property, let us come back to the closed, composite system shown in figure 1.2, and consider another derivative:

$$\frac{dS}{dt} = \frac{dS_1}{dt} + \frac{dS_2}{dt} = \frac{dS_1}{dE_1}\frac{dE_1}{dt} + \frac{dS_2}{dE_2}\frac{dE_2}{dt}. \tag{1.10}$$

If the internal state of each part is very close to equilibrium (as was assumed from the very beginning) at each moment of time, we can use Eq. (1.9) to replace the derivatives $dS_{1,2}/dE_{1,2}$ for $1/T_{1,2}$ and get

$$\frac{dS}{dt} = \frac{1}{T_1}\frac{dE_1}{dt} + \frac{1}{T_2}\frac{dE_2}{dt}. \tag{1.11}$$

Since in a closed system $E = E_1 + E_2 = \text{const}$, these time derivatives are related as $dE_2/dt = -dE_1/dt$, and Eq. (1.11) yields

$$\frac{dS}{dt} = \left(\frac{1}{T_1} - \frac{1}{T_2}\right)\frac{dE_1}{dt}. \tag{1.12}$$

But in accordance with the 2nd law of thermodynamics, the derivative cannot be negative: $dS/dt \geqslant 0$. Hence,

$$\left(\frac{1}{T_1} - \frac{1}{T_2}\right)\frac{dE_1}{dt} \geqslant 0. \tag{1.13}$$

For example, if $T_1 > T_2$ (i.e. $1/T_1 < 1/T_2$), then $dE_1/dt \leqslant 0$, i.e. the warmer part gives energy to the colder counterpart.

Note also that such a heat exchange, at fixed volumes $V_{1,2}$, and $T_1 \neq T_2$, increases the total system's entropy, without performing any 'useful' mechanical work—see Eq. (1.1).

1.3 The 1st and 3rd laws of thermodynamics, and heat capacity

Now let us consider a *thermally insulated* system whose volume V may be changed by a deterministic force—see, for example, figure 1.1. Such a system is different from the fully closed one, because its energy E may be changed by the external force's work—see Eq. (1.1):

$$dE = d\mathscr{W} = -PdV. \tag{1.14}$$

Let the volume change be so slow ($dV/dt \to 0$) that the system is virtually at equilibrium at any instant. Such a slow process is called *reversible*, and in this

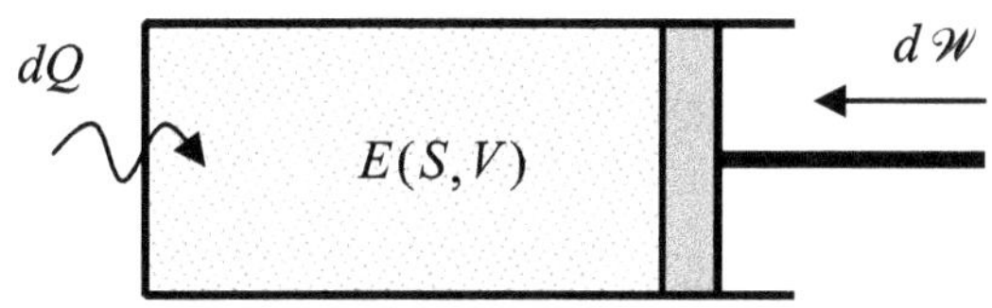

Figure 1.3. An example of the thermodynamic process involving both the mechanical work by the environment, and the heat exchange with it.

particular case of a thermally insulated system, it is also called *adiabatic*. If the pressure P (or any generalized external force $\mathscr{F}_j$) is deterministic, i.e. is a predetermined function of time, independent on the state of the system under analysis, it may be considered as coming from a fully ordered system, i.e. the one having zero entropy, with the total system (the system under our analysis plus the source of the force) completely closed. Since the entropy of the total closed system should stay constant (see the second of Eqs. (1.5) above), S of the system under analysis should stay constant on its own. Thus we arrive at a very important conclusion: at an adiabatic process, the entropy of a system cannot change[20]. This means that we may use Eq. (1.14) to write

$$P = -\left(\frac{\partial E}{\partial V}\right)_S. \tag{1.15}$$

Now let us consider a more general thermodynamic system that may also exchange thermal energy ('heat') with its environment (figure 1.3). For such a system, our previous conclusion about the entropy's constancy is not valid, so that S, in equilibrium, may be a function of not only the system's energy E, but also of its volume V: $S = S(E, V)$. Let us resolve this relation for energy: $E = E(S, V)$, and write the general mathematical expression for the full differential of E as a function of these two independent arguments:

$$dE = \left(\frac{\partial E}{\partial S}\right)_V dS + \left(\frac{\partial E}{\partial V}\right)_S dV. \tag{1.16}$$

This formula, based on the stationary relation $E = E(S, V)$, is evidently valid not only in equilibrium, but also for all very slow, reversible[21] processes. Now, using Eqs. (1.9) and (1.15), we may rewrite Eq. (1.16) as

$$dE = TdS - PdV. \tag{1.17}$$

According to Eq. (1.1), the second term on the right-hand side of this equation is just the work of the external force, so that due to the conservation of energy[22], the first

[20] A process conserving entropy is sometimes called *isentropic*.

[21] Let me emphasize again that any adiabatic process is reversible, but not vice versa.

[22] Such conservation, expressed by Eqs. (1.18) and (1.19), is commonly called the *1st law of thermodynamics*. While it (in contrast with the 2nd law) does not present any new law of nature on the top of mechanics, and in particular was already used de-facto to write the first of Eqs. (1.5) and also Eq. (1.14), such a grand name was quite justified in the mid-19th century when the mechanical nature of the internal energy (thermal motion) was not at all clear. In this context, the names of two great scientists, J von Mayer (who was first to conjecture the conservation of the sum of the thermal and macroscopic mechanical energies in 1841), and J Joule (who proved this conservation experimentally two years later), have to be reverently mentioned.

term has to be equal to the *heat dQ* transferred from the environment to the system (see figure 1.3):

$$dE = dQ + d\mathscr{W}, \tag{1.18}$$

$$dQ = TdS. \tag{1.19}$$

The last relation, divided by T and then integrated along an arbitrary (but reversible!) process,

$$S = \int \frac{dQ}{T} + \text{const}, \tag{1.20}$$

is sometimes used as an alternative definition of entropy S—provided that temperature is defined not by Eq. (1.9), but in some independent way. It is useful to recognize that the entropy (like energy) may be defined to an arbitrary constant, which does not affect any other thermodynamic observables. The common convention is to take

$$S \to 0, \quad \text{at } T \to 0. \tag{1.21}$$

This condition is sometimes called the *3rd law of thermodynamics*, but it is important to realize that this is just a convention rather than a real law[23]. Indeed, the convention corresponds well to the notion of the full order at $T = 0$ in some systems (e.g. separate atoms or perfect crystals), but creates an ambiguity for other systems, e.g. amorphous solids (like the usual glasses) that may remain, for 'astronomic' times, highly disordered even at $T \to 0$.

Now let us discuss the notion of *heat capacity* that, by definition, is the ratio dQ/dT, where dQ is the amount of heat that should be given to a system to raise its temperature by a small amount dT.[24] (This notion is very important, because the heat capacity may be most readily measured experimentally.) The heat capacity depends, naturally, on whether the heat dQ goes only into an increase of the internal energy dE of the system (as it does if its volume V is constant), or also into the mechanical work ($-d\mathscr{W}$) that is performed by the system at its expansion—as happens, for example, if the pressure P, rather than the volume V, is fixed (the so-called *isobaric* process—see figure 1.4). Hence we should discuss at least two different quantities, the *heat capacity at fixed volume*,

$$C_V \equiv \left(\frac{\partial Q}{\partial T}\right)_V \tag{1.22}$$

[23] Actually, the 3rd law (also called the *Nernst theorem*) as postulated by W Nernst in 1912 was different—and really meaningful: 'It is impossible for any procedure to lead to the isotherm $T = 0$ in a finite number of steps.' I will discuss this theorem at the end of section 1.6.

[24] By this definition, the full heat capacity of a system is an extensive variable, but it may be used to form such intensive variables as the heat capacity per particle, called the *specific heat capacity*, or just the *specific heat*. (Please note that this term is very ambiguous: it is also used for the heat capacity per unit mass, per unit volume, and sometimes even for the heat capacity of the system as the whole, so that some caution is in order.)

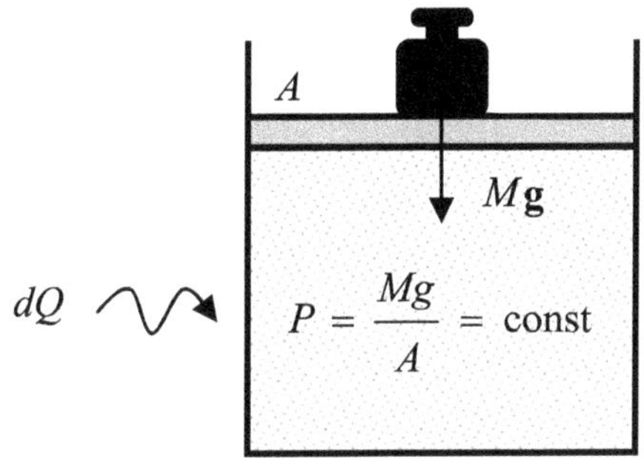

Figure 1.4. The simplest example of an isobaric process.

and heat capacity at fixed pressure

$$C_P \equiv \left(\frac{\partial Q}{\partial T}\right)_P, \tag{1.23}$$

and expect that for all 'normal' (mechanically stable) systems, $C_P \geqslant C_V$.[25] The difference between C_P and C_V is rather minor for most liquids and solids, but may be very substantial for gases—see section 1.4.

1.4 Thermodynamic potentials

Since for a fixed volume, $d\mathscr{W} = -PdV = 0$, and Eq. (1.18) yields $dQ = dE$, we may rewrite Eq. (1.22) in another convenient form

$$C_V = \left(\frac{\partial E}{\partial T}\right)_V. \tag{1.24}$$

so that in order to calculate C_V from a certain statistical-physics model, we only need to calculate E as a function of temperature and volume. If we want to obtain a similarly convenient expression for C_P, the best way is to introduce a new notion of so-called *thermodynamic potentials*—whose introduction and effective use is perhaps one of the most impressive formalisms of thermodynamics. For that, let us combine Eqs. (1.1) and (1.18) to write the 1st law of thermodynamics in its most common form

$$dQ = dE + PdV. \tag{1.25}$$

At an isobaric process (figure 1.4), i.e. at $P = \text{const}$, this expression is reduced to

$$(dQ)_P = dE + d(PV) = d(E + PV). \tag{1.26}$$

Thus, if we introduce a new function with the dimensionality of energy[26]:

---

[25] Dividing both sides of Eq. (1.19) by dT, we get the general relation $dQ/dT = TdS/dT$, which may be used to rewrite the definitions (1.22) and (1.23) in the following forms

$$C_V = T\left(\frac{\partial S}{\partial T}\right)_V, \qquad C_P = T\left(\frac{\partial S}{\partial T}\right)_P,$$

more convenient for some applications.

[26] From the point of view of mathematics, Eq. (1.27) is a particular case of the so-called *Legendre transformations*.

$$H \equiv E + PV, \tag{1.27}$$

called *enthalpy* (or, more rarely, the 'heat function' or 'heat contents')[27], we may rewrite Eq. (1.23) as

$$C_P = \left(\frac{\partial H}{\partial T}\right)_P. \tag{1.28}$$

Comparing Eqs. (1.28) and (1.24) we see that for the heat capacity, the enthalpy H plays the same role at fixed pressure as the internal energy E plays at fixed volume.

Now let us explore properties of the enthalpy at an arbitrary reversible process, i.e. lifting the restriction P = const, but still keeping the definition (1.27). Differentiating this equality, we get

$$dH = dE + PdV + VdP. \tag{1.29}$$

Plugging into this relation Eq. (1.17) for dE, we see that the terms $\pm PdV$ cancel, yielding a very simple expression

$$dH = TdS + VdP, \tag{1.30}$$

whose right-hand side differs from Eq. (1.17) only by the swap of P and V in the second term, with the simultaneous change of its sign. Eq. (1.30) shows that if H has been found (say, experimentally measured or calculated for a certain microscopic model) as a function of the entropy S and the pressure P of a system, we can calculate its temperature T and volume V by simple partial differentiation:

$$T = \left(\frac{\partial H}{\partial S}\right)_P, \qquad V = \left(\frac{\partial H}{\partial P}\right)_S. \tag{1.31}$$

The comparison of the first of these relations with Eq. (1.9) shows that not only for the heat capacity, but for the temperature as well, the enthalpy plays the same role at fixed pressure, as is played by the internal energy at fixed volume.

This success immediately raises the question whether we could develop it further on, by defining other useful thermodynamic potentials—the variables with the dimensionality of energy that would have similar properties, first of all a potential that would enable a similar swap of T and S in its full differential. We already know that an adiabatic processes is the reversible process with fixed entropy, inviting an analysis of a reversible process with fixed temperature. Such *isothermal process* may be implemented, for example, by placing the system under consideration into a thermal contact with a much larger system (called either the *heat bath*, or 'heat reservoir', or 'thermostat') that remains in thermodynamic equilibrium at all times—see figure 1.5.

Due to its very large size, the heat bath temperature T does not depend on what is being done with our system, and if the change is being done sufficiently slowly (i.e. reversibly), this temperature is also the temperature of our system—see Eq. (1.8) and its discussion. Let us calculate the elementary mechanical work $d\mathscr{W}$ (1.1) at such a

[27] This function (as well as the Gibbs free energy G, see below), had been introduced in 1875 by J Gibbs, though the term 'enthalpy' was coined (much later) by H Onnes.

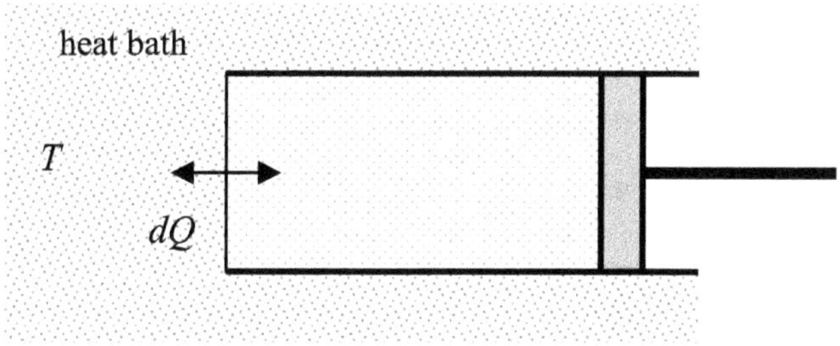

Figure 1.5. The simplest example of an isothermal process.

reversible isothermal process. According to the general Eq. (1.18), $d\mathscr{W} = dE - dQ$. Plugging dQ from Eq. (1.19) into this equality, for $T = \text{const}$ we get

$$(d\mathscr{W})_T = dE - TdS = d(E - TS) = dF, \tag{1.32}$$

where the following combination,

$$F \equiv E - TS, \tag{1.33}$$

is called the *free energy* (or the 'Helmholtz free energy', or just the 'Helmholtz energy'[28]). Just as we have done for the enthalpy, let us establish properties of this new thermodynamic potential for an arbitrary small, reversible (now not necessarily isothermal!) variation of variables, while keeping the definition (1.33). Differentiating this relation and using Eq. (1.17), we get

$$dF = -SdT - PdV. \tag{1.34}$$

Thus, if we know the function $F(T, V)$, we can calculate S and P by simple differentiation:

$$S = -\left(\frac{\partial F}{\partial T}\right)_V, \quad P = -\left(\frac{\partial F}{\partial V}\right)_T. \tag{1.35}$$

Now we may notice that the system of all partial derivatives may be made full and symmetric if we introduce one more thermodynamic potential. Indeed, we have already seen that each of the three already introduced thermodynamic potentials (E, H, and F) has an especially simple full differential if it is considered as a function of its two *canonical* arguments: one of the 'thermal variables' (either S or T) and one of the 'mechanical variables' (either P or V):[29]

$$E = E(S, V), \quad H = H(S, P), \quad \text{and} \quad F = F(T, V). \tag{1.36}$$

---

[28] It was named after H von Helmholtz (1821–94). The last of the listed terms for F was recommended by the most recent (1988) IUPAC's decision, but I will use the first term, which prevails is physics literature. The origin of the adjective 'free' stems from Eq. (1.32): F is may be interpreted as the internal energy's part that is 'free' to be transferred to the mechanical work, at the (most common) reversible, isothermal process.

[29] Note the similarity of this situation with that is analytical mechanics (see, e.g. *Part CM* chapters 2 and 10): the Lagrangian function may be used to derive equations of motion if it is expressed as a function of generalized coordinates and *velocities*, while in order to use the Hamiltonian function in a similar way, it has to be expressed as a function of the generalized coordinates and *momenta*.

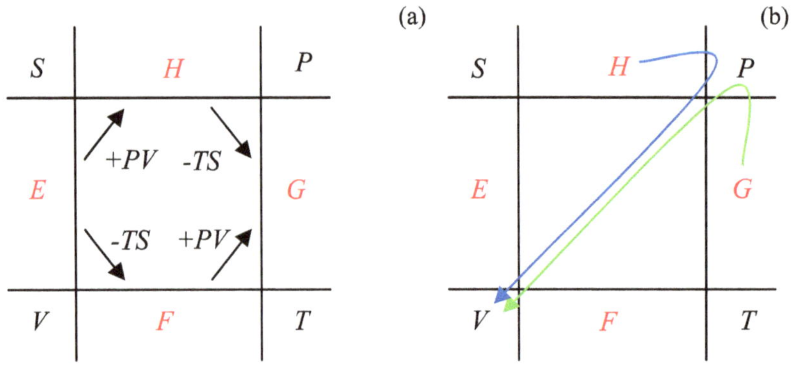

Figure 1.6. (a) The circular diagram and (b) an example of its use for variable calculation. The thermodynamic potentials are typeset in red, each flanked with its two canonical arguments.

In this list of pairs of four arguments, only one pair is missing: $\{T, P\}$. The thermodynamic function of this pair, which gives the two remaining variables (S and V) by simple differentiation, is called the *Gibbs energy* (or sometimes the 'Gibbs free energy'): $G = G(T, P)$. The way to define it in a symmetric way is evident from the so-called *circular diagram* shown in figure 1.6.

In this diagram, each thermodynamic potential is placed between its two canonical arguments—see Eq. (1.36). The left two arrows in figure 1.6a show the way the potentials H and F have been obtained from the energy E—see Eqs. (1.27) and (1.33). This diagram hints that G has to be defined as shown by the right two arrows on that panel, i.e. as

$$G \equiv F + PV = H - TS = E - TS + PV. \tag{1.37}$$

In order to verify this idea, let us calculate the full differential of this new thermodynamic potential, using, e.g. the first form of Eq. (1.37) together with Eq. (1.34):

$$\begin{aligned} dG &= dF + d(PV) = (-SdT - PdV) + (PdV + VdP) \\ &\equiv -SdT + VdP, \end{aligned} \tag{1.38}$$

so that if we know the function $G(T, P)$, we can indeed readily calculate both entropy and volume:

$$S = -\left(\frac{\partial G}{\partial T}\right)_P, \quad V = \left(\frac{\partial G}{\partial P}\right)_T. \tag{1.39}$$

The circular diagram completed in this way is a good mnemonic tool for describing Eqs. (1.9), (1.15), (1.31), (1.35), and (1.39), which express thermodynamic variables as partial derivatives of thermodynamic potentials. Indeed, the variable in any corner of the diagram may be found as a derivative of any of two potentials that are not its immediate neighbors, over the variable in the opposite corner. For example, the green line in figure 1.6b corresponds to the second of Eqs. (1.39), while the blue line, to the second of Eqs. (1.31). At this procedure, all the derivatives giving the variables of the upper row (S and P) have to be taken with negative signs, while those giving the variables of the bottom row (V and T), with positive signs[30].

Now I have to justify the collective name 'thermodynamic potentials' used for E, H, F, and G. For that, let us consider an *irreversible* process, for example, a direct thermal contact of two bodies with different initial temperatures. As was discussed in section 1.2, at such a process, the entropy may grow even without the external heat flow: $dS \geqslant 0$ at $dQ = 0$—see Eq. (1.12). This means that at a more general process with $dQ \neq 0$, the entropy may grow faster than predicted by Eq. (1.19), which has been derived for a reversible process, so that

$$dS \geqslant \frac{dQ}{T}, \tag{1.40}$$

with the equality approached in the reversible limit. Plugging Eq. (1.40) into Eq. (1.18) (which, being just the energy conservation law, remains valid for irreversible processes as well), we get

$$dE \leqslant TdS - PdV. \tag{1.41}$$

We can use this relation to have a look at the behavior of other thermodynamic potentials in irreversible situations, still keeping their definitions given by Eqs. (1.27), (1.33), and (1.37). Let us start from the (very common) case when both the temperature T and the volume V of a system are kept constant. If the process is reversible, then according to Eq. (1.34), the full time derivative of the free energy F would equal zero. Eq. (1.41) says that at an irreversible process it is not necessarily so: if $dT = dV = 0$, then

$$\frac{dF}{dt} = \frac{d}{dt}(E - TS) = \frac{dE}{dt} - T\frac{dS}{dt} \leqslant T\frac{dS}{dt} - T\frac{dS}{dt} \equiv 0. \tag{1.42}$$

Hence, in the general (irreversible) situation, F can only decrease, but not increase in time. This means that F eventually approaches its minimum value $F(T, S)$, given by the equations of reversible thermodynamics. To re-phrase this important conclusion, in the case $T =$ const, $V =$ const, the free energy F, i.e. the difference $E - TS$, plays the role of the potential energy in the classical mechanics of dissipative processes: its minimum corresponds to the (in the case of F, thermodynamic) equilibrium of the system. This is one of the key results of thermodynamics, and I invite the reader to give it some thought. One of its possible handwaving interpretations of this fact is that the heat bath with fixed $T > 0$, i.e. with a substantial thermal agitation of its components, 'wants' to impose thermal disorder in the system immersed into it, by 'rewarding' it by the lowering F for any increase of disorder.

---

[30] There is also a wealth of other relations between thermodynamic variables that may be represented as second derivatives of the thermodynamic potentials, including four *Maxwell relations* such as $(\partial S/\partial V)_T = (\partial P/\partial T)_V$, etc. (They may be readily recovered from the well-known property of a function of two independent arguments, say, $f(x, y)$: $\partial(\partial f/\partial x)/\partial y = \partial(\partial f/\partial y)/\partial x$.) In this chapter, I will list only the thermodynamic relations that will be used later in the course; a more complete list may found, e.g. in section 16 of the book by L Landau and E Lifshitz [4].

Repeating the calculation for the different case, $T = \text{const}$, $P = \text{const}$, it is easy to see that in this case the same role is played by the Gibbs energy:

$$\begin{aligned}\frac{dG}{dt} = \frac{d}{dt}(E - TS + PV) &= \frac{dE}{dt} - T\frac{dS}{dt} + P\frac{dV}{dt} \\ &\leqslant \left(T\frac{dS}{dt} - P\frac{dV}{dt}\right) - T\frac{dS}{dt} + P\frac{dV}{dt} \equiv 0,\end{aligned} \tag{1.43}$$

so that the thermal equilibrium now corresponds to the minimum of G rather than F.

For two remaining thermodynamic potentials, E and H, the calculations similar to Eqs. (1.42) and (1.43) make less sense, because that would require taking $S =$ const (with $V =$ const for E, and $P =$ const for H), but it is hard to prevent the entropy from growing if initially it had been lower than its equilibrium value, at least on the long-term basis[31]. Thus the circular diagram is not so symmetric after all: G and F are somewhat more useful for most practical calculations than E and H.

Note that the difference $G - F = PV$ between the two 'more useful' potentials has very little to do with thermodynamics at all, because this difference exists (although not much advertised) in classical mechanics as well[32]. Indeed, the difference may be generalized as $G - F = -\mathscr{F}q$, where q is a generalized coordinate and $\mathscr{F}$ is the corresponding generalized force. The minimum of F corresponds to the equilibrium of an autonomous system (with $\mathscr{F} = 0$), while the equilibrium position of the same system under the action of external force $\mathscr{F}$ is given by the minimum of G. Thus the external force 'wants' the system to subdue to its effect, 'rewarding' it by lowering its G.

Moreover, the difference between F and G becomes a bit ambiguous (approach-dependent) when the product $\mathscr{F}q$ may be partitioned into single-particle components—just as is done in Eqs. (1.2*b*) and (1.3*b*) for the electric and magnetic fields. Here the applied field may be taken into account on the microscopic level, including its effect directly into the energy ε_k of each particle. In this case, the field contributes to the total internal energy E directly, and hence the thermodynamic equilibrium (at $T = \text{const}$) is described as the minimum of F. (We may say that in this case $F = G$, unless a difference between these thermodynamic potentials is created by the actual mechanical pressure P.) However, in some cases, typically for condensed systems, with their strong interparticle interactions, it is easier (and sometimes the only one practicable[33]) to account for the field on the macroscopic level, taking $G - F = -\mathscr{F}q$. In this case, the

---

[31] There are a few practicable systems, notably including the so-called *adiabatic magnetic refrigerators* (to be discussed in chapter 2), where the unintentional growth of S is so slow that the condition $S =$ const may be closely approached.

[32] It is convenient to describe it as the difference between the 'usual' (internal) potential energy U of the system to its 'Gibbs potential energy' U_G—see *Part CM* section 1.4. For the readers who skipped that discussion: my pet example is the usual elastic spring with $U = kx^2/2$, under the effect of an external force $\mathscr{F}$, whose equilibrium position, $x_0 = \mathscr{F}/k$, evidently corresponds to the minimum of $U_G = U - \mathscr{F}x$, rather than U.

[33] An example of such an extreme situation is the case when an external magnetic field $\mathscr{H}$ is applied to a superconductor in its so-called *intermediate state*, in which the sample partitions into domains of the 'normal' phase with $\mathscr{B} = \mu_0\mathscr{H}$, and the superconducting phase with $\mathscr{B} = 0$. In this case, the field is effectively applied to the interfaces between the domains, very similarly to the mechanical pressure applied to a gas via the piston—see figure 1.1 again.

same equilibrium state is described as the minimum of G. Several examples of this dichotomy will be given later in this course. Whatever the choice, one should be careful not to take the same field effect into account twice.

One more important conceptual question I would like to discuss here is why usually statistical physics pursues the calculation of thermodynamic potentials, rather than just a relation between P, V, and T. (Such a relation is called the *equation of state* of the system.) Let us explore this issue on the example of an *ideal classical gas* in thermodynamic equilibrium, for which the equation of state should be well known to the reader from undergraduate physics[34]:

$$PV = NT, \tag{1.44}$$

where N is the number of particles in volume V. (In chapter 3, we will derive Eq. (1.44) from statistics.) Let us try to use it for the calculation of all thermodynamic potentials, and all other thermodynamic variables discussed above. We may start, for example, from the calculation of the free energy F. Indeed, integrating the second of Eqs. (1.35) with the pressure calculated from Eq. (1.44), $P = NT/V$, we get

$$\begin{aligned} F &= -\int PdV\,|_{T=\text{const}} = -NT\int\frac{dV}{V} \equiv -NT\int\frac{d(V/N)}{(V/N)} \\ &= -NT\ln\frac{V}{N} + Nf(T), \end{aligned} \tag{1.45}$$

where V has been divided by N in both instances just to represent F as a manifestly extensive variable, in this uniform system proportional to N. The integration 'constant' $f(T)$ is some function of temperature, which cannot be recovered from the equation of state. This function affects all other thermodynamic potentials, and the entropy as well. Indeed, using the first of Eqs. (1.35) together with Eq. (1.45), we get

$$S = -\left(\frac{\partial F}{\partial T}\right)_V = N\left[\ln\frac{V}{N} - \frac{df(T)}{dT}\right], \tag{1.46}$$

and now may combine Eqs. (1.33) with (1.46) to calculate the (internal) energy of the gas[35],

$$\begin{aligned} E = F + TS &= \left[-NT\ln\frac{V}{N} + Nf(T)\right] + T\left[N\ln\frac{V}{N} - N\frac{df(T)}{dT}\right] \\ &\equiv N\left[f(T) - T\frac{df(T)}{dT}\right], \end{aligned} \tag{1.47}$$

[34] The long history of the gradual discovery of this relation includes contributions by R Townely, H Power, R Boyle, E Mariotte, J Charles, J Dalton, and J Gay-Lussac. It was fully formulated by E Clapeyron (in 1834) in the form $PV = nRT_K$, where n is the number of moles in the gas sample, and $R \approx 8.31$ J mole$^{-1}\cdot$K^{-1} is the so-called *gas constant*. This form is equivalent to Eq. (1.44), taking into account that $R = k_B N_A$, where $N_A = 6.022\ 140\ 76 \times 10^{23}$ mole^{-1} is the *Avogadro number*, i.e. the number of molecules per mole. (By the mole's definition, N_A is just the reciprocal mass, in grams, of a baryon—more exactly, by convention, a 1/12 part of the ^{12}C atom's mass.)

[35] In particular, Eq. (1.47) describes a very important property of the ideal classical gas: its energy depends only on temperature, and not on volume or pressure.

then use Eqs. (1.27), (1.44) and (1.47) to calculate its enthalpy,

$$H = E + PV = E + NT = N\left[f(T) - T\frac{df(T)}{dT} + T\right], \tag{1.48}$$

and, finally, plug Eqs. (1.44) and (1.45) into Eq. (1.37) to calculate the Gibbs energy

$$G = F + PV = N\left[-T\ln\frac{V}{N} + f(T) + T\right]. \tag{1.49}$$

One might ask whether the function $f(T)$ is physically significant, or is something like the inconsequential, arbitrary constant that may be always added to the potential energy in non-relativistic mechanics. In order to address this issue, let us calculate, from Eqs. (1.24) and (1.28), both heat capacities, which are evidently measurable quantities:

$$C_V = \left(\frac{\partial E}{\partial T}\right)_V = -NT\frac{d^2f}{dT^2}, \tag{1.50}$$

$$C_P = \left(\frac{\partial H}{\partial T}\right)_P = N\left(-T\frac{d^2f}{dT^2} + 1\right) = C_V + N. \tag{1.51}$$

We see that the function $f(T)$, or at least its second derivative, is measurable[36]. (In chapter 3, we will calculate this function for two simple 'microscopic' models of the ideal classical gas.) The meaning of this function is evident from the physical picture of the ideal gas: the pressure P exerted on the walls of the containing volume is produced only by the translational motion of the gas molecules, while their internal energy E (and hence other thermodynamic potentials) may be also contributed by the internal dynamics of the molecules—their rotations, vibrations, etc. Thus, the equation of state does *not* give us the full thermodynamic description of a system, while the thermodynamic potentials do.

1.5 Systems with variable number of particles

Now we have to consider one more important case when the number N of particles in a system is not rigidly fixed, but may change as a result of a thermodynamic process. A typical example of such a system is a gas sample separated from the environment by a penetrable partition—see figure 1.7.[37] Let us analyze this situation for the simplest case when all the particles are similar. (In section 4.1, this analysis will be extended to systems with particles of several sorts.) In this case we can consider N as an independent thermodynamic variable whose variation may change the energy E of the system, so that (for a slow, reversible process) Eq. (1.17) should be now generalized as

---

[36] Note, however, that the difference $C_P - C_V = N$ is independent of $f(T)$. (If temperature is measured in kelvins, this relation takes a more familiar form $C_P - C_V = nR$.) It is straightforward (and hence left for the reader's exercise) to show that the difference $C_P - C_V$ of *any* system is fully determined by its equation of state.

[37] Another important example is a gas in contact with the open-surface liquid of similar molecules.

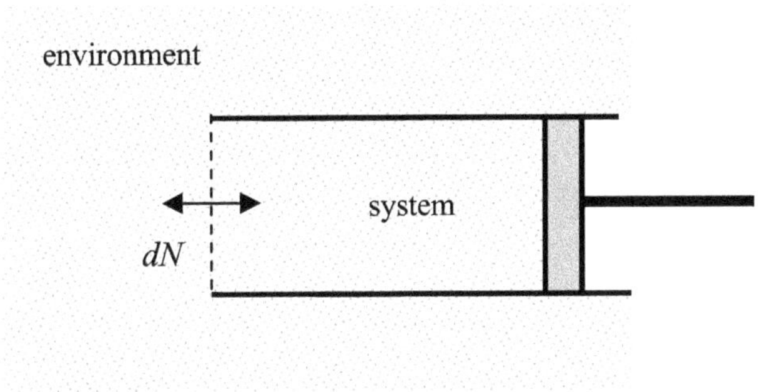

Figure 1.7. An example of a system with a variable number of particles.

$$dE = TdS - PdV + \mu dN, \tag{1.52}$$

where μ is a new function of state, called the *chemical potential*[38]. Keeping the definitions of other thermodynamic potentials, given by Eqs. (1.27), (1.33), and (1.37), intact, we see that the expressions for their differentials should be generalized as

$$dH = TdS + VdP + \mu dN, \tag{1.53a}$$

$$dF = -SdT - PdV + \mu dN, \tag{1.53b}$$

$$dG = -SdT + VdP + \mu dN, \tag{1.53c}$$

so that the chemical potential may be calculated as either of the following partial derivatives[39]:

$$\mu = \left(\frac{\partial E}{\partial N}\right)_{S,V} = \left(\frac{\partial H}{\partial N}\right)_{S,P} = \left(\frac{\partial F}{\partial N}\right)_{T,V} = \left(\frac{\partial G}{\partial N}\right)_{T,P}. \tag{1.54}$$

Despite the formal similarity of all Eqs. (1.54), one of them is more consequential than the others. Indeed, the Gibbs energy G is the only thermodynamic potential that is a function of two *intensive* parameters, T and P. However, as all thermodynamic potentials, G has to be extensive, so that in a system of similar particles it has to be proportional to N:

$$G = Ng, \tag{1.55}$$

where g is some function of T and P. Plugging this expression into the last of Eqs. (1.54), we see that μ equals exactly this function; in other words,

$$\mu = \frac{G}{N}, \tag{1.56}$$

so that the chemical potential is just *the Gibbs energy per particle*.

---

[38] This name, of a historic origin, is a bit misleading: as evident from Eq. (1.52), μ has a clear *physical* sense of the average energy cost of adding one more particle to the system of $N \gg 1$ particles.

[39] Note that strictly speaking, Eqs. (1.9), (1.15), (1.31), (1.35) and (1.39) should be now generalized by adding one more lower index, N, to the corresponding derivatives; I will just imply this.

In order to demonstrate how vital the notion of chemical potential may be, let us consider the situation (parallel to that shown in figure 1.2) when a system consists of two parts, with equal pressure and temperature, that can exchange particles at a relatively slow rate (much slower than the speed of the internal relaxation of each of the parts). Then we can write two equations similar to Eqs. (1.5):

$$N = N_1 + N_2, \qquad G = G_1 + G_2, \tag{1.57}$$

where $N = \text{const}$, and Eq. (1.56) may be used to describe each component of G:

$$G = \mu_1 N_1 + \mu_2 N_2. \tag{1.58}$$

Plugging the N_2 expressed from the first of Eqs. (1.57), $N_2 = N - N_1$, into Eq. (1.58), we see that

$$\frac{dG}{dN_1} = \mu_1 - \mu_2, \tag{1.59}$$

so that the minimum of G is achieved at $\mu_1 = \mu_2$. Hence, in the conditions of fixed temperature and pressure, i.e. when G is the appropriate thermodynamic potential, the chemical potentials of the system parts should be equal—the so-called *chemical equilibrium*.

Later in the course we will also run into several cases when the volume V of a system, its temperature T, and the chemical potential μ are all fixed. (The last condition may be readily implemented by allowing the system of our interest to exchange particles with their storage so large that its μ stays constant.) The thermodynamic potential appropriate for this case may be obtained from the free energy F by subtraction of the product μN, resulting in the so-called *grand thermodynamic potential* (or the 'Landau potential'):

$$\Omega \equiv F - \mu N = F - \frac{G}{N}N \equiv F - G = -PV. \tag{1.60}$$

Indeed, for a reversible process, the full differential of this potential is

$$\begin{aligned} d\Omega &= dF - d(\mu N) = (-SdT - PdV + \mu dN) - (\mu dN + Nd\mu) \\ &= -SdT - PdV - Nd\mu, \end{aligned} \tag{1.61}$$

so that if Ω has been calculated as a function of T, V and μ, other thermodynamic variables may be found as

$$S = -\left(\frac{\partial \Omega}{\partial T}\right)_{V,\mu}, \quad P = -\left(\frac{\partial \Omega}{\partial V}\right)_{T,\mu}, \quad N = -\left(\frac{\partial \Omega}{\partial \mu}\right)_{T,V}. \tag{1.62}$$

Now acting exactly as we have done for other potentials, it is straightforward to prove that for an irreversible process with fixed T, V, and μ, $d\Omega/dt \leqslant 0$, so that the system's equilibrium indeed corresponds to the minimum of the grand potential Ω. We will repeatedly use this fact later in this course.

1.6 Thermal machines

In order to complete this brief review of thermodynamics, I cannot completely pass the topic of *thermal machines*—not because it will be used much in this course, but mostly because of its practical and historic significance. (After all, the whole field of thermodynamics was spurred by the famous 1824 work by S Carnot, where he in particular gave an alternative, indirect form of the 2nd law of thermodynamics—see below.)

Figure 1.8a shows the generic scheme of a thermal machine that may perform mechanical work on its environment (in the notation of Eq. (1.1), equal to $-\mathscr{W}$) during each cycle of the expansion/compression of the 'working gas', by transferring different amounts of heat from a high-temperature heat bath (Q_H) and to the low-temperature bath (Q_L).

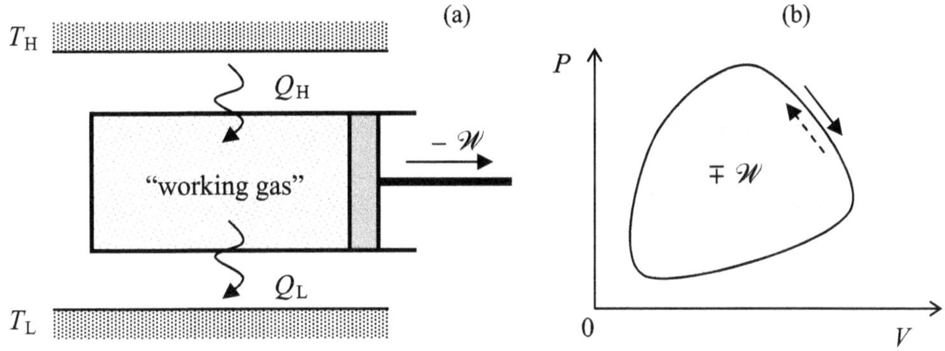

Figure 1.8. (a) The simplest implementation of a thermal machine, and (b) the graphic representation of the mechanical work it performs. On panel (b), the solid arrow indicates the heat engine cycle direction, while the dashed arrow indicates the refrigerator cycle direction.

One relation between the three amounts Q_H, Q_L, and $\mathscr{W}$ is immediately given by the energy conservation (i.e. by the 1st law of thermodynamics):

$$Q_H - Q_L = -\mathscr{W}. \tag{1.63}$$

From Eq. (1.1), the mechanical work during the cycle may be calculated as

$$-\mathscr{W} = \oint PdV, \tag{1.64}$$

and hence represented by the area circumvented by the state-representing point on the $[P, V]$ plane—see figure 1.8b. Note that the sign of this circular integral depends on the direction of rotation of the point; in particular, the work ($-\mathscr{W}$) done by the working gas is positive at its clockwise rotation (pertinent to *heat engines*) and negative in the opposite case (implemented in *refrigerators* and *heat pumps*). Evidently, the work depends on the exact form of the cycle, which in turn may depend not only on T_H and T_L, but also on the working gas' properties.

An exception from this rule is the famous *Carnot cycle*, consisting of two isothermal and two adiabatic processes (all reversible!). In its heat engine's form, the cycle may start, for example, from an isothermic expansion of the gas in contact with the hot bath (i.e. at $T = T_H$). It is followed by its additional adiabatic expansion (without working gas' contact with any of heat baths) until its temperature drops to T_L. Then an isothermal compression of the gas is performed in its contact with the cold bath (at $T = T_L$), followed by its additional adabatic compression to raise T to T_H again, after which the cycle is repeated again and again. (Note that during this cycle the working gas is never in contact with both heat baths simultaneously, thus avoiding the irreversible heat transfer between them.) The cycle's shape on the $[V, P]$ plane (figure 1.9a) depends on exact properties of the working gas and may be rather complicated. However, since the entropy is constant at any adabatic process, the Carnot cycle's shape on the $[S, T]$ plane is always rectangular—see figure 1.9b.

Since during each isotherm, the working gas is brought into thermal contact only with the corresponding heat bath, i.e. its temperature is constant, the relation (1.19), $dQ = TdS$, may be immediately integrated to yield

$$Q_H = T_H(S_2 - S_1), \qquad Q_L = T_L(S_2 - S_1). \tag{1.65}$$

Hence the ratio of these two heat flows is completely determined by their temperature ratio:

$$\frac{Q_H}{Q_L} = \frac{T_H}{T_L}, \tag{1.66}$$

regardless of the working gas properties. Eqs. (1.63) and (1.66) are sufficient to find the ratio of the work $-\mathscr{W}$ to any of Q_H and Q_L. For example, the main figure-of-merit of a thermal machine used as a heat engine ($Q_H > 0$, $Q_L > 0$, $-\mathscr{W} = |\mathscr{W}| > 0$), is its *efficiency*

$$\eta \equiv \frac{|\mathscr{W}|}{Q_H} = \frac{Q_H - Q_L}{Q_H} \equiv 1 - \frac{Q_L}{Q_H} \leqslant 1. \tag{1.67}$$

For the Carnot cycle, this definition, together with Eq. (1.66), immediately yield the famous relation,

$$\eta_{\text{Carnot}} = 1 - \frac{T_L}{T_H}, \tag{1.68}$$

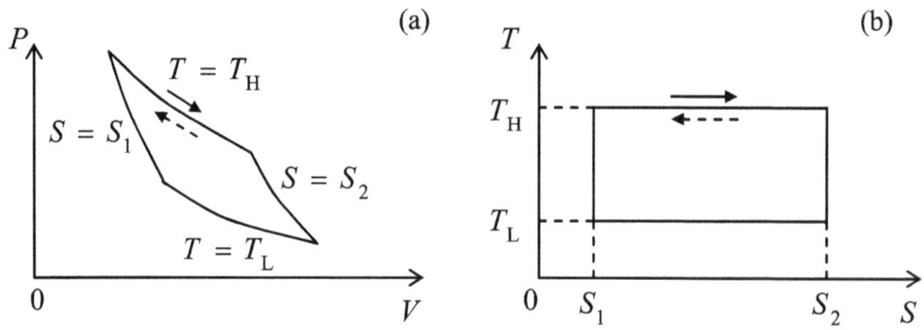

Figure 1.9. Representation of the Carnot cycle: (a) the $[V, P]$ plane (schematically), and (b) on the $[S, T]$ plane. The meaning of the arrows is the same as in figure 1.8.

which shows that at given T_L (that is typically the ambient temperature ~300 K), the efficiency may be increased, ultimately to 1, by raising the temperature T_H of the heat source[40]. The unique reversible nature of the Carnot cycle (see figure 1.9b again) makes its efficiency (1.68) the upper limit for any heat engine[41].

On the other hand, if the cycle is reversed (see the dashed arrows in figures 1.8 and 1.9), the same thermal machine may serve as a *refrigerator*, providing heat removal from the low-temperature bath ($Q_L < 0$) at the cost of consuming external mechanical work: $\mathscr{W} > 0$. This reversal does not affect the basic relation (1.63), which now may be used to calculate the relevant figure-of-merit, called the *cooling coefficient of performance* ($\text{COP}_{\text{cooling}}$):

$$\text{COP}_{\text{cooling}} \equiv \frac{|Q_L|}{\mathscr{W}} = \frac{Q_L}{Q_H - Q_L}. \tag{1.69}$$

Notice that this coefficient may be above unity; in particular, for the Carnot cycle we may use Eq. (1.66) (which is also unaffected by the cycle reversal) to get

$$\left(\text{COP}_{\text{cooling}}\right)_{\text{Carnot}} = \frac{T_L}{T_H - T_L}, \tag{1.70}$$

so that the $\text{COP}_{\text{cooling}}$ is larger than 1 at $T_H < 2T_L$, and even may be much larger than that when the temperature difference ($T_H - T_L$), sustained by the refrigerator, tends to zero. For example, in a typical air-conditioning system, this difference is of the order of 10 K, while $T_L \sim 300$ K, so that $(T_H - T_L) \sim T_L/30$, i.e. the Carnot value of $\text{COP}_{\text{cooling}}$ is as high as ~30. (In the state-of-the-art commercial HVAC systems it is within the range of 3–4.) This is why the term 'cooling efficiency', used in some textbooks instead of $(\text{COP})_{\text{cooling}}$, may be misleading.

Since in the reversed cycle $Q_H = -\mathscr{W} + Q_L < 0$, i.e. the system also provides heat flow into the high-temperature heat bath, it may be used as a *heat pump* for heating purposes. The figure-of-merit appropriate for this application is different from Eq. (1.69):

$$\text{COP}_{\text{heating}} \equiv \frac{|Q_H|}{\mathscr{W}} = \frac{Q_H}{Q_H - Q_L}, \tag{1.71}$$

so that for the Carnot cycle, using Eq. (1.66) again, we get

$$\left(\text{COP}_{\text{heating}}\right)_{\text{Carnot}} = \frac{T_H}{T_H - T_L}. \tag{1.72}$$

[40] Semi-qualitatively, such a trend is valid for other, less efficient but more practicable heat engine cycles—see problems 1.13–1.16. This is the leading reason why internal combustion engines, with T_H of the order of 1500 K, are more efficient than steam engines, with the difference $T_H - T_L$ of at most a few hundred K.

[41] In some alternative axiomatic systems of thermodynamics, this fact is postulated, and serves the role of the 2nd law. This is why it is under persisting (dominantly, theoretical) attacks by suggestions of more efficient heat engines—recently, mostly quantum systems using sophisticated protocols such as the so-called *shortcut-to-adiabaticity*—see, e.g. the recent paper [5] and references therein. So far, reliable analyses of all these suggestions have confirmed that the Carnot efficiency (1.68) is the highest possible.

Note that this COP is *always* larger than 1, meaning that the Carnot heat pump is always more efficient than the direct conversion of work into heat (when $Q_H = -\mathscr{W}$, so that $\text{COP}_{\text{heating}} = 1$), though practical electricity-driven heat pumps are substantially more complex, and hence more expensive than, say, simple electric heaters. Such heat pumps, with the typical $\text{COP}_{\text{heating}}$ values around 4 in summer and 2 in winter, are frequently used for heating large buildings.

I have dwelled so long on the Carnot cycle, because it has a remarkable property: the highest possible efficiency of all heat-engine cycles. Indeed, in the Carnot cycle the transfer of heat between any heat bath and the working gas is performed reversibly, when their temperatures are equal. If this is not so, heat might flow from a hotter to colder bath without performing any work. Hence the result (1.68) also yields the maximum efficiency of *any* heat engine. In particular, it shows that $\eta_{\max} = 0$ at $T_H = T_L$, i.e. no heat engine can perform mechanical work in the absence of temperature gradients[42].

Note also that according to Eq. (1.70), the $\text{COP}_{\text{cooling}}$ of the Carnot cycle tends to zero at $T_L \to 0$, making it impossible to reach the absolute zero of temperature, and hence illustrating the meaningful (Nernst's) formulation of the 3rd law of thermodynamics. Indeed, let us prescribe a finite, but very large heat capacity $C(T)$ to the low-temperature bath, and use the definition of this variable to write the following expression for the (very small) change of its temperature as a result of a relatively large number dn of similar refrigeration cycles:

$$C(T_L)dT_L = Q_L dn. \tag{1.73}$$

Together with Eq. (1.66), this relation yields

$$\frac{C(T_L)dT_L}{T_L} = -\frac{|Q_H|}{T_H}dn. \tag{1.74}$$

The right-hand side of this equality does not depend on T_L, so that if we integrate it to describe the result of many (n) cycles, with constant Q_H and T_H, we get the following simple relation for the initial and final values of T_L:

$$\int_{T_{\text{ini}}}^{T_{\text{fin}}} \frac{C(T)dT}{T} = -\frac{|Q_H|}{T_H}n. \tag{1.75}$$

For example, if $C(T)$ is a constant, Eq. (1.75) yields an exponential law,

$$T_{\text{fin}} = T_{\text{ini}} \exp\left\{-\frac{|Q_H|}{CT_H}n\right\}, \tag{1.76}$$

[42] Such a hypothetical heat engine, which would violate the 2nd law of thermodynamics, is called the 'perpetual motion machine of the 2nd kind'—in contrast to the 'perpetual motion machine of the 1st kind' which would violate the 1st law, i.e. the energy conservation.

with the absolute zero of temperature not reached as any finite n. Even for an arbitrary function $C(T)$, which does not vanish at $T \to 0$, Eq. (1.74) proves the Nernst theorem, because dn diverges at $T_L \to 0$.[43]

1.7 Problems

Problem 1.1. Two bodies, with temperature-independent heat capacities C_1 and C_2, and different initial temperatures T_1 and T_2, are placed into a weak thermal contact. Calculate the change of the total entropy of the system before it reaches the thermal equilibrium.

Problem 1.2. A gas portion has the following properties:

(i) its heat capacity $C_V = aT^b$, and
(ii) the work $\mathscr{W}_T$ needed for its isothermal compression from V_2 to V_1 equals $cT\ln(V_2/V_1)$,

where a, b, and c are some constants. Find the equation of state of the gas, and calculate the temperature dependence of its entropy S and thermodynamic potentials E, H, F, G and Ω.

Problem 1.3. A close volume with an ideal classical gas of similar molecules is separated with a partition in such a way that the number N of molecules in both parts is the same, but their volumes are different. The gas is initially in thermal equilibrium, and its pressure in one part is P_1, and in another, P_2. Calculate the change of entropy resulting from a fast removal of the partition, and analyze the result.

Problem 1.4. An ideal classical gas of N particles is initially confined to volume V, and is in thermal equilibrium with a heat bath of temperature T. Then the gas is allowed to expand to volume $V' > V$ in one the following ways:

(i) The expansion is slow, so that due to the sustained thermal contact with the heat bath, the gas temperature remains equal to T all the times.
(ii) The partition separating the volumes V and $(V' - V)$ is removed very fast, allowing the gas to expand rapidly.

For each process, calculate the eventual changes of pressure, temperature, energy, and entropy of the gas, resulting from its expansion.

Problem 1.5. For an ideal classical gas with temperature-independent specific heat, derive the relation between P and V at the adiabatic expansion/compression.

[43] Note that for such metastable systems as glasses the situation is more complicated. (For a detailed discussion of this issue see, e.g. [6].) Fortunately, this issue does not affect other aspects of statistical physics—at least those to be discussed in this course.

Problem 1.6. Calculate the speed and the wave impedance of acoustic waves propagating in an ideal classical gas with temperature-independent specific heat, in the limits when the propagation may be treated as:

(i) the isothermal process, and
(ii) the adiabatic process.

Which of these limits is achieved at higher wave frequencies?

Problem 1.7. As will be discussed in section 3.5, the so-called 'hardball' models of classical particle interaction yield the following equation of state of a gas of such particles:

$$P = T\varphi(n),$$

where $n = N/V$ is the particle density, and the function $\varphi(n)$ is generally different from that ($\varphi_{\text{ideal}}(n) = n$) of the ideal gas, but still independent of temperature. For such a gas, with temperature-independent c_V, calculate:

(i) the energy of the gas, and
(ii) its pressure as a function of n at the adiabatic compression.

Problem 1.8. For an arbitrary thermodynamic system with a fixed number of particles, prove the following *Maxwell relations* (already mentioned in section 1.4):

$$\text{(i)}: \left(\frac{\partial S}{\partial V}\right)_T = \left(\frac{\partial P}{\partial T}\right)_V, \qquad \text{(ii)}: \left(\frac{\partial V}{\partial S}\right)_P = \left(\frac{\partial T}{\partial P}\right)_S,$$

$$\text{(iii)}: \left(\frac{\partial S}{\partial P}\right)_T = -\left(\frac{\partial V}{\partial T}\right)_P, \qquad \text{(iv)}: \left(\frac{\partial P}{\partial S}\right)_V = -\left(\frac{\partial T}{\partial V}\right)_S,$$

and also the following relation:

$$\left(\frac{\partial E}{\partial V}\right)_T = T\left(\frac{\partial P}{\partial T}\right)_V - P.$$

Problem 1.9. Express the heat capacity difference, $C_P - C_V$, via the equation of state $P = P(V, T)$ of the system.

Problem 1.10. Prove that the *isothermal compressibility*, defined as[44]

$$\kappa_T \equiv -\frac{1}{V}\left(\frac{\partial V}{\partial P}\right)_{T,N},$$

in a single-phase system may be expressed in two different ways:

[44] The compressibility is just the reciprocal bulk modulus, $\kappa = 1/K$—see, e.g. *Part CM* section 7.3.

$$\kappa_T = \frac{V^2}{N^2}\left(\frac{\partial^2 P}{\partial \mu^2}\right)_T = \frac{V}{N^2}\left(\frac{\partial N}{\partial \mu}\right)_{T,V}.$$

Problem 1.11. A reversible process, performed with a fixed portion of an ideal gas, may be represented on the [*P*, *V*] plane with the straight line shown in the figure below. Find the point at which the heat flow into/out of the gas changes its direction.

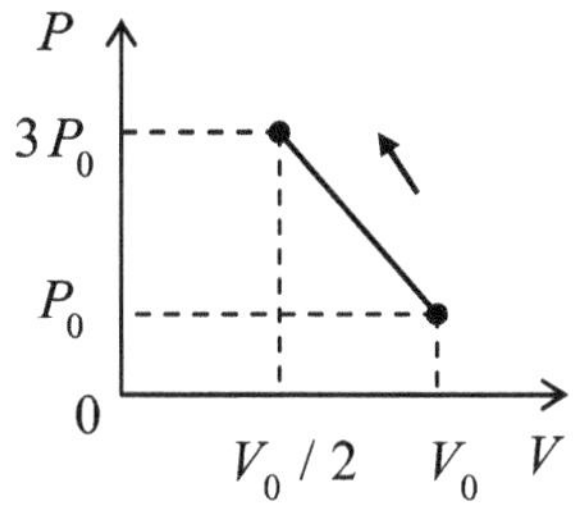

Problem 1.12. Two bodies have equal, temperature-independent heat capacities *C*, but different temperatures, T_1 and T_2. Calculate the maximum mechanical work obtainable from this system, using a heat engine.

Problem 1.13. Express the efficiency η of a heat engine that uses the so-called *Joule cycle*, consisting of two adiabatic and two isobaric processes (see the figure below), via the minimum and maximum values of pressure, and compare the result with η_{Carnot}. Assume an ideal classical working gas with temperature-independent C_P and C_V.

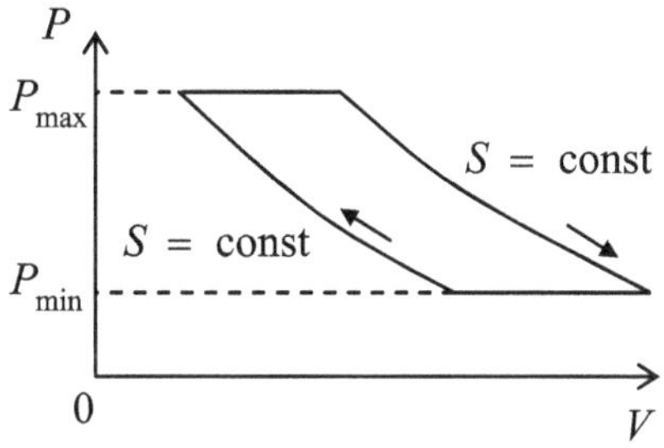

Problem 1.14. Calculate the efficiency of a heat engine using the *Otto cycle*[45], which consists of two adiabatic and two isochoric (constant-volume) reversible processes—see the figure below. Explore how does the efficiency depend on the ratio $r \equiv V_{\max}/V_{\min}$, and compare it with the Carnot cycle's efficiency. Assume an ideal classical working gas with temperature-independent heat capacity.

[45] This name stems from the fact that the cycle is an approximate model of operation of the four-stroke internal-combustion engine, which was improved and made practicable (though not invented!) by N Otto in 1876.

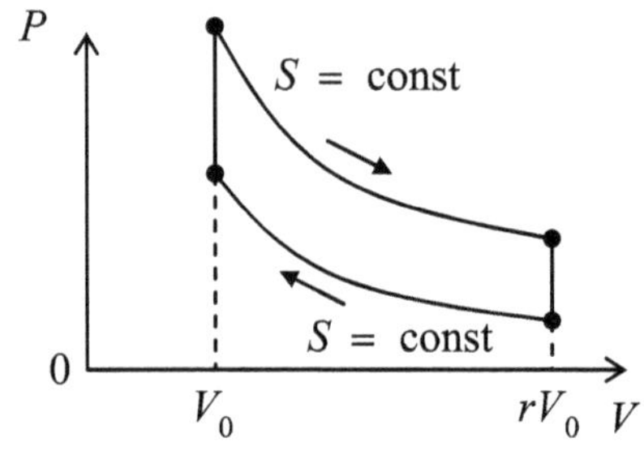

Problem 1.15. A heat engine's cycle consists of two isothermal (T = const) and two isochoric (V = const) reversible processes—see the figure below[46].

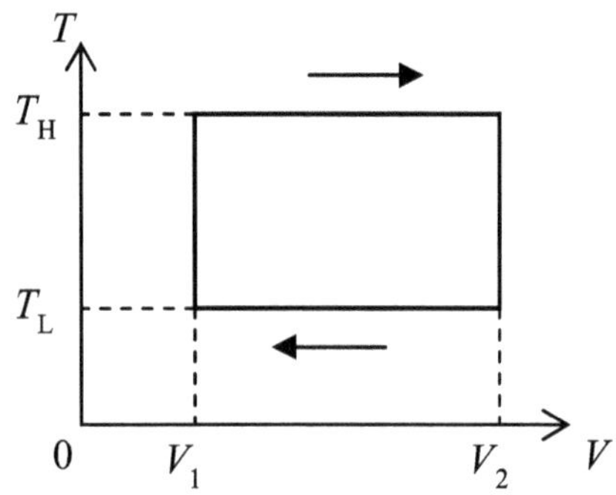

(i) Assuming that the working gas is an ideal classical gas of N particles, calculate the mechanical work performed by the engine during one cycle.
(ii) Are the specified conditions sufficient to calculate the engine's efficiency?

Problem 1.16. The *Diesel cycle* (an approximate model of the Diesel internal combustion engine's operation) consists of two adiabatic processes, one isochoric process, and one isobaric process—see the figure below. Assuming an ideal working gas with temperature-independent C_V and C_P, express the efficiency η of the heat engine using this cycle via the gas temperatures in its transitional states, corresponding to the corners of the cycle diagram.

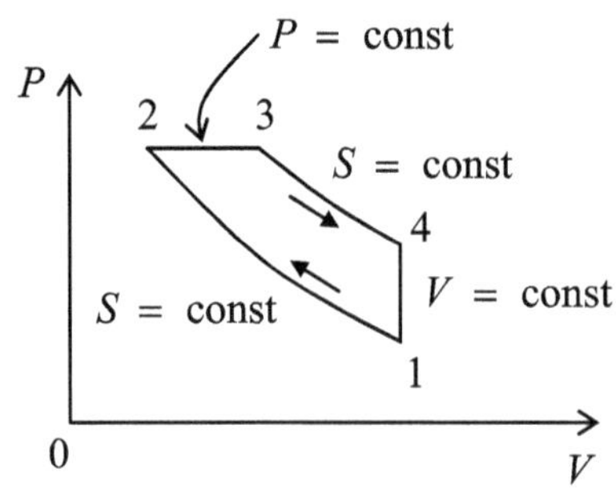

[46] The reversed cycle of this type is a reasonable approximation for the operation of *Stirling* and *Gifford-McMahon* (GM) refrigerators, broadly used for *cryocooling*—for a recent review see, e.g. [7].

References

[1] Kittel C and Kroemer H 1980 *Thermal Physics* 2nd edn (W H Freeman)
[2] Reif F 2008 *Fundamentals of Statistical and Thermal Physics* (Waveland)
[3] Schroeder D 1999 *Introduction to Thermal Physics* (Addison Wesley)
[4] Landau L and Lifshitz E 1980 *Statistical Physics, Part 1* 3rd edn (Pergamon)
[5] Abah O and Lutz E 2017 *Europhysics Lett.* **118** 40005
[6] Wilks J 1961 *The Third Law of Thermodynamics* (Oxford University Press)
[7] de Waele A 2011 *J. Low Temp. Phys.* **164** 179

Chapter 2

Principles of physical statistics

This chapter is the key part of this course. It starts with a brief discussion of such basic notions of statistical physics as statistical ensembles, probability, and ergodicity. Then the so-called microcanonical distribution postulate is formulated, simultaneously with the statistical definition of the entropy. This allows a derivation of the famous Gibbs ('canonical') distribution—the most frequently used tool of statistical physics. Then we will discuss one more, grand canonical distribution, which is more convenient for some tasks, in particular for the derivation of the most important Boltzmann, Fermi–Dirac, and Bose–Einstein statistics for systems of independent particles, which will be repeatedly used in the following chapters.

2.1 Statistical ensembles and probability

As has been already discussed in section 1.1, statistical physics deals with situations when either unknown initial conditions, or a system's complexity, or the laws of its motion (as in the case of quantum mechanics) do not allow a definite prediction of measurement results. The main formalism for the analysis of such systems is the probability theory, so let me start with a very brief review of its basic concepts, using informal 'physical' language—less rigorous but (hopefully) more transparent than standard mathematical treatments[1].

Consider $N \gg 1$ independent similar experiments carried out with *apparently* similar systems (i.e. systems with identical *macroscopic* parameters such as volume, pressure, etc), but still giving, by any of the reasons outlined above, different results of measurements. Such a collection of experiments, together with a fixed method of result processing, is a good example of a *statistical ensemble*. Let us start from the

[1] For the reader interested in reviewing a more rigorous approach, I can recommend, for example, chapter 18 of the handbook by G Korn and T Korn—see section A.16(ii).

doi:10.1088/2053-2563/aaf503ch2

case when the experiments may have M different *discrete* outcomes, and the number of experiments giving the corresponding different results is N_1, N_2,…, N_M, so that

$$\sum_{m=1}^{M} N_m = N. \tag{2.1}$$

The *probability* of each outcome, for the given statistical ensemble, is then defined as

$$W_m \equiv \lim_{N\to\infty}\frac{N_m}{N}. \tag{2.2}$$

Though this definition is so close to our everyday experience that it is almost self-evident, a few remarks may still be relevant.

First, the probabilities W_m depend on the exact statistical ensemble they are defined for, notably including the method of result processing. As the simplest example, consider throwing the standard cubic-shaped dice. For the ensemble of all thrown and counted dice, the probability of each outcome (say, '1') is 1/6. However, nothing prevents us from defining another statistical ensemble of dice-throwing experiments in which all outcomes '1' are discounted. Evidently, the probability of finding outcomes '1' in this modified (but legitimate) ensemble is 0, while those for all five other outcomes ('2' to '6') is 1/5 rather than 1/6.

Second, a statistical ensemble does not necessarily require N different physical systems, e.g. N different dice. It is intuitively clear that tossing the same die N times constitutes an ensemble with similar statistical properties. More generally, a set of N experiments with the same system gives a statistical ensemble equivalent to the set of experiments with N different systems, provided that the experiments are kept independent, i.e. that outcomes of past experiments do not affect those of the experiments to follow. Moreover, for many physical systems of interest, no special preparation of each experiment is necessary, and N different experiments, separated by sufficiently long time intervals, form a 'good' statistical ensemble—the property called *ergodicity*[2].

Third, the reference to infinite N in Eq. (2.2) does not strip the notion of probability from its practical relevance. Indeed, it is easy to prove (see chapter 5) that, at very general conditions, at finite but sufficiently large N, the numbers N_m are approaching their *average* (or *expectation*) *values*[3]

[2] The most popular counter-example is an energy-conserving system. Consider, for example, a system of particles placed in a potential which is a quadratic form of its coordinates. Theory of oscillations tells us (see, e.g. *Part CM* section 6.2) that this system is equivalent to a set of non-interacting harmonic oscillators. Each of these oscillators conserves its own initial energy E_j forever, so that the statistics of N measurements of one such system may differ from that of N different systems with random distribution of E_j, even if the total energy of the system, $E = \Sigma_j E_j$, is the same. Such non-ergodicity, however, is a rather feeble phenomenon, and is readily destroyed by any of many mechanisms, such as weak interaction with environment (leading, in particular, to oscillation damping), potential anharmonicity (see, e.g. *Part CM* chapter 5), and chaos (*Part CM* chapter 9), all of them strongly enhanced by increasing the number of particles in the system, i.e. the number of its degrees of freedom. This is why an overwhelming part of real-life systems are ergodic; for readers interested in non-ergodic exotics, I can recommend the monograph by V Arnold and A Avez [1].

[3] Here, and everywhere in this series, angle brackets $\langle\ldots\rangle$ mean averaging over a statistical ensemble, which is generally different from averaging over time—as will be the case in quite a few examples considered below.

$$\langle N_m \rangle \equiv W_m N, \tag{2.3}$$

with the relative deviations decreasing as $\sim 1/\langle N_m \rangle^{1/2}$.

Now let me list those properties of probabilities that we will immediately need. First, dividing Eq. (2.1) by N and following the limit $N \to \infty$, we get the well-known *normalization condition*

$$\sum_{m=1}^{M} W_m = 1; \tag{2.4}$$

just remember that it is true only if each experiment definitely yields one of the outcomes $N_1, N_2, \ldots, N_M$. Next, if we have an additive function of the results,

$$f = \frac{1}{N} \sum_{m=1}^{M} N_m f_m, \tag{2.5}$$

where f_m are some definite (deterministic) coefficients, the *statistical average* (also called the *expectation value*) of the function is naturally defined as

$$\langle f \rangle \equiv \lim_{N\to\infty} \frac{1}{N} \sum_{m=1}^{M} \langle N_m \rangle f_m, \tag{2.6}$$

so that using Eq. (2.3) we get

$$\langle f \rangle = \sum_{m=1}^{M} W_m f_m. \tag{2.7}$$

Notice that Eq. (2.3) may be considered as the particular form of this general result, when all $f_m = 1$.

Next, the spectrum of possible experimental outcomes is frequently continuous for all practical purposes. (Think, for example, about the positions of the marks left by bullets fired into a target from afar.) The above formulas may be readily generalized to this case; let us start from the simplest situation when all different outcomes may be described by just one continuous scalar variable q—which replaces the discrete index m in Eqs. (2.1)–(2.7). The basic relation for this case is the self-evident fact that the probability dW of having an outcome within a small interval dq near point q is proportional to the magnitude of that interval:

$$dW = w(q)dq, \tag{2.8}$$

where $w(q)$ is some function of q, which does not depend on dq. Such a function is called the *probability density*. Now all the above formulas may be recast by replacing the probabilities W_m with products (2.8), and the summation over m, with the integration over q. In particular, instead of Eq. (2.4) the normalization condition now becomes

$$\int w(q)dq = 1, \tag{2.9}$$

where the integration should be extended over the whole range of possible values of q. Similarly, instead by Eq. (2.5), it is natural to consider a function $f(q)$. Then instead of Eq. (2.7), the expectation value of the function may be calculated as

$$\langle f \rangle = \int w(q)f(q)dq. \tag{2.10}$$

It is also straightforward to generalize these formulas to the case of more variables. For example, results of measurements of a classical particle with three degrees of freedom may be described by the probability density w defined in the 6D space of its generalized radius-vector $\mathbf{q}$ and momentum $\mathbf{p}$. As a result, the expectation value of a function of these variables may be expressed as a 6D integral

$$\langle f \rangle = \int w(\mathbf{q}, \mathbf{p})f(\mathbf{q}, \mathbf{p})d^3q d^3p. \tag{2.11}$$

Some systems considered in this course consist of components whose quantum properties cannot be ignored, so let us discuss how $\langle f \rangle$ should be calculated in this case. If by f_m we mean measurement results, Eq. (2.7) (and its generalizations) of course remains valid, but since these numbers themselves may be affected by the intrinsic quantum-mechanical uncertainty, it may make sense to have a bit of a deeper look into this situation. Quantum mechanics tells us[4] that the most general expression for the expectation value of an observable f in a certain ensemble of macroscopically similar systems is

$$\langle f \rangle = \sum_{m,m'} W_{mm'} f_{m'm} \equiv \mathrm{Tr}(\mathrm{Wf}). \tag{2.12}$$

Here $f_{mm'}$ are the matrix elements of the quantum-mechanical operator $\hat{f}$ corresponding to the observable f, in a full basis of orthonormal states m,

$$f_{mm'} = \langle m|\hat{f}\,|m'\rangle, \tag{2.13}$$

while the coefficients $W_{mm'}$ are elements of the so-called *density matrix* W, which represents, in the same basis, the *density operator* $\hat{W}$ describing properties of this ensemble. Eq. (2.12) is evidently more general than Eq. (2.7), and is reduced to it only if the density matrix is diagonal:

$$W_{mm'} = W_m \delta_{mm'}, \tag{2.14}$$

(where $\delta_{mm'}$ is the Kronecker symbol), when the diagonal elements W_m play the role of probabilities of the corresponding states.

Thus the largest difference between the quantum and classical description is the presence, in Eq. (2.12), of the off-diagonal elements of the density matrix. They have

[4] See, e.g. *Part QM* section 7.1.

largest values in the *pure* (also called 'coherent') ensemble, in which the state of the system may be described with *state vectors*, e.g. the *ket-vector*

$$|\alpha\rangle = \sum_m \alpha_m |m\rangle, \tag{2.15}$$

where α_m are some complex coefficients. In this simple case, the density matrix elements are merely

$$W_{mm'} = \alpha_m^* \alpha_{m'}, \tag{2.16}$$

so that the off-diagonal elements are of the same order as the diagonal elements. For example, in the very important particular case of a *two-level system*, the pure-state density matrix is

$$\mathrm{W} = \begin{pmatrix} \alpha_1^* \alpha_1 & \alpha_1^* \alpha_2 \\ \alpha_2^* \alpha_1 & \alpha_2^* \alpha_2 \end{pmatrix}, \tag{2.17}$$

so that the product of its off-diagonal components is as large as that of the diagonal components.

In the most important basis of stationary states, i.e. the eigenstates of the system's time-independent Hamiltonian, the coefficients α_m oscillate in time as[5]

$$\alpha_m(t) = \alpha_m(0)\exp\left\{-i\frac{E_m}{\hbar}t\right\} \equiv |\alpha_m|\exp\left\{-i\frac{E_m}{\hbar}t + i\varphi_m\right\}, \tag{2.18}$$

where E_m are the corresponding eigenenergies, and φ_m are constant phase shifts. This means that while the diagonal terms of the density matrix (2.16) remain constant, its off-diagonal components are oscillating functions of time:

$$W_{mm'} = \alpha_{m'}^* \alpha_m = |\alpha_{m'}\alpha_m|\exp\left\{i\frac{E_m - E_{m'}}{\hbar}t\right\} \exp\{i(\varphi_{m'} - \varphi_m)\}. \tag{2.19}$$

Due to the extreme smallness of the Planck constant (on the human scale of things), miniscule random perturbations of eigenenergies are equivalent to substantial random changes of the phase multiplier, so that the time average of any off-diagonal matrix element tends to zero. Moreover, even if our statistical ensemble consists of systems with exactly the same E_m, but different values φ_m (which are typically hard to control at the initial preparation of the system), the average values of all $W_{mm'}$ (with $m \neq m'$) vanish again.

This is why, besides some very special cases, typical statistical ensembles of quantum particles are far from being pure, and in most cases (certainly including the thermodynamic equilibrium), a good approximation for their description is given by the opposite limit of the so-called *classical mixture* in which all off-diagonal matrix elements of the density matrix equal zero, and its diagonal elements W_{mm} are merely

[5] Here I use the Schrödinger picture of quantum mechanics in which the matrix elements $f_{nn'}$, representing quantum-mechanical operators, do not evolve in time. The final result of the discussion does not depend on the particular picture—see, e.g. *Part QM* section 4.6.

the probabilities W_m of the corresponding eigenstates. In this case, for the observables compatible with energy, Eq. (2.12) is reduced to Eq. (2.7), with f_m being the eigenvalues of the variable f.

2.2 Microcanonical ensemble and distribution

Let us start with the discussion of physical statistics with the simplest, *microcanonical statistical ensemble*[6] that is defined as a set of macroscopically similar closed (isolated) systems with *virtually* the same total energy E. Since in quantum mechanics the energy of a closed system is quantized, in order to make the forthcoming discussion suitable for quantum systems as well, it is convenient to include in the ensemble all systems with energies E_m within a narrow interval $\Delta E \ll E$ (see figure 2.1) that is nevertheless much larger than the average distance δE between the energy levels, so that the number M of different quantum states within the interval ΔE is large, $M \gg 1$. Such a choice of ΔE is only possible if $\delta E \ll E$; however, the reader should not worry too much about this condition, because the most important applications of the microcanonical ensemble are for very large systems (and/or very high energies) when the energy spectrum is very dense[7].

This ensemble serves as the basis for the formulation of the postulate which is most frequently called the *microcanonical distribution* (or sometimes the 'main statistical hypothesis'): *in the thermodynamic equilibrium, all possible states of the microcanonical ensemble have equal probability,*

$$W_m = \frac{1}{M} = \text{const.} \tag{2.20}$$

Though in some constructs of statistical mechanics this equality is derived from other axioms, which look more plausible to their authors, I believe that Eq. (2.20) may be taken as the starting point of the statistical physics, supported 'just' by the compliance of all its corollaries with experimental observations[8].

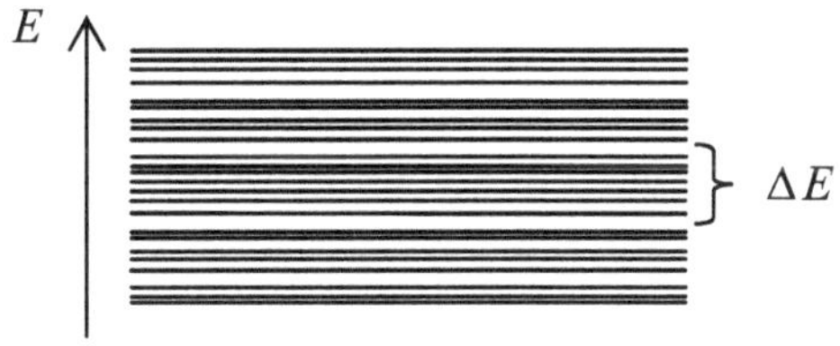

Figure 2.1. A *very* schematic image of the microcanonical ensemble. (Actually, the ensemble deals with quantum *states* rather than energy *levels*. An energy level may be degenerate, i.e. correspond to several states.)

[6] The terms 'microcanonical', as well as 'canonical' (see section 2.4 below) are apparently due to J Gibbs, and I could not find out his motivation for these names. ('Canonical' in the sense of 'standard' or 'common' is quite appropriate, but why 'micro'? Perhaps to reflect the smallness of ΔE?)

[7] Formally, the main result of this section, Eq. (2.20), is valid for any M (including $M = 1$), it is just less informative for small M—and trivial for $M = 1$.

[8] Though I have to move on, let me note that the microcanonical distribution (2.20) is a very nontrivial postulate, and my advice to the reader to give some thought to this foundation of the whole building of statistical mechanics.

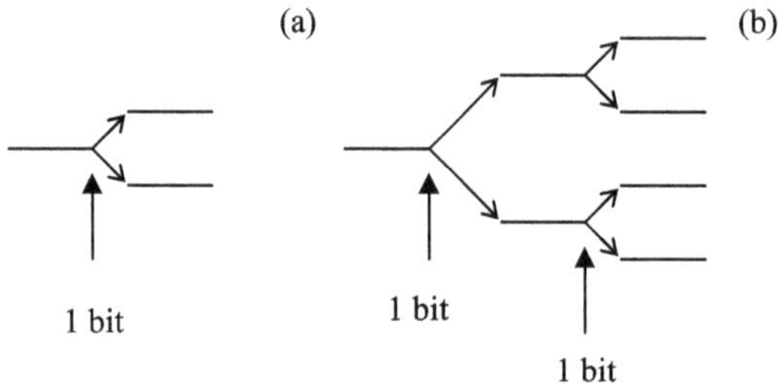

Figure 2.2. 'Logarithmic trees' of binary decisions for making a choice between (a) $M = 2$, and (b) $M = 4$ opportunities with equal probabilities.

Note that the postulate (2.20) sheds some light on the nature of the macroscopic irreversibility of microscopically reversible (closed) systems: if such a system was initially in a certain state, its time evolution with just miniscule interactions with environment (which is necessary for reaching the thermodynamic equilibrium) would eventually lead to the uniform distribution of its probability among all states with the essentially same energy. Each of these states is not 'better' than the initial one; rather, in a macroscopic system, there are just so many of these states that the chance to find the system in the initial state is practically nil—again, think about the ink drop diffusion into a glass of water.

Now let us find a suitable definition of entropy S of a microcanonical ensemble's member—for now, in the thermodynamic equilibrium only. Since S is a measure of *disorder*, it should be related to the amount of information *lost* when the system goes from the full order to the full disorder, i.e. into the microcanonical distribution (2.20), or, in other words, the amount of information[9] *necessary* to find the exact state of your system in a microcanonical ensemble.

In information theory, the amount of information necessary to make a definite choice between two options with equal probabilities (figure 2.2a) is defined as

$$I(2) \equiv \log_2 2 = 1. \tag{2.21}$$

This unit of information is called a *bit*. Now, if we need to make a choice between four equally probable opportunities, it can be made in two similar steps (figure 2.2b), each requiring one bit of information, so that the total amount of information necessary for the choice is

$$I(4) = 2I(2) = 2 = \log_2 4. \tag{2.22}$$

An obvious extension of this process to the choice between $M = 2^m$ states gives

$$I(M) = mI(2) = m = \log_2 M. \tag{2.23}$$

[9] I will rely on the reader's common sense and intuitive understanding of what information is, because in formal information theory this notion is also essentially postulated—see, e.g. the wonderfully clear text by J Pierce [2].

This measure, if extended naturally to any integer M, is quite suitable for the definition of entropy at equilibrium, with the only difference that, following tradition, the binary logarithm is replaced with the natural one[10]:

$$S \equiv \ln M. \tag{2.24a}$$

Using Eq. (2.20), we may recast this definition in the most frequently used form

$$S = \ln \frac{1}{W_m} \equiv - \ln W_m. \tag{2.24b}$$

(Again, please note that Eq. (2.24) is valid in the thermodynamic equilibrium only!)

Eq. (2.24) satisfies the major condition for the entropy definition in thermodynamics, i.e. to be a unique characteristics of disorder. Indeed, according to Eq. (2.20), M is the only possible measure characterizing the microcanonical distribution, and so is its unique function $\ln M$. This function also satisfies another thermodynamic requirement to the entropy, of being an extensive variable. Indeed, mathematics tells us that for two independent systems, the joint probability is just a product of their partial probabilities, and hence, according to Eq. (2.24), their entropies just add up.

Now let us see whether Eqs. (2.20) and (2.24) are compatible with the 2nd law of thermodynamics. For that, we need to generalize Eq. (2.24) for S to an arbitrary state of the system (generally, out of thermodynamic equilibrium), with arbitrary state probabilities W_m. For that, let us first recognize that M in Eq. (2.24) is just the number of possible ways to commit a particular system to a certain state m ($m = 1, 2,\ldots M$), in a statistical ensemble where each state is equally probable. Now let us consider a more general ensemble, still consisting of a large number $N \gg 1$ of similar systems, but with a certain number $N_m = W_m N \gg 1$ of systems in each of M states, with W_m not necessarily equal. In this case the evident generalization of Eq. (2.24) is that the entropy S_N of the whole ensemble is

$$S_N \equiv \ln M(N_1, N_2, ..), \tag{2.25}$$

where $M(N_1,N_2,\ldots)$ is the number of ways to commit a particular system to a certain state m, while keeping all numbers N_m fixed. Such number $M(N_1,N_2,\ldots)$ is clearly equal to the number of ways to distribute N distinct balls between M different boxes, with the fixed number N_m of balls in each box, but in no particular order within it.

---

[10] This is of course just the change of a constant factor: $S(M) = \ln M = \ln 2 \times \log_2 M = \ln 2 \times I(M) \approx 0.693\, I(M)$. A review of chapter 1 shows that nothing in thermodynamics prevents us from choosing such a coefficient arbitrarily, with the corresponding change of the temperature scale—see Eq. (1.9). In particular, in the SI units, where Eq. (2.24b) becomes $S = -k_B \ln W_m$, one bit of information corresponds to the entropy change $\Delta S = k_B \ln 2 \approx 0.693\, k_B \approx 0.965\times10^{-23}\,\mathrm{J\,K^{-1}}$. By the way, the formula '$S = k \log W$' is engraved on the tombstone of L Boltzmann (1844–1906), who was the first one to recognize this intimate connection between entropy and probability.

Comparing this description with the definition of the so-called *multinomial coefficients*[11], we get

$$M(N_1, N_2, \ldots) = {}^N C_{N_1, N_2, \ldots, N_M} \equiv \frac{N!}{N_1!N_2!\ldots N_M!}, \quad \text{with } N = \sum_{m=1}^{M} N_m. \tag{2.26}$$

In order to simplify the resulting expression for S_N, we can use the famous *Stirling formula*, in its crudest de Moivre's form[12], whose accuracy is suitable for most purposes of statistical physics:

$$\ln (N!)_{|N\to\infty} \to N(\ln N - 1). \tag{2.27}$$

When applied to our current problem, this formula gives the following average entropy per system[13],

$$\begin{aligned} S \equiv \frac{S_N}{N} &= \frac{1}{N}\left[\ln(N!) - \sum_{m=1}^{M} \ln(N_m!)\right] \\ &\to_{N_m\to\infty} \frac{1}{N}\left[N(\ln N - 1) - \sum_{m=1}^{M} N_m(\ln N_m - 1)\right] \\ &\equiv - \sum_{m=1}^{M} \frac{N_m}{N} \ln \frac{N_m}{N}. \end{aligned} \tag{2.28}$$

and since this result is only valid in the limit $N_m \to \infty$ anyway, we may use Eq. (2.2) to represent it as

$$S = -\sum_{m=1}^{M} W_m \ln W_m = \sum_{m=1}^{M} W_m \ln \frac{1}{W_m}. \tag{2.29}$$

This extremely important result[14] may be interpreted as the average of the entropy values given by Eq. (2.24), weighed with specific probabilities W_m in accordance with the general formula (2.7).[15]

Now let us find what distribution of probabilities W_m provides the largest value of the entropy (2.29). The answer is almost evident from a good glance at Eq. (2.29).

---

[11] See, e.g. Eq. (A.6). Despite the intimidating name, Eq. (2.26) may be very simply derived. Indeed, $N!$ is just the number of all possible permutations of N balls, i.e. the ways to place them in *certain* positions—say, inside M boxes. Now in order to take into account that the particular order of the balls in each box is not important, that number should be divided by all numbers $N_m!$ of possible permutations of balls within each box—that's it.

[12] See, e.g. Eq. (A.13).

[13] Strictly speaking, I should use the notation $\langle S \rangle$ here. However, following the style accepted in thermodynamics, I will drop the averaging signs until we will really need them to avoid confusion. Again, this shorthand is not too bad because the relative fluctuations of entropy (as those of any macroscopic variable) are very small at $N \gg 1$.

[14] With the replacement of $\ln W_m$ with $\log_2 W_m$ (i.e. division of both sides by ln 2), Eq. (2.29) is the famous *Shannon* (or 'Boltzmann–Shannon') *formula* for the average information I per symbol in a long communication string using M different symbols, with probability W_m each.

[15] In some textbooks, this simple argument is even accepted as the derivation of Eq. (2.29); however, it is evidently less strict than the one outlined above.

For example, if for a subgroup of $M' \leqslant M$ states the coefficients W_m are constant (and hence equal to $1/M'$), while $W_m = 0$ for all other states, all M' nonvanishing terms in the sum (2.29) are equal to each other, so that

$$S = M' \frac{1}{M'} \ln M' = \ln M', \tag{2.30}$$

so that the closer M' to its maximum value M the larger S. Hence, the maximum of S is reached at the uniform distribution given by Eq. (2.24).

In order to prove this important fact more strictly, let us find the maximum of the function given by Eq. (2.29). If its arguments $W_1, W_2, \ldots W_M$ were completely independent, this could be done by finding the point (in the M-dimensional space of the coefficients W_m) where all partial derivatives $\partial S/\partial W_m$ equal zero. However, since the probabilities are constrained by the condition (2.4), the differentiation has to be carried out more carefully, taking into account this interdependence:

$$\left[\frac{\partial}{\partial W_m} S(W_1, W_2, \ldots)\right]_{\text{cond}} = \frac{\partial S}{\partial W_m} + \sum_{m' \neq m} \frac{\partial S}{\partial W_{m'}} \frac{\partial W_{m'}}{\partial W_m}. \tag{2.31}$$

At the maximum of the function S, all such expressions should be equal to zero simultaneously. This condition may be represented as $\partial S/\partial W_m = \lambda$, where the so-called *Lagrange multiplier* λ is independent of m. Indeed, at such point Eq. (2.31) becomes

$$\begin{aligned}\left[\frac{\partial}{\partial W_m} S(W_1, W_2, \ldots)\right]_{\text{cond}} &= \lambda + \sum_{m' \neq m} \lambda \frac{\partial W_{m'}}{\partial W_m} \\ &= \lambda \left(\frac{\partial W_m}{\partial W_m} + \sum_{m' \neq m} \frac{\partial W_{m'}}{\partial W_m} \right) \\ &= \lambda \frac{\partial}{\partial W_m}(1) = 0.\end{aligned} \tag{2.32}$$

For our particular expression (2.29), the condition $\partial S/\partial W_m = \lambda$ yields

$$\frac{\partial S}{\partial W_m} \equiv \frac{d}{dW_m}[-W_m \ln W_m] \equiv -\ln W_m - 1 = \lambda. \tag{2.33}$$

This equality may hold for all m (and hence the entropy reach its maximum value) only if W_m is independent on m. Thus the entropy (2.29) indeed reaches its maximum value (2.24) at equilibrium.

To summarize, we see that the definition (2.24) of the entropy in statistical physics does fit all the requirements imposed on this variable by thermodynamics. In particular, we have been able to prove the 2nd law of thermodynamics, starting from that definition, together with a more fundamental postulate (2.20).

Now let me discuss one possible point of discomfort with that definition: the values of M, and hence W_m, depend on the accepted energy interval ΔE of the microcanonical ensemble, for whose choice no exact guidance is offered. However, if the interval ΔE contains many states, $M \gg 1$, then with a very small relative error (vanishing in the limit $M \to \infty$), M may be represented as

$$M = g(E)\Delta E, \tag{2.34}$$

where $g(E)$ is the *density of states* of the system:

$$g(E) \equiv \frac{d\Sigma(E)}{dE}, \tag{2.35}$$

$\Sigma(E)$ being the total number of states with energies *below* E. (Note that the average interval δE between energy levels, mentioned in the beginning of this section, is just $\delta E = \Delta E/M = 1/g(E)$.) Plugging Eq. (2.34) into Eq. (2.24), we get

$$S = \ln M = \ln g(E) + \ln \Delta E, \tag{2.36}$$

so that the only effect of a particular choice of ΔE is an offset of entropy by a constant, and in chapter 1 we have seen that such a constant shift does not affect any measurable quantity. Of course, Eq. (2.34), and hence Eq. (2.36) are only precise in the limit when the density of states $g(E)$ is so large that the range available for the appropriate choice of ΔE,

$$g^{-1}(E) \ll \Delta E \ll E, \tag{2.37}$$

is sufficiently broad: $g(E)E = E/\delta E \gg 1$.

In order to get some feeling of the functions $g(E)$ and $S(E)$ and the feasibility of the condition (2.37), and also to see whether the microcanonical distribution may be directly used for calculations of thermodynamic variables in particular systems, let us apply it to a microcanonical ensemble of many sets of $N \gg 1$ independent, similar harmonic oscillators with frequency ω. (Please note that the requirement of a virtually fixed energy is applied, in this case, to the total energy E_N of each *set* of oscillators, rather to a single oscillator—whose energy E may be virtually arbitrary, though certainly less than $E_N \sim NE \gg E$.) Basic quantum mechanics tells us[16] that the eigenenergies of such an oscillator form a discrete, equidistant spectrum:

$$E_m = \hbar\omega\left(m + \frac{1}{2}\right), \quad \text{where } m = 0, 1, 2, \ldots \tag{2.38}$$

If ω is kept constant, the ground-state energy $\hbar\omega/2$ does not contribute to any thermodynamic properties of the system[17], so that for the sake of simplicity we may take that point as the energy origin, and replace Eq. (2.38) with $E_m = m\hbar\omega$. Let us carry out an approximate analysis of the system for the case when its average energy per oscillator,

[16] See, e.g. *Part QM* sections 2.9 and 5.4.

[17] Let me hope that the reader knows that the ground-state energy *is* experimentally measurable—for example using the famous *Casimir effect*—see, e.g. *Part QM* section 9.1. (In section 5.5 below I will briefly discuss another method of experimental observation of that energy.)

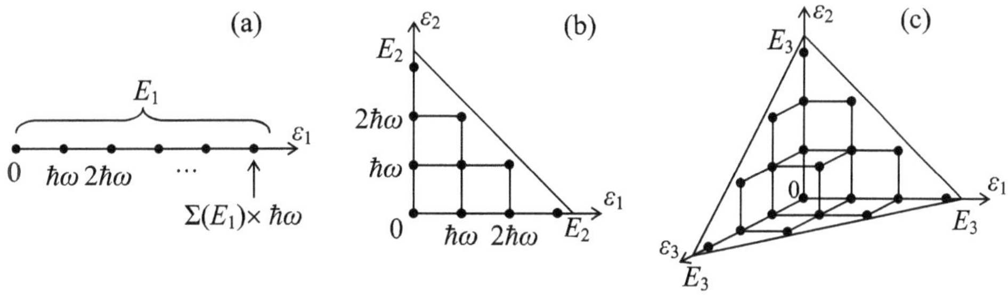

Figure 2.3. Calculating functions $\Sigma(E_N)$ for systems of (a) one, (b) two and (c) three harmonic oscillators.

$$E = \frac{E_N}{N}, \tag{2.39}$$

is much larger than the energy quantum $\hbar\omega$.

For one oscillator, the number of states with an energy ε_1 below a certain value $E_1 \gg \hbar\omega$ is evidently $\Sigma(E_1) \approx E_1/\hbar\omega \equiv (E_1/\hbar\omega)/1!$—see figure 2.3a. For two oscillators, all possible values of the total energy $(\varepsilon_1 + \varepsilon_2)$ below some level E_2 correspond to the points of a 2D square grid within the right triangle shown in figure 2.3b, giving $\Sigma(E_2) \approx (1/2)(E_2/\hbar\omega)^2 \equiv (E_2/\hbar\omega)^2/2!$. For three oscillators, the possible values of the total energy $(\varepsilon_1 + \varepsilon_2 + \varepsilon_3)$ correspond to those points of the 3D cubic grid that fit inside the right pyramid shown in figure 2.3c, giving $\Sigma(E_3) \approx (1/3)[(1/2)(E_3/\hbar\omega)^3] \equiv (E_3/\hbar\omega)^3/3!$, etc.

An evident generalization of these formulas to arbitrary N gives the number of states[18]

$$\Sigma(E_N) \approx \frac{1}{N!}\left(\frac{E_N}{\hbar\omega}\right)^N. \tag{2.40}$$

Differentiating this expression over the energy, we get

$$g(E_N) = \frac{d\Sigma(E_N)}{dE_N} = \frac{1}{(N-1)!}\frac{E_N^{N-1}}{(\hbar\omega)^N}, \tag{2.41}$$

so that

$$\begin{aligned} S_N(E_N) &= \ln g(E_N) + \text{const} \\ &= -\ln[(N-1)!] + (N-1)\ln E_N - N\ln(\hbar\omega) + \text{const}. \end{aligned} \tag{2.42}$$

For $N \gg 1$ we can ignore the difference between N and $(N-1)$ in both instances, and use the Stirling formula (2.27) to simplify this result as

[18] The coefficient $1/N!$ has the geometrical meaning of the (hyper)volume of the N-dimensional right pyramid with unit sides.

$$S_N(E) - \text{const} \approx N\left(\ln \frac{E_N}{N\hbar\omega} + 1\right) \approx N\left(\ln \frac{E}{\hbar\omega}\right) \equiv \ln\left[\left(\frac{E}{\hbar\omega}\right)^N\right]. \tag{2.43}$$

(The second, approximate step is only valid at very high $E/\hbar\omega$ ratios, when the logarithm in Eq. (2.43) is substantially larger than 1.) Returning for a second to the density of states, we see that in the limit $N \to \infty$, it is exponentially large:

$$g(E_N) = e^{S_N} \approx \left(\frac{E}{\hbar\omega}\right)^N, \tag{2.44}$$

so that the conditions (2.37) may be satisfied within a very broad range of ΔE.

Now we can use Eq. (2.43) to find all thermodynamic properties of the system, though only in the limit $E \gg \hbar\omega$. Indeed, according to thermodynamics, if the system's volume and number of particles are fixed, the derivative dS/dE is nothing else than the reciprocal temperature in thermal equilibrium—see Eq. (1.9). In our current case, we imply that the harmonic oscillators are distinct, for example by their spatial positions. Hence, even if we can speak of some volume of the system, it is certainly fixed[19]. Differentiating Eq. (2.43) over energy E, we get

$$\frac{1}{T} = \frac{dS_N}{dE_N} = \frac{N}{E_N} = \frac{1}{E}. \tag{2.45}$$

Reading this result backwards, we see that the average energy E of a harmonic oscillator equals T (i.e. $k_B T_K$ is SI units). At this point, a first-time student of thermodynamics should be very relieved to see that the counter-intuitive thermodynamic definition (1.9) of temperature does indeed correspond to what we all have known about this notion from our kindergarten physics courses.

The result (2.45) may be readily generalized. Indeed, in quantum mechanics a harmonic oscillator with eigenfrequency ω may by described by the Hamiltonian operator

$$\hat{H} = \frac{\hat{p}^2}{2\mathscr{m}} + \frac{\kappa\hat{q}^2}{2}, \tag{2.46}$$

where q is some generalized coordinate, and p the corresponding generalized momentum, $\mathscr{m}$ is oscillator's mass[20], and κ is the spring constant, so that $\omega = (\kappa/\mathscr{m})^{1/2}$. Since in the thermodynamic equilibrium the density matrix is always diagonal (see section 2.1 above) in the basis of stationary states m, the quantum-mechanical averages of the kinetic and potential energies may be found from Eq. (2.7):

$$\left\langle \frac{p^2}{2\mathscr{m}} \right\rangle = \sum_{m=0}^{\infty} W_m \langle m| \frac{\hat{p}^2}{2\mathscr{m}} |m\rangle, \qquad \left\langle \frac{\kappa q^2}{2} \right\rangle = \sum_{m=0}^{\infty} W_m \langle m| \frac{\kappa\hat{q}^2}{2} |m\rangle, \tag{2.47}$$

---

[19] By the same reason, the notion of pressure P in such a system is not clearly defined, and neither are any thermodynamic potentials but E and F.

[20] I am using this fancy font for the mass to avoid any chance of its confusion with the state number.

where W_m is the probability to occupy m-th energy level, and bra- and ket-vectors describe the stationary state corresponding to that level[21]. However, both classical and quantum mechanics teach us that for any m, the bra-kets under the sums in Eqs. (2.47), which represent the average kinetic and mechanical energies of the oscillator on its mth energy level, are equal to each other, and hence each of them is equal to $E_m/2$. Hence, even though we do not know the probability distribution W_m yet (it will be calculated in section 2.5 below), we may conclude that in the 'classical limit' $T \gg \hbar\omega$,

$$\left\langle \frac{p^2}{2m} \right\rangle = \left\langle \frac{\kappa q^2}{2} \right\rangle = \frac{T}{2}. \tag{2.48}$$

Now let us consider a system with an arbitrary number of degrees of freedom, described by a more general Hamiltonian[22]:

$$\hat{H} = \sum_j \hat{H}_j, \quad \text{with} \quad \hat{H}_j = \frac{\hat{p}_j^2}{2m_j} + \frac{\kappa_j \hat{q}_j^2}{2}, \tag{2.49}$$

with (generally, different) frequencies $\omega_j = (\kappa_j/m_j)^{1/2}$. Since the 'modes' (effective harmonic oscillators), contributing to this Hamiltonian, are independent, the result (2.48) is valid for each of the modes. This is the famous *equipartition theorem*: at thermal equilibrium with $T \gg \hbar\omega_j$, the average energy of each so-called *half-degree of freedom* (which is defined as either of variables p_j or q_j, giving a quadratic contribution to the system's Hamiltonian), is equal to $T/2$.[23] In particular, for each of three Cartesian component contributions to the kinetic energy of a free-moving particle, this theorem is valid for *any* temperature, because such components may be considered as 1D harmonic oscillators with vanishing potential energy, i.e. $\omega_j = 0$, so that condition $T \gg \hbar\omega_j$ is fulfilled at any temperature.

I believe that this case study of harmonic oscillator systems was a fair illustration of both the strengths and the weaknesses of the microcanonical ensemble approach[24]. On one hand, we could readily calculate virtually everything we wanted in the classical limit $T \gg \hbar\omega$, but calculations for an arbitrary $T \sim \hbar\omega$, though possible, are rather unpleasant, because for that, all vertical steps of the function $\Sigma(E_N)$ have to be carefully counted. In section 2.4, we will see that other statistical ensembles are much more convenient for such calculations.

---

[21] Note again that though we have committed the energy E_N of N oscillators to be fixed (in order to apply Eq. (2.36), valid only for a microcanonical ensemble at thermodynamic equilibrium), single oscillator's energy E in our analysis may be arbitrary—within limits $\hbar\omega \ll E < E_N \sim NT$.

[22] As a reminder, the Hamiltonian of any system whose classical Lagrangian function is an arbitrary quadratic form of its generalized coordinates and the corresponding generalized velocities, may be brought to the form (2.49) by an appropriate choice of 'normal coordinates' q_j which are certain linear combinations of the original coordinates—see, e.g. *Part CM* section 5.2.

[23] This also means that in the classical limit, the heat capacity of a system is equal to one half of the number of its half-degrees of freedom (in SI units, multiplied by k_B).

[24] The reader is strongly urged to solve problem 2.2, whose task is to do a similar calculation for another key ('two-level') physical system, and compare the results.

Let me conclude this discussion of entropy with a short notice on deterministic classical systems with just a few degrees of freedom (and even simpler mathematical objects called 'maps') that may exhibit essentially disordered behavior, called the *deterministic chaos*[25]. Such chaotic system may be approximately characterized by an entropy defined similarly to Eq. (2.29), where W_m are the probabilities to find it in different small regions of phase space, at well separated small time intervals. On the other hand, one can use an expression slightly more general than Eq. (2.29) to define the so-called *Kolmogorov* (or 'Kolmogorov–Sinai') *entropy* K that characterizes the speed of loss of information about the initial state of the system, and hence what is called the 'chaos' depth'. In the definition of K, the sum over m is replaced with the summation over all possible permutations $\{m\} = m_0, m_1, \ldots, m_{N-1}$ of small space regions, and W_m is replaced with $W_{\{m\}}$, the probability of finding the system in the corresponding regions m at time moment t_m, with $t_m = m\tau$, in the limit $\tau \to 0$, with $N\tau = \text{const}$. For chaos in the simplest objects, 1D maps, K is equal to the Lyapunov exponent $\lambda > 0$.[26] For systems of higher dimensionality, which are characterized by several Lyapunov exponents λ, the Kolmogorov entropy is equal to the phase-space average of the sum of all positive λ. These facts provide a much more practicable way of (typically, numerical) calculation of the Kolmogorov entropy than the direct use of its definition[27].

2.3 Maxwell's Demon, information, and computation

Before proceeding to other statistical distributions, I would like to address one more popular concern about Eq. (2.24), the direct relation between the entropy and information. Some physicists are still uneasy with entropy being nothing other than the (deficit of) information, though to the best of my knowledge, nobody has yet been able to suggest any experimentally verifiable difference between these two notions. Let me give one example of their direct relation[28]. Consider a volume containing just one molecule (considered as a point particle), and separated to two equal halves by a movable partition with a door that may be opened and closed at will, at no energy cost—see figure 2.4a. If the door is open and the system is in

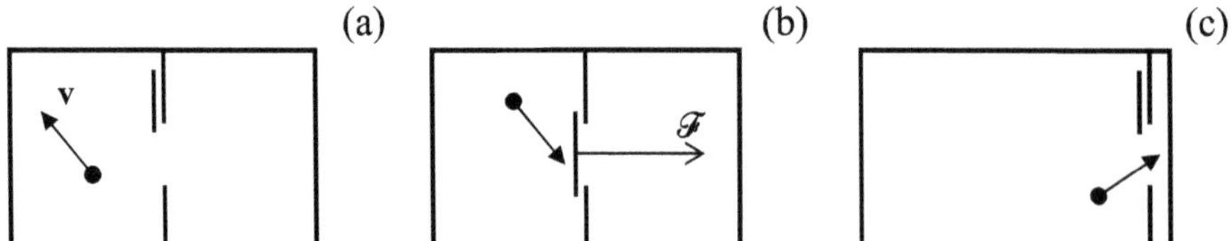

Figure 2.4. The Szilard engine: the volume with a single molecule (a) before and (b) after closing the door, and (c) after opening the door in the end of the expansion stage.

[25] See, e.g. *Part CM* chapter 9 and literature therein.

[26] For the definition of λ, see, e.g. *Part CM* Eq. (9.9).

[27] For more discussion, see, e.g. either section 6.2 of the monograph [3], or the monograph by Arnold and Avez, cited in section 2.1.

[28] This system is frequently called the *Szilard engine*, after L Szilard who performed its detailed theoretical analysis in 1929, but is essentially a straightforward extension of the thought experiment suggested by J Maxwell as early as in 1867.

thermodynamic equilibrium, we do not know on which side of the partition the molecule is. Here the disorder, i.e. the entropy has the largest value, and there is no way to get, from a large ensemble of such systems, any useful mechanical energy.

Now, let us consider that we know (as instructed by, in Lord Kelvin's formulation, an omniscient *Maxwell's Demon*) on which side of the partition the molecule is currently located. Then we may close the door trapping the molecule, so that its repeated impacts on the partition create, on the average, a pressure force $\mathscr{F}$ directed toward the empty part of the volume (in figure 2.4b, the right one). Now we can get from the molecule some mechanical work, say by allowing the force $\mathscr{F}$ to move the partition to the right, and picking up the resulting mechanical energy by some deterministic (zero-entropy) external mechanism. After the partition has been moved to the right end of the volume, we can open the door again (figure 2.4c), equalizing the molecule's average pressure on both sides of the partition, and then slowly move the partition back to the middle of the volume—without its resistance, i.e. without doing any substantial work. With the continuing help by the Maxwell's Demon, we can repeat the cycle again and again, and hence make the system to do unlimited mechanical work, fed 'only' by the molecule's thermal motion, and the information about its position—thus implementing the perpetual motion machine of the 2nd kind—see section 1.6. The fact that such heat engines do not exist means that getting any *new* information, at non-zero temperature (i.e. at a substantial thermal agitation of particles) has a non-zero energy cost.

In order to evaluate this cost, let us calculate the maximum work per cycle made by the Szilard engine (figure 2.4), assuming that it is constantly in the thermal equilibrium with a heat bath of temperature T. Eq. (2.21) tells us that the information supplied by the demon (on what exactly half of the volume contains the molecule) is exactly one bit, $I(2) = 1$. According to Eq. (2.24), this means that by getting this information we are changing the entropy of our system by

$$\Delta S_I = -\ln 2. \tag{2.50}$$

Now, it would be a mistake to plug this (negative) entropy change into Eq. (1.19). First, that relation is only valid for slow, reversible processes. Moreover (and more importantly), this equation, as well as its irreversible version (1.41), is only valid for a fixed statistical ensemble. The change ΔS_I does not belong to this category, and may be formally described by the change of the statistical ensemble—from the one consisting of all similar systems (experiments) with an unknown location of the molecule, to a new ensemble consisting of the systems with the molecule in its certain (in figure 2.4, left) half[29].

Now let us consider the slow expansion of the 'gas' after the door had been closed. At this stage, we do not need the Demon's help any longer (i.e. the statistical ensemble may be fixed), and we can indeed use the relation (1.19). At the assumed isothermal conditions (T = const), this relation may be integrated over the whole

[29] This procedure of *redefining the statistical ensemble* is the central point of the connection between the information theory and physics, and is crucial in particular for any (or rather any meaningful :-) discussion of measurements in quantum mechanics—see, e.g. *Part QM* sections 2.5 and 10.1.

expansion process, getting $\Delta Q = T\Delta S$. At the final position shown in figure 2.4c, the system's entropy should be the same as initially, i.e. before the door had been opened, because we again do not know where in the volume the molecule is. This means that the entropy was replenished, during the reversible expansion, from the heat bath, by $\Delta S = -\Delta S_I = +\ln 2$, so that $\Delta Q = T\Delta S = T \ln 2$. Since by the end of the whole cycle the internal energy E of the system is the same as before, all this heat should have gone into the mechanical energy obtained during the expansion. Thus the obtained work per cycle (i.e. for each obtained information bit) is $T \ln 2$ ($k_B T_K \ln 2$ in SI units), about 4×10^{-21} J at room temperature. This is exactly the energy cost of getting one bit of *new* information about a system at temperature T. The smallness of that amount on the everyday human scale has left the Szilard engine an academic exercise for almost a century. However, recently several such devices, of various physical nature, were implemented experimentally (with the Maxwell's Demon role played by an instrument measuring the position of the particle without a substantial effect on its motion), and the relation $\Delta Q = T \ln 2$ was proved, with a gradually increasing precision[30].

Actually, the discussion of the Maxwell's Demon resumed even earlier, in the 1960s in the context of energy consumption at numerical calculations. It was motivated by the exponential (*Moore's-law*) progress of digital integrated circuits, which has led in particular, to a fast reduction of the energy ΔE 'spent' (turned into heat) per one binary logic operation. In the current generations of semiconductor digital integrated circuits, the average ΔE is still above 10^{-17} J, i.e. still exceeds the room-temperature value of $T \ln 2 \approx 4\times10^{-21}$ J by several orders of magnitude. Still, some engineers still believe that thermodynamics imposes an important lower limit on ΔE and hence presents an insurmountable obstacle to the future progress of computation[31], so that the issue deserves at least a brief discussion.

Let me believe that the reader of these notes understands that, in contrast to naïve popular talk, computers do not *create* any new information; all they can do is to reshape ('process') it, *losing* most input information on the go. Indeed, any digital computation algorithm may be decomposed into simple, binary logical operations, each of them performed by a circuit called the *logic gate*. Some of these gates (e.g. logical NOT performed by inverters, as well as memory READ and WRITE operations) do not change the amount of information in the computer. On the other hand, such *information-irreversible* logic gates as two-input NAND (or NOR, or XOR, etc) actually erase one bit at each operation, because they turn two input bits into one output bit (figure 2.5a).

In 1961, R Landauer arrived at the conclusion that each logic operation should turn into heat at least energy

$$\Delta E_{\min} = T \ln 2 \equiv k_B T_K \ln 2. \tag{2.51}$$

[30] See, for example, [4–7].

[31] Unfortunately, in the 2000s this delusion resulted in a substantial and unjustified shift of electron device research resources toward using 'non-charge degrees of freedom' (such as spin)—as if they do not obey the general laws of statistical physics!

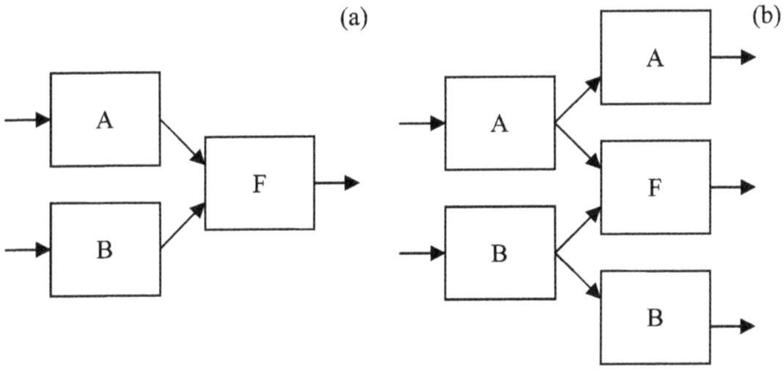

Figure 2.5. Simple examples of (a) irreversible and (b) potentially reversible logic circuits. Each rectangle presents a circuit storing one bit of information.

This result may be illustrated with the Szilard engine (figure 2.4), operated in a reversed cycle. At the first stage, with the door closed, it uses external mechanical work $\Delta E = T \ln 2$ to reduce the volume in which of the molecule is confined from V to $V/2$, pumping heat $\Delta Q = \Delta E$ into the heat bath. To model a logically-irreversible logic gate, let us now open the door in the partition, and thus lose 1 bit of information about the molecule's position. Then we will never get the work $T \ln 2$ back, because moving the partition back to the right, with the door open, takes place at zero average pressure. Hence, Eq. (2.51) gives a fundamental limit for energy loss (per bit) at the logically irreversible computation.

However, in 1973 C Bennett came up with convincing arguments that it is possible to avoid such energy loss by using only operations that are *reversible* not only physically, but also logically[32]. For that, one has to avoid any loss of information, i.e. any erasure of intermediate results, for example in the way shown in figure 2.5b.[33] In the end of all calculations, after the result has been copied into a memory, the intermediate results may be 'rolled back' through reversible gates to be eventually merged into a copy of input data, again without erasing a single bit. The minimal energy dissipation at such reversible calculation tends to zero as the operation speed is decreased, so that the average energy loss per bit may be less than the perceived 'fundamental thermodynamic limit' (2.51). The price to pay for this ultralow dissipation is a very high complexity of the hardware necessary for storage of all intermediate results. However, using irreversible gates sparsely, it may be possible to reduce the complexity dramatically, so that in future such *mostly* reversible computation may be able to reduce energy consumption in practical digital electronics[34].

[32] [8]; see also [9].

[33] For that, all gates have to be *physically reversible*, with no static power consumption. Such logic devices do exist, though they are still not very practicable—see, e.g. [10]. (Another reason for citing, rather reluctantly, my own paper is that it also gave a constructive proof that the reversible computation may also beat the perceived 'fundamental quantum limit', $\Delta E \Delta t > \hbar$, where Δt is the time of the binary logic operation.)

[34] Some types of *quantum computing* are also reversible—see, e.g. *Part QM* section 8.5 and references therein.

Before we leave the Maxwell's Demon behind, let me use it to revisit, for one more time, the relation between the reversibility of the classical and quantum mechanics of Hamiltonian systems and the irreversibility possible in thermodynamics and statistical physics. In the gedanken experiment shown in figure 2.4, the laws of mechanics governing the motion of the molecule are reversible at all times. Still, at partition's motion to the right, driven by molecule's impacts, the entropy grows, because the molecule picks up the heat $\Delta Q > 0$, and hence the entropy $\Delta S = \Delta Q/T > 0$, from the heat bath. The physical mechanism of this irreversible entropy (read: disorder) growth is the interaction of the molecule with uncontrollable components of the heat bath, and the resulting loss of information about the motion of the molecule. Philosophically, such emergence of irreversibility in large systems is a strong argument against the *reductionism*—a naïve belief that knowing the exact laws of Nature at the lowest, most fundamental level of its complexity, we can readily understand all the phenomena on the higher levels of its organization. In reality, the macroscopic irreversibility of large systems is a wonderful example of a new law (in this case, the 2nd law of thermodynamics) that becomes relevant on a substantially new level of complexity—without defying the lower-level laws. Without such new laws, very little of the higher-level organization of Nature may be understood.

2.4 Canonical ensemble and the Gibbs distribution

As was shown in section 2.2 (see also a few problems of the list given at the end of this chapter), the microcanonical distribution may be directly used for solving some simple problems. However, its further development, also due to J Gibbs, turns out to be much more convenient for calculations.

Let us consider a statistical ensemble of similar systems, each in thermal equilibrium with a much larger heat bath of temperature T (figure 2.6a). Such an ensemble is called *canonical*.

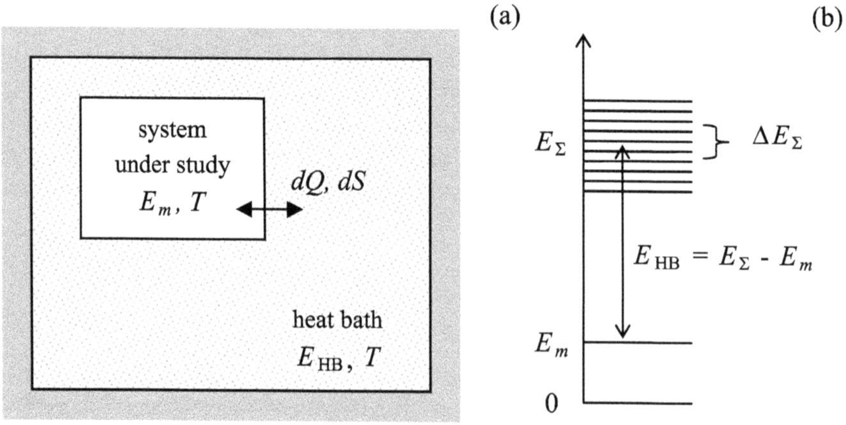

Figure 2.6. (a) System in a heat bath (a canonical ensemble member) and (b) the energy spectrum of the composite system (including the heat bath).

It is intuitively evident that if the heat bath is sufficiently large, any thermodynamic variables characterizing the system under study should not depend on the *heat bath's* environment. In particular, we may assume that the heat bath is thermally insulated; then the total energy E_Σ of the *composite system* (consisting of the system of our interest, plus the heat bath) does not change in time. For example, if the system of our interest is on its certain (say, mth) energy level, then the sum

$$E_\Sigma = E_m + E_{HB} \tag{2.52}$$

does not depend on m. Now let us partition this *canonical* ensemble into much smaller sub-ensembles, each being a *microcanonical* ensemble of composite systems whose total energy E_Σ is the same—as discussed in section 2.2, within a certain small energy interval $\Delta E_\Sigma \ll E_\Sigma$. Due to the very large size of each heat bath in comparison with that of the system under study, the heat bath' density of states g_{HB} is very high, and ΔE_Σ may be selected so that

$$\frac{1}{g_{HB}} \ll \Delta E_\Sigma \ll |E_m - E_{m'}| \ll E_{HB}, \tag{2.53}$$

where m and m' are any states of the system of our interest.

According to the microcanonical distribution, the probabilities to find the composite system, within this new ensemble, in any state are equal. Still, the heat bath energies $E_{HB} = E_\Sigma - E_m$ (figure 2.6b) of members of this microcanonical sub-ensemble may be different due to the difference in E_m. The probability $W(E_m)$ to find the system of our interest (within the selected sub-ensemble) on some energy level E_m is proportional to the number ΔM of the corresponding heat baths in the sub-ensemble. As figure 2.6b shows, in this case we may write $\Delta M = g_{HB}(E_{HB})\Delta E_\Sigma$. As a result, within the microcanonical ensemble with the total energy E_Σ,

$$W_m \propto \Delta M = g_{HB}(E_{HB})\Delta E_\Sigma = g_{HB}(E_\Sigma - E_m)\Delta E_\Sigma. \tag{2.54}$$

Let us simplify this expression further, using the Taylor expansion with respect to relatively small $E_m \ll E_\Sigma$. However, here we should be careful. As we have seen in section 2.2, the density of states of a large system is an extremely rapidly growing function of energy, so that if we applied the Taylor expansion directly to Eq. (2.54), the Taylor series would converge for very small E_m only. A much broader applicability range may be obtained by taking logarithm of both parts of Eq. (2.54) first:

$$\ln W_m = \text{const} + \ln [g_{HB}(E_\Sigma - E_m)] + \ln \Delta E_\Sigma = \text{const} + S_{HB}(E_\Sigma - E_m), \tag{2.55}$$

where the second equality results from the application of Eq. (2.36) to the heat bath, and $\ln \Delta E_\Sigma$ has been incorporated into the (inconsequential) constant. Now, we can Taylor-expand the (much more smooth) function of energy on the right-hand side, and limit ourselves to the two leading terms of the series:

$$\ln W_m \approx \text{const} + S_{HB}|_{E_m=0} - \frac{dS_{HB}}{dE_{HB}}|_{E_m=0} E_m. \tag{2.56}$$

But according to Eq. (1.9), the derivative participating in this expression is nothing else than the reciprocal heat bath temperature, which (due to the large bath size) does not depend on whether E_m is equal to zero or not. Since our system of interest is in the thermal equilibrium with the bath, this is also the temperature T of the system—see Eq. (1.8). Hence Eq. (2.56) is merely

$$\ln W_m = \text{const} - \frac{E_m}{T}. \tag{2.57}$$

This equality describes a substantial decrease of W_m as E_m is increased by $\sim T$, and hence our linear approximation (2.56) is virtually exact as soon as E_{HB} is much larger than T—the condition that is quite easy to satisfy, because as we have seen in section 2.2, the average energy of one degree of freedom of the system of interest is of the order of T.

Now we should be careful again, because so far Eq. (2.57) has been only derived for a sub-ensemble with a certain fixed E_Σ. However, since the right-hand side of Eq. (2.57) includes only E_m and T, which are independent of E_Σ, this relation is valid for all sub-ensembles of the canonical ensemble, and hence for the ensemble as the whole. Hence for the total probability to find our system of interest in state with energy E_m, in the canonical ensemble with temperature T, we can write

$$W_m = \text{const} \times \exp\left\{-\frac{E_m}{T}\right\} \equiv \frac{1}{Z}\exp\left\{-\frac{E_m}{T}\right\}. \tag{2.58}$$

This is the famous *Gibbs distribution* (sometimes called the 'canonical distribution')[35], which is arguably the summit of statistical physics[36], because it may be used for a straightforward (or at least conceptually straightforward :-) calculation of all statistical and thermodynamic variables of a vast range of systems. Before illustrating this, let us first calculate the coefficient Z participating in Eq. (2.58) for the general case. Requiring, in accordance with Eq. (2.4), the sum of all W_m to equal 1, we get

$$Z = \sum_m \exp\left\{-\frac{E_m}{T}\right\}, \tag{2.59}$$

where the summation is formally extended to all quantum states of the system, though in practical calculations, the sum may be truncated to include only the states that are noticeably occupied. The apparently humble normalization coefficient Z turns out to be so important for the relation between the Gibbs distribution (i.e. statistics) and

[35] The temperature dependence of the type $\exp\{-\text{const}/T\}$, especially when showing up in rates of certain events, e.g. chemical reactions, is also frequently called the *Arrhenius law*—after chemist S Arrhenius who has noticed this law in numerous experimental data. In all cases I am aware of, the Gibbs distribution is the underlying reason of the Arrhenius law; we will see several examples of that later in the course.

[36] This opinion is shared by several authoritative colleagues, including R Feynman—who climbs on this summit as early as the first page of his brilliant book [11]. (Despite its title, this monograph is a collection of lectures on a few diverse, mostly advanced topics of statistical physics, rather than a systematic course of the discipline, so that unfortunately it can hardly be used as the first textbook on the subject. However, I certainly recommend reading its first chapter to everybody.)

thermodynamics that it has a special name—or actually, two names: either the *statistical sum* or the *partition function*.

To demonstrate how important Z is, let us use the general Eq. (2.29) for the entropy to calculate it for the particular case of the canonical ensemble, i.e. the Gibbs distribution of the probabilities W_n:

$$S = -\sum_m W_m \ln W_m = \frac{\ln Z}{Z}\sum_m \exp\left\{-\frac{E_m}{T}\right\} + \frac{1}{ZT}\sum_m E_m \exp\left\{-\frac{E_m}{T}\right\}. \tag{2.60}$$

On the other hand, according to the general rule (2.7), the thermodynamic (i.e. ensemble-average) value E of the internal energy of the system is

$$E = \sum_m W_m E_m = \frac{1}{Z}\sum_m E_m \exp\left\{-\frac{E_m}{T}\right\}, \tag{2.61a}$$

so that the second term on the right-hand side of Eq. (2.60) is just E/T, while the first term equals just $\ln Z$, due to Eq. (2.59). (As a parenthetic remark, using the notion of *reciprocal temperature* $\beta \equiv 1/T$, Eq. (2.61*a*), with the account of Eq. (2.59), may be also rewritten as

$$E = -\frac{\partial(\ln Z)}{\partial\beta}. \tag{2.61b}$$

This formula is very convenient for calculations if our prime interest is the average energy E rather than F or W_n.) With these substitutions, Eq. (2.60) yields a very simple relation between the statistical sum and the entropy of the system:

$$S = \frac{E}{T} + \ln Z. \tag{2.62}$$

Now using Eq. (1.33), we see that Eq. (2.62) gives a straightforward way to calculate the free energy F of the system in the thermodynamic equilibrium from nothing other than its statistical sum (and temperature):

$$F \equiv E - TS = -T\ln Z. \tag{2.63}$$

The relations (2.61*b*) and (2.63) play the key role in the connection of statistics to thermodynamics, because they enable the calculation, from Z alone, of all thermodynamic potentials of the system in equilibrium, and hence of all other variables of interest, using the general thermodynamic relations—see especially the circular diagram shown in figure 1.6, and its discussion in section 1.4. Let me only note that in order to calculate the pressure P, e.g. from the second of Eqs. (1.35), we would need to know the explicit dependence of F, and hence of the statistical sum Z on the system's volume V. This would require the calculation, by appropriate methods of either classical or quantum mechanics, of the dependence of the eigenenergies E_m on the volume. Numerous examples of such calculations will be given later in the course.

As the final note of this section, Eqs. (2.59) and (2.63) may be readily combined to give a very elegant expression,

$$\exp\left\{-\frac{F}{T}\right\} = \sum_m \exp\left\{-\frac{E_m}{T}\right\}, \tag{2.64}$$

which offers a convenient interpretation of the free energy as a (rather specific, temperature-weighed) average of the eigenenergies of the system. One more convenient formula may be obtained by using Eq. (2.64) to rewrite the Gibbs distribution (2.58) in the form

$$W_m = \exp\left\{\frac{F - E_m}{T}\right\}. \tag{2.65}$$

In particular, this expression shows that that since all probabilities W_m are below 1, F is always lower than the lowest energy level. Also, Eq. (2.65) clearly shows that the probabilities W_m do not depend on the energy reference choice, i.e. on an arbitrary constant added to all E_m—and hence to E and F.

2.5 Harmonic oscillator statistics

The last property may be immediately used in our first example of the Gibbs distribution application to a particular, but very important system—the harmonic oscillator, for the more general case than was done in section 2.2, namely for an arbitrary relation between T and $\hbar\omega$.[37] Let us consider a canonical ensemble of similar oscillators, each in contact with a heat bath of temperature T. Selecting the ground-state energy $\hbar\omega/2$ for the origin of E, the oscillator eigenenergies (2.38) become $E_m = m\hbar\omega$ ($m = 0, 1,\ldots$), so that the Gibbs distribution (2.58) for probabilities of these states is

$$W_m = \frac{1}{Z}\exp\left\{-\frac{E_m}{T}\right\} = \frac{1}{Z}\exp\left\{-\frac{m\hbar\omega}{T}\right\}, \tag{2.66}$$

with the statistical sum

$$Z = \sum_{m=0}^{\infty} \exp\left\{-\frac{m\hbar\omega}{T}\right\} \equiv \sum_{m=0}^{\infty} \lambda^m, \quad \text{where } \lambda \equiv \exp\left\{-\frac{\hbar\omega}{T}\right\} \leqslant 1. \tag{2.67}$$

This is just the well-known infinite geometric progression (the 'geometric series')[38], with the sum

$$Z = \frac{1}{1-\lambda} \equiv \frac{1}{1 - e^{-\hbar\omega/T}}, \tag{2.68}$$

so that Eq. (2.66) yields

[37] A simpler task of making a similar calculation for another key quantum-mechanical object, the two-level system, is left for the reader's exercise.

[38] See, e.g. Eq. (A.11*b*).

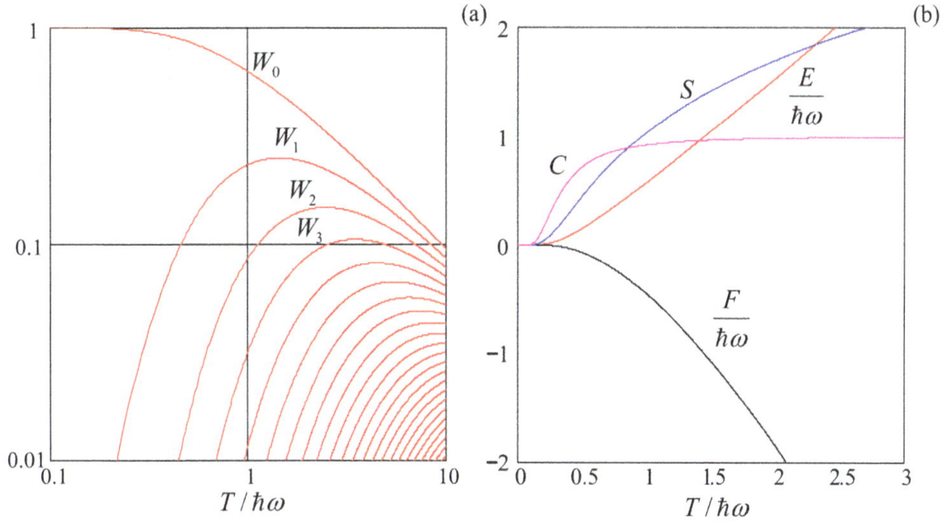

Figure 2.7. Statistical and thermodynamic parameters of a harmonic oscillator, as functions of temperature.

$$W_m = (1 - e^{-\hbar\omega/T})\ e^{-m\hbar\omega/T}. \tag{2.69}$$

Figure 2.7a shows these probabilities W_m, for several lowest energy levels, as functions of temperature, or rather of the $T/\hbar\omega$ ratio. It shows that the probability to find the oscillator in each particular state (except for the ground one, with $m = 0$) vanishes in both low- and high-temperature limits, and reaches its maximum value $W_m \sim 0.3/m$ at $T \sim m\hbar\omega$, so that the contribution $m\hbar\omega W_m$ of each excited level into the average oscillator energy E is always smaller than $\hbar\omega$.

This average energy may be calculated in any of two ways: either using Eq. (2.61*a*) directly:

$$E = \sum_{m=0}^{\infty} E_m W_m = \ (1 - e^{-\hbar\omega/T}) \sum_{m=0}^{\infty} m\hbar\omega\ e^{-m\hbar\omega/T}, \tag{2.70}$$

or (simpler) using Eq. (2.61*b*), as

$$E = -\frac{\partial}{\partial\beta} \ln Z = \frac{\partial}{\partial\beta} \ln(1 - \exp\{-\beta\hbar\omega\}), \quad \text{where } \beta \equiv \frac{1}{T}. \tag{2.71}$$

Both methods give (of course) the same result[39],

$$E = E(\omega, T) = \hbar\omega \frac{1}{e^{\hbar\omega/T} - 1}, \tag{2.72}$$

which is valid for arbitrary temperature and plays the key role in many fundamental problems of physics. The red line in figure 2.7b shows this result as a function of the

[39] It was first obtained in 1924 by S Bose, and is sometimes called the *Bose distribution*—a particular case of the *Bose–Einstein distribution* to be discussed in section 2.8 below.

normalized temperature. At low temperatures, $T \ll \hbar\omega$, the oscillator is predominantly in its lowest (ground) state, and its energy (on top of the constant energy $\hbar\omega/2$, which was used in our calculation as the reference) is exponentially small: $E \approx \hbar\omega \exp\{-\hbar\omega/T\} \ll T, \hbar\omega$. On the other hand, in the high-temperature limit the energy tends to T. This is exactly the result (a particular case of the equipartition theorem) that was obtained in section 2.2 from the microcanonical distribution. Please note how much simpler is the calculation starting from the Gibbs distribution, even for an arbitrary ratio $T/\hbar\omega$.

To complete the discussion of thermodynamic properties of the harmonic oscillator, we can calculate its free energy using Eq. (2.63):

$$F = T \ln \frac{1}{Z} = T \ln (1 - e^{-\hbar\omega/T}). \tag{2.73}$$

Now the entropy may be found from thermodynamics: either from the first of Eqs. (1.35), $S = -(\partial F/\partial T)_V$, or (even more easily) from Eq. (1.33): $S = (E - F)/T$. Both relations give, of course, the same result:

$$S = \frac{\hbar\omega}{T}\frac{1}{e^{\hbar\omega/T} - 1} - \ln (1 - e^{-\hbar\omega/T}). \tag{2.74}$$

Finally, since in the general case the dependence of the oscillator properties (essentially, of ω) on volume V in this problem is not specified, such variables as P, μ, G, W, and Ω are not defined, and we may calculate only the average heat capacity C per one oscillator:

$$C = \frac{\partial E}{\partial T} = \left(\frac{\hbar\omega}{T}\right)^2 \frac{e^{\hbar\omega/T}}{(e^{\hbar\omega/T} - 1)^2} \equiv \left[\frac{\hbar\omega/2T}{\sinh(\hbar\omega/2T)}\right]^2. \tag{2.75}$$

The calculated thermodynamic variables are shown in figure 2.7b. In the low-temperature limit ($T \ll \hbar\omega$), they all tend to zero. On the other hand, in the high temperature limit ($T \gg \hbar\omega$), $F \to -T \ln(T/\hbar\omega) \to -\infty$, $S \to \ln(T/\hbar\omega) \to +\infty$, and $C \to 1$ (in the SI units, $C \to k_B$). Note that the last limit is the direct corollary of the equipartition theorem: each of the two 'half-degrees of freedom' of the oscillator gives, in the classical limit, the same contribution $C = ½$ into its heat capacity.

Now let us use Eq. (2.69) to discuss the statistics of the quantum oscillator described by Hamiltonian (2.46), in the coordinate representation. Again using the density matrix' diagonality in thermodynamic equilibrium, we may use a relation similar to Eqs. (2.47) to calculate the probability density to find the oscillator at coordinate q:

$$\begin{aligned} w(q) &= \sum_{m=0}^{\infty} W_m w_m(q) = \sum_{m=0}^{\infty} W_m \,|\psi_m(q)|^2 \\ &= (1 - e^{-\hbar\omega/T}) \sum_{m=0}^{\infty} e^{-m\hbar\omega/T} \,|\psi_m(q)|^2, \end{aligned} \tag{2.76}$$

where $\psi_m(q)$ is the normalized eigenfunction of mth stationary state of the oscillator. Since each $\psi_m(q)$ is proportional to the Hermite polynomial[40] that requires at least m elementary functions for its representation, working out the sum in Eq. (2.76) is a bit tricky[41], but the final result is rather simple: $w(q)$ is just a normalized *Gaussian distribution* (the 'bell curve'),

$$w(q) = \frac{1}{(2\pi)^{1/2}\delta q}\exp\left\{-\frac{q^2}{2(\delta q)^2}\right\}, \tag{2.77}$$

with $\langle q\rangle = 0$, and

$$\langle q^2\rangle = (\delta q)^2 = \frac{\hbar}{2m\omega}\coth\frac{\hbar\omega}{2T}. \tag{2.78}$$

Since the function $\coth\xi$ tends to 1 at $\xi \to \infty$, and diverges as $1/\xi$ at $\xi \to 0$, Eq. (2.78) shows that the width δq of the coordinate distribution is constant (and equal to that, $(\hbar/2m\omega)^{1/2}$, of the ground-state wavefunction ψ_0) at $T \ll \hbar\omega$, and grows as $(T/m\omega^2)^{1/2} \equiv (T/\kappa)^{1/2}$ at $T/\hbar\omega \to \infty$.

As a sanity check, we may use Eq. (2.78) to write the following expression,

$$U \equiv \left\langle\frac{\kappa q^2}{2}\right\rangle = \frac{\hbar\omega}{4}\coth\frac{\hbar\omega}{2T} \to \begin{cases}\hbar\omega/4, & \text{at } T \ll \hbar\omega,\\ T/2, & \text{at } \hbar\omega \ll T,\end{cases} \tag{2.79}$$

for the average potential energy of the oscillator. In order to comprehend this result, let us recall that Eq. (2.72) for the average full energy E was obtained by counting it from the ground state energy $\hbar\omega/2$ of the oscillator. If we add this reference energy to that result, we get

$$E = \frac{\hbar\omega}{e^{\hbar\omega/T} - 1} + \frac{\hbar\omega}{2} \equiv \frac{\hbar\omega}{2}\coth\frac{\hbar\omega}{2T}. \tag{2.80}$$

We see that for arbitrary temperature, $U = E/2$, as was already discussed in section 2.2. This means that the average kinetic energy, equal to $E - U$, is also the same:

$$\left\langle\frac{p^2}{2m}\right\rangle = \left\langle\frac{\kappa q^2}{2}\right\rangle = \frac{E}{2} = \frac{\hbar\omega}{4}\coth\frac{\hbar\omega}{2T}. \tag{2.81}$$

In the classical limit $T \gg \hbar\omega$, both energies equal $T/2$, reproducing the equipartition theorem result (2.48).

2.6 Two important applications

The results of the previous section, especially Eq. (2.72), have enumerable applications in physics and related disciplines, though I have time for a brief discussion of only two of them.

[40] See, e.g. *Part QM* section 2.10.

[41] The calculation may be found, e.g. in *Part QM* section 7.2.

(i) *Blackbody radiation.* Let us consider a free-space volume V limited by non-absorbing (i.e. ideally reflecting) walls. Electrodynamics tells us[42] that the electromagnetic field in such a 'resonance cavity' may be represented as a sum of 'modes' with time evolution similar to that of the usual harmonic oscillator. If the volume V is large enough[43], the number of these modes within a small range dk of the wavevector magnitude k is

$$dN = \frac{gV}{(2\pi)^3}d^3k = \frac{gV}{(2\pi)^3}4\pi k^2 dk, \tag{2.82}$$

where for electromagnetic waves, the degeneracy factor g is equal to 2, due to their two different independent (e.g. linear) polarizations for the same wave vector $\mathbf{k}$. With the isotropic dispersion relation for waves in vacuum, $k = \omega/c$, the elementary volume d^3k corresponding to a small interval $d\omega$ is a spherical shell of small thickness $dk = d\omega/c$, and Eq. (2.82) yields

$$dN = \frac{2V}{(2\pi)^3}4\pi\frac{\omega^2 d\omega}{c^3} = V\frac{\omega^2}{\pi^2 c^3}d\omega. \tag{2.83}$$

On the other hand, quantum mechanics says[44] that the energy of such a 'field oscillator' is quantized in accordance with Eq. (2.38), so that at thermal equilibrium its average energy is described by Eq. (2.72). Plugging that result into Eq. (2.83), we see that the spectral density of the electromagnetic field's energy, per unit volume, is

$$u(\omega) \equiv \frac{E}{V}\frac{dN}{d\omega} = \frac{\hbar\omega^3}{\pi^2 c^3}\frac{1}{e^{\hbar\omega/T} - 1}. \tag{2.84}$$

This is the famous *Planck's blackbody radiation law*[45]. To understand why its common name mentions radiation, let us consider a small planar part, of area dA, of a surface that completely absorbs electromagnetic waves incident from any direction. (Such 'perfect black body' approximation may be closely approached using special experimental structures, especially in limited frequency intervals.) Figure 2.8 shows that if the arriving wave was planar, with the incidence angle θ, then the power $d\mathscr{P}_\theta(\omega)$ absorbed by the surface within a small frequency interval $d\omega$, would be equal to the radiation energy within the same frequency interval, contained inside an imaginary cylinder of the height c, base area $dA\cos\theta$, and hence volume $dV = c\ dA\cos\theta$:

$$d\mathscr{P}_\theta(\omega) = u(\omega)d\omega\ dV = u(\omega)d\omega\ c\ dA\cos\theta. \tag{2.85}$$

[42] See, e.g. *Part EM* section 7.8.

[43] In our current context, the volume should be much larger than $(c\hbar/T)^3$, where $c \approx 3\times10^8$ m s^{-1} is the speed of light. For room temperature ($T \approx k_B \times 300$ K $\approx 4 \times 10^{-21}$ J), this lower bound is of the order of 10^{-16} m^3.

[44] See, e.g. *Part QM* section 9.1.

[45] Let me hope the reader knows that this law was first suggested in 1900 by M Planck as an empirical fit for the experimental data on blackbody radiation, and this was the historic point at which the Planck constant $\hbar$ (or rather $h \equiv 2\pi\hbar$) was introduced—see, e.g. *Part QM* section 1.1.

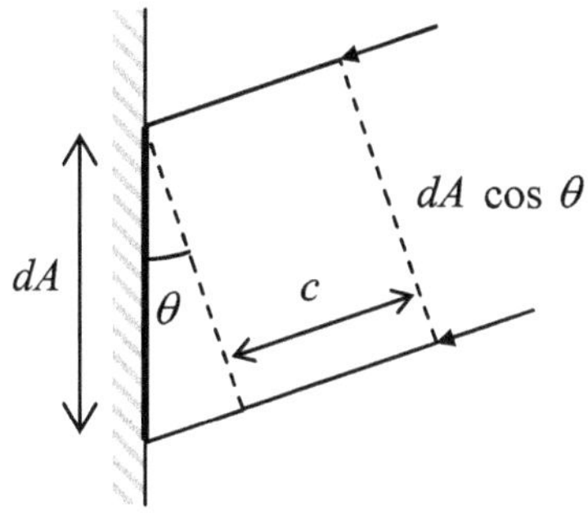

Figure 2.8. Calculating the relation between $d\mathscr{P}(w)$ and $u(\omega)d\omega$.

Since the thermally-induced field is isotropic, i.e. propagates equally in all directions, this result should be averaged over all solid angles within the polar angle interval $0 \leqslant \theta \leqslant \pi/2$:

$$\begin{aligned}\frac{d\mathscr{P}(\omega)}{dAd\omega} &= \frac{1}{4\pi}\int \frac{d\mathscr{P}_\theta(\omega)}{dAd\omega}d\Omega \\ &= cu(\omega)\frac{1}{4\pi}\int_0^{\pi/2} \sin\theta d\theta \int_0^{2\pi} d\varphi \cos\theta \\ &= \frac{c}{4}u(\omega).\end{aligned} \tag{2.86}$$

Hence the Planck's expression (2.84), multiplied by $c/4$, gives the power absorbed by such 'blackbody' surface. But at thermal equilibrium, this absorption has to be exactly balanced by the surface's own radiation, due to its finite temperature T.

I am confident that the reader is familiar with the main features of the Planck law (2.84), including its general shape (figure 2.9), with the low-frequency asymptote $u(\omega) \propto \omega^2$ (due to its historic significance bearing the special name of the *Rayleigh–Jeans law*), the exponential drop at high frequencies (the *Wien law*), and the resulting maximum of the function $u(\omega)$, reached at the frequency ω_{max} with

$$\hbar\omega_{\text{max}} \approx 2.82\ T, \tag{2.87}$$

i.e. at the wavelength $\lambda_{\text{max}} = 2\pi/k_{\text{max}} = 2\pi c/\omega_{\text{max}} \approx 2.22\ c\hbar/T$.

Still, I cannot help mentioning two particular values: one corresponding to the visible light ($\lambda_{\text{max}} \sim 500$ nm) for the Sun's visible surface temperature $T_K \approx 6000$ K, and another one corresponding to the mid-infrared range ($\lambda_{\text{max}} \sim 10$ μm) for the Earth's surface temperature $T_K \approx 300$ K. The balance of these two radiations, absorbed and emitted by the Earth, determines its surface temperature, and hence has the key importance for all life on our planet. This is why it is at the front and center of the current climate change discussions. As one more example, the cosmic microwave background (CMB) radiation, closely following the Planck law with $T_K = 2.726$ K (and hence having the maximum density at $\lambda_{\text{max}} \approx 1.9$ mm), and in particular its (slight) anisotropy, is a major source of data for all modern cosmology.

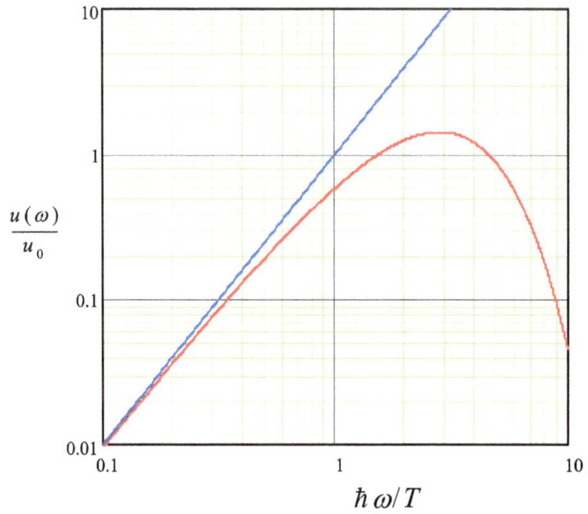

Figure 2.9. The frequency dependence of the blackbody radiation density, normalized by $u_0 \equiv T^3/\pi^2\hbar^2c^3$, according to the Planck law (red line) and the Rayleigh–Jeans law (blue line).

Now let us calculate the total energy E of this radiation in some volume V. It may be found from Eq. (2.84) by its integration over all frequencies[46,47]:

$$\begin{aligned} E &= V\int_0^\infty u(\omega)d\omega = V\int_0^\infty \frac{\hbar\omega^3}{\pi^2c^3}\frac{d\omega}{e^{\hbar\omega/T}-1} \\ &= \frac{VT^4}{\pi^2\hbar^3c^3}\int_0^\infty \frac{\xi^3 d\xi}{e^\xi - 1} = V\frac{\pi^2}{15\hbar^3c^3}T^4. \end{aligned} \tag{2.88}$$

Using Eq. (2.86) to recast Eq. (2.88) into the total power radiated by a blackbody surface, we get the well-known *Stefan* (or 'Stefan–Boltzmann') *law*

$$\frac{d\mathscr{P}}{dA} = \frac{\pi^2}{60\hbar^3c^2}T^4 \equiv \sigma T_K^4, \tag{2.89a}$$

where σ is the *Stefan–Boltzmann constant*

$$\sigma \equiv \frac{\pi^2}{60\hbar^3c^2}k_B^4 \approx 5.67\times 10^{-8}\frac{\text{W}}{\text{m}^2\text{K}^4}. \tag{2.89b}$$

By this time, the thoughtful reader should have an important concern ready: Eq. (2.84) and hence Eq. (2.88) are based on Eq. (2.72) for the average energy of each

[46] The last step in Eq. (2.88) uses a table integral, equal to $\Gamma(4)\zeta(4) = (3!)(\pi^4/90) = \pi^4/15$—see, e.g. Eqs. (A.35*b*), (A.10*b*), and (A.34*c*).

[47] Note that the heat capacity $C_V \equiv (\partial E/\partial T)_V$, following from Eq. (2.88), is proportional to T^3 at any temperature, and hence does not obey the trend $C_V \to$ const at $T \to \infty$. This is the result of the unlimited growth, with temperature, of the number of thermally-exited field oscillators with frequencies ω below $T/\hbar$.

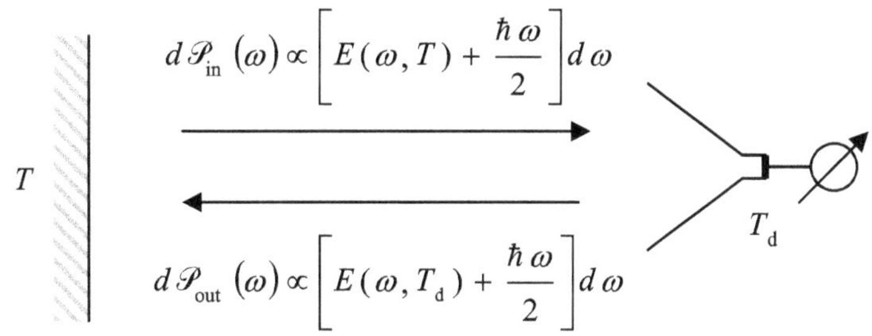

Figure 2.10. The power balance at the electromagnetic radiation power measurement.

oscillator, referred to as its ground-state energy $\hbar\omega/2$. However, the radiation power should not depend on the energy origin; why have we not included the ground energy of each oscillator into the integration (2.88), as we have done in Eq. (2.80)? The answer is that usual radiation detectors only measure the *difference* between the power $\mathscr{P}_{in}$ of the incident radiation (say, that of a blackbody surface with temperature T) and their own back-radiation power $\mathscr{P}_{out}$, corresponding to some effective temperature T_d of the detector—see figure 2.10. But however low T_d is, the temperature-independent contribution $\hbar\omega/2$ of the ground-state energy to the back radiation is always there. Hence, the term $\hbar\omega/2$ drops out from the balance, and cannot be detected—at least in this simple way. This is the reason why we had the right to ignore this contribution in Eq. (2.88)—very fortunately, because it would lead to the integral's divergence at its upper limit. However, let me repeat that the ground-state energy of the electromagnetic field oscillators is physically real—and important (see section 5.5 below).

One more interesting result may be deduced from the free energy F of the electromagnetic radiation, which may be also calculated by integration of Eq. (2.73) over all the modes, with the appropriate weight (2.83):

$$\begin{aligned} F &= \sum_{\omega} T \ln(1 - e^{-\hbar\omega/T}) \rightarrow \int_0^{\infty} T \ln(1 - e^{-\hbar\omega/T}) \frac{dN}{d\omega} d\omega \\ &= \int_0^{\infty} T \ln(1 - e^{-\hbar\omega/T}) \left(V \frac{\omega^2}{\pi^2 c^3} \right) d\omega. \end{aligned} \tag{2.90}$$

Representing $\omega^2 d\omega$ as $d(\omega^3)/3$, we can readily work out this integral by parts, reducing it to a table integral similar to that in Eq. (2.88), and getting a surprisingly simple result:

$$F = -V \frac{\pi^2}{45\hbar^3 c^3} T^4 \equiv -\frac{E}{3}. \tag{2.91}$$

Now we can use the second of the general thermodynamic relations (1.35) to calculate the pressure exerted by the radiation on the walls of the containing volume V:[48]

[48] This formula may be also derived from the expression for the forces exerted by the electromagnetic radiation on the walls (see, e.g. *Part EM* section 9.8), but the above calculation is much simpler.

$$P = -\left(\frac{\partial F}{\partial V}\right)_T = \frac{\pi^2}{45\hbar^3 c^3} T^4 = \frac{E}{3V}. \tag{2.92a}$$

Rewritten in the form,

$$PV = \frac{E}{3}, \tag{2.92b}$$

this result may be considered as the equation of state of the electromagnetic field, i.e. from the quantum-mechanical point of view, of the *photon gas*. Note that the equation of state (1.44) of the ideal classical gas may be represented in a similar form, but with a coefficient generally different from Eq. (2.92). Indeed, according to the equipartition theorem, for an ideal gas of non-relativistic atoms whose internal degrees of freedom are in their ground state, the whole energy is that of the three translational 'half-degrees of freedom', $E = 3N(T/2)$. Solving this equality for the product $NT = (2E/3)$, and plugging it into Eq. (1.44), we get a relation similar to Eq. (2.92), but with a twice larger factor before E. On the other hand, a *relativistic* treatment of the classical gas shows that Eq. (2.92) is valid for any gas in the ultra-relativistic limit, $T \gg mc^2$, where m is the rest mass of the gas particle. Evidently, photons (i.e. particles with $m = 0$) satisfy this condition[49].

Finally, let me note that Eq. (2.92) allows the following interesting interpretation. The last of Eqs. (1.60), being applied to Eq. (2.92), shows that in this particular case the grand potential Ω equals $(-E/3)$, so that according to Eq. (2.91), it is equal to F. But according to the definition of Ω, i.e. the first of Eqs. (1.60), this means that the chemical potential of the electromagnetic field excitations vanishes:

$$\mu = \frac{F - \Omega}{N} = 0. \tag{2.93}$$

In section 2.8 below, we will see that the same result follows from the comparison of Eq. (2.72) and the general Bose–Einstein distribution for arbitrary bosons. As a result, from the statistical point of view, photons may be considered as bosons with zero chemical potential.

(ii) *Specific heat of solids*. The heat capacity of solids is readily measurable, and in the early 1900s its experimentally observed temperature dependence served as an important test for emerging quantum theories. However, the theoretical calculation of C_V is not simple[50]—even for insulators, whose specific heat is due to thermally-induced vibrations of their crystal lattice alone[51]. Indeed, a solid may be treated as an elastic continuum only at low relatively frequencies. Such a continuum supports

[49] Note that according to Eqs. (1.44), (2.88), and (2.92), the difference between the equations of state of the photon gas and an ideal gas of non-relativistic particles, expressed in the more usual form $P = P(V, T)$, is much more dramatic: $P \propto T^4 V^0$ vs $P \propto T^1 V^{-1}$.

[50] Due to low temperature expansion of solids, the difference between their C_V and C_P is small.

[51] In good conductors (e.g. metals), specific heat is contributed (and at low temperatures, dominated) by free electrons—see section 3.3 below.

three different modes of mechanical waves with the same frequency ω, that obey similar, linear dispersion laws, $\omega = vk$, but the velocity $v = v_l$ for one of these modes (the *longitudinal sound*) is higher than that (v_t) of two other modes (the *transverse sound*)[52]. At such frequencies the wave mode density may be described by an evident generalization of Eq. (2.83):

$$dN = V\frac{1}{(2\pi)^3}\left(\frac{1}{v_l^3} + \frac{2}{v_t^3}\right)4\pi\omega^2 d\omega. \qquad (2.94a)$$

For what follows, it is convenient to rewrite this relation in a form similar to Eq. (2.83):

$$dN = \frac{3V}{(2\pi)^3}4\pi\frac{\omega^2 d\omega}{v^3}, \qquad \text{with } v \equiv \left[\frac{1}{3}\left(\frac{1}{v_l^3} + \frac{2}{v_t^3}\right)\right]^{-1/3}. \qquad (2.94b)$$

However, the basic wave theory shows[53] that as frequency ω of a sound wave in a periodic structure is increased so that its half-wavelength π/k approaches the crystal period d, the dispersion law $\omega(k)$ becomes nonlinear before the frequency reaches its maximum at $k = \pi/d$. To make things even more complex, 3D crystals are generally anisotropic, so that the dispersion law is different in different directions of the wave propagation. As a result, the exact statistics of thermally excited sound waves, and hence the heat capacity of crystals, is rather complex and specific for each particular crystal type.

In 1912, P Debye suggested an approximate theory of the temperature dependence of the specific heat, which is in surprisingly good agreement with experiment for many insulators, including polycrystalline and amorphous materials. In his model, the linear (*acoustic*) dispersion law $\omega = vk$, with the effective sound velocity v defined by the latter of Eqs. (2.94*b*), is assumed to be exact all the way up to some cutoff frequency ω_D, the same for all three wave modes. This *Debye frequency* may be defined by the requirement that the total number of acoustic modes, calculated within this model from Eq. (2.94*b*),

$$N = V\frac{1}{(2\pi)^3}\frac{3}{v^3}\int_0^{\omega_D} 4\pi\omega^2 d\omega = \frac{V\omega_D^3}{2\pi^2 v^3}, \qquad (2.95)$$

is equal to the universal number $N = 3nV$ of the degrees of freedom (and hence of independent oscillation modes) in a system of nV elastically coupled particles, where n is the atomic density of the crystal, i.e. the number of atoms per unit volume. Within this model, Eq. (2.72) immediately yields the following expression for the average energy and specific heat (in thermal equilibrium at temperature T):

$$E = V\frac{1}{(2\pi)^3}\frac{3}{v^3}\int_0^{\omega_D}\frac{\hbar\omega}{e^{\hbar\omega/T} - 1}4\pi\omega^2 d\omega = 3nVT \cdot D(x)_{x=T_D/T}, \qquad (2.96)$$

[52] See, e.g. *Part CM* section 7.7.

[53] See, e.g. *Part CM* section 6.3, in particular figure 6.5 and its discussion.

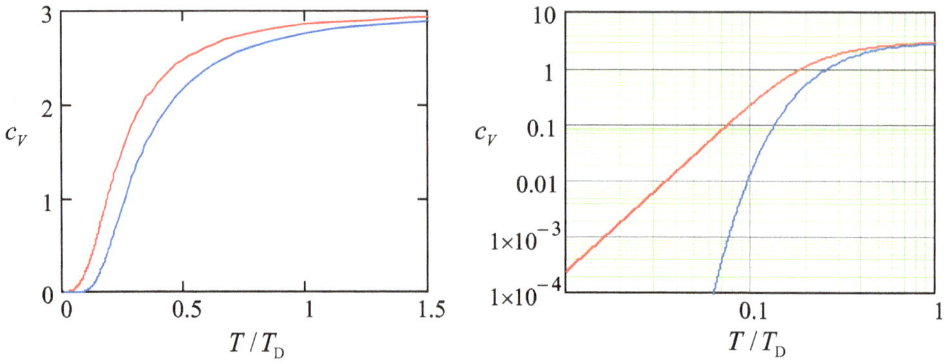

Figure 2.11. The specific heat as a function of temperature in the Debye (red lines) and Einstein (blue lines) models.

$$c_V \equiv \frac{C_V}{nV} = \frac{1}{nV}\left(\frac{\partial E}{\partial T}\right)_V = 3\left[D(x) - x\frac{dD(x)}{dx}\right]_{x=T_D/T}, \tag{2.97}$$

where $T_D \equiv \hbar\omega_D$ is called the *Debye temperature*[54], and

$$D(x) \equiv \frac{3}{x^3}\int_0^x \frac{\xi^3 d\xi}{e^\xi - 1} \to \begin{cases} 1, & \text{at } x \to 0, \\ \pi^4/5x^3, & \text{at } x \to \infty, \end{cases} \tag{2.98}$$

the *Debye function*. Red lines in figure 2.11 show the temperature dependence of the specific heat c_V (per particle) within the Debye model. At high temperatures, it approaches a constant value of 3, corresponding to the energy $E = 3nVT$, in accordance with the equipartition theorem for each of three degrees of freedom (i.e. six half-degrees of freedom) of each mode. (This value of c_V is known as the *Dulong–Petit law*.) In the opposite limit of low temperatures, the specific heat is much smaller:

$$c_V \approx \frac{12\pi^4}{5}\left(\frac{T}{T_D}\right)^3 \ll 1, \tag{2.99}$$

reflecting the reduction of the number of excited waves with $\hbar\omega < T$ as temperature is decreased.

As a historic curiosity, P Debye's work followed that by A Einstein, who had suggested (in 1907) a simpler model of crystal vibrations. In this model, all $3nV$ independent oscillatory modes of nV atoms of the crystal have approximately the same frequency, say ω_E, and Eq. (2.72) immediately yields

$$E = 3nV\frac{\hbar\omega_E}{e^{\hbar\omega_E/T} - 1}, \tag{2.100}$$

[54] In SI units, the Debye temperature T_D is of the order of a few hundred K for most simple solids (e.g. close to 430 K for aluminum and 340 K for copper), with somewhat lower values for crystals with heavy atoms (~105 K for lead), and reaches the highest value ~2200 K for diamond, with its relatively light atoms and very stiff lattice.

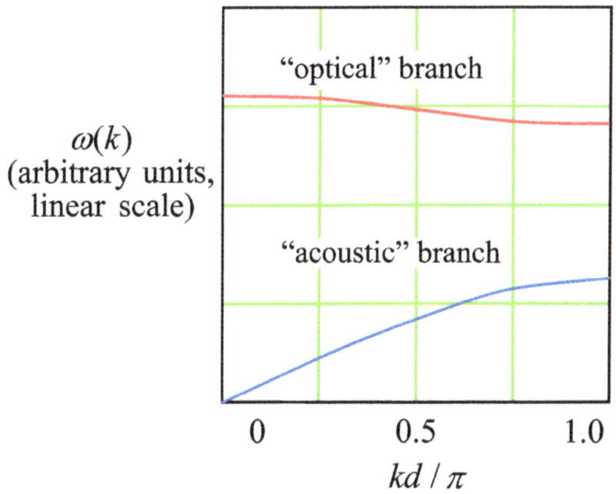

Figure 2.12. The dispersion relation for mechanical waves in a simple 1D model of a solid, with similar interparticle distances d, but alternating particle masses, plotted for a particular mass ratio $r = 5$—see *Part CM* chapter 6.

so that the specific heat is functionally similar to Eq. (2.75):

$$c_V \equiv \frac{1}{nV}\left(\frac{\partial E}{\partial T}\right)_V = 3\left[\frac{\hbar\omega_E/2T}{\sinh(\hbar\omega_E/2T)}\right]^2. \tag{2.101}$$

This dependence $c_V(T)$ is shown with blue lines in figure 2.11 (assuming, for the sake of simplicity, $\hbar\omega_E = T_D$). At high temperatures, this result does satisfy the universal Dulong–Petit law ($c_V = 3$), but at low temperatures the Einstein's model predicts a much faster (exponential) drop of the specific heart as temperature is reduced. (The difference between the Debye and Einstein models is not too spectacular on the linear scale, but in the log–log plot, shown on the right panel of figure 2.11, it is rather dramatic[55].) The Debye model is in much better agreement with experimental data for simple, monoatomic crystals, thus confirming the conceptual correctness of his wave-based approach.

Note, however, that when a genius such as A Einstein makes an error, there is usually some deep and important background behind it. Indeed, crystals with the basic cell consisting of atoms of two or more types (such as NaCl, etc), feature two or more separate branches of the dispersion law $\omega(k)$—see, e.g. figure 2.12. While the lower, 'acoustic' branch is virtually similar to those for monoatomic crystals, and may be approximated by the Debye model, $\omega = vk$, reasonably well, the upper ('optical'[56]) branch does not approach $\omega = 0$ at any k. Moreover, for large values of the atomic mass ratio r, the optical branches are almost flat, with virtually k-independent frequencies ω_0, which correspond to simple oscillations of each light

[55] This is why there is a general 'rule of thumb' in quantitative sciences: if you plot your data on a linear rather than log scale, you had better have a good excuse ready. (An example of a valid excuse: the variable you are plotting changes sign within the range you want to exhibit.)

[56] This term stems from the fact that at $k \to 0$, the mechanical waves corresponding to these branches have phase velocities $v_{ph} \equiv \omega(k)/k$ that are much higher than that of the acoustic waves, and may approach the speed of light. As a result, these waves can strongly interact with electromagnetic (practically, optical) waves of the same frequency, while the acoustic waves cannot.

atom between its heavy counterparts. For thermal excitations of such oscillations, and their contribution to the specific heat, the Einstein model (with $\omega_E = \omega_0$) gives a very good approximation, so that the specific heat may be well described by a sum of the Debye and Einstein laws (2.97) and (2.101), with appropriate weights.

2.7 Grand canonical ensemble and distribution

As we have seen, the Gibbs distribution is a very convenient way to calculate statistical and thermodynamic properties of systems with a fixed number N of particles. However, for systems in which N may vary, another distribution is preferable for some applications. Several examples of such situations (as well as the basic thermodynamics of such systems) have already been discussed in section 1.5. Perhaps even more importantly, statistical distributions for systems with variable N are also applicable to the ensembles of independent particles on a certain single-particle energy level—see the next section.

With this motivation, let us consider what is called the *grand canonical ensemble* (figure 2.13). It is similar to the canonical ensemble discussed in the previous section (see figure 2.6) in all aspects, besides that now the system under study and the heat bath (in this case typically called the *environment*) may exchange not only heat but also particles. In all system members of the ensemble, the environments are in both the thermal and chemical equilibrium, and their temperatures T and chemical potentials μ are equal.

Now let us assume that the system of interest is also in the chemical and thermal equilibrium with its environment. Then using exactly the same arguments as in section 2.4 (including the specification of microcanonical sub-ensembles with fixed E_Σ and N_Σ), we may generalize Eq. (2.55), taking into account that entropy S_{env} of the environment is now a function of not only its energy $E_{env} = E_\Sigma - E_{m,N}$,[57] but also of the number of particles $N_{env} = N_\Sigma - N$, with E_Σ and N_Σ fixed:

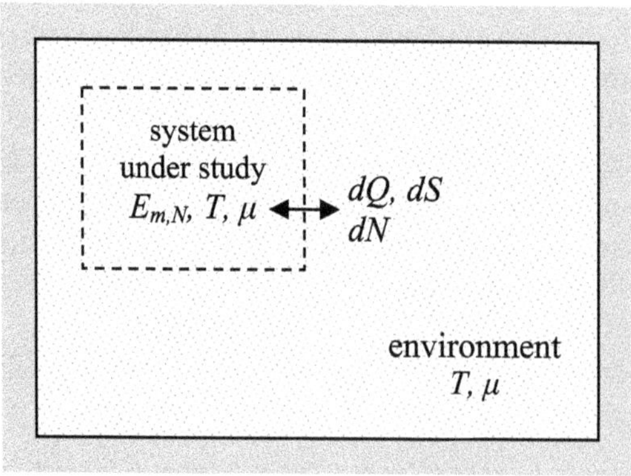

Figure 2.13. A member of the grand canonical ensemble.

[57] The additional index in the new notation $E_{m,N}$ for the energy of the system of interest reflects the fact that its spectrum is generally dependent on the number N of particles in it.

$$\begin{aligned}\ln W_{m,N} \propto \ln M &= \ln g_{\text{env}}(E_\Sigma - E_{m,N}, N_\Sigma - N) + \ln \Delta E_\Sigma \\ &= S_{\text{env}}(E_\Sigma - E_{m,N}, N_\Sigma - N) + \text{const} \\ &\approx S_{\text{env}}\Big|_{E_\Sigma, N_\Sigma} - \frac{\partial S_{\text{env}}}{\partial E_{\text{env}}}\Big|_{E_\Sigma, N_\Sigma} E_{m,N} - \frac{\partial S_{\text{env}}}{\partial N_{\text{env}}}\Big|_{E_\Sigma, N_\Sigma} N + \text{const.}\end{aligned} \quad (2.102)$$

In order to simplify this relation, we may rewrite Eq. (1.52) in the following equivalent form:

$$dS = \frac{1}{T}dE + \frac{P}{T}dV - \frac{\mu}{T}dN. \quad (2.103)$$

Hence, if the entropy S of a system is expressed as a function of E, V, and N, then

$$\left(\frac{\partial S}{\partial E}\right)_{V,N} = \frac{1}{T}, \quad \left(\frac{\partial S}{\partial V}\right)_{E,N} = \frac{P}{T}, \quad \left(\frac{\partial S}{\partial N}\right)_{E,V} = -\frac{\mu}{T}. \quad (2.104)$$

Applying the first one and the last one of these relations to Eq. (2.102), and using the equality of the temperatures T and chemical potentials μ in the system under study and its environment, at their equilibrium (as was discussed in section 1.5), we get

$$\ln W_{m,N} = S_{\text{env}}(E_\Sigma, N_\Sigma) - \frac{1}{T}E_{m,N} + \frac{\mu}{T}N + \text{const.} \quad (2.105)$$

Again, exactly as at the derivation of the Gibbs distribution in section 2.4, we may argue that since $E_{m,N}$, T and μ do not depend on the choice of environment's size, i.e. on E_Σ and N_Σ, the probability $W_{m,N}$ for a system to have N particles and be in mth quantum state in the whole grand canonical ensemble should also obey Eq. (2.105). As a result, we get the so-called *grand canonical distribution*:

$$W_{m,N} = \frac{1}{Z_G}\exp\left\{\frac{\mu N - E_{m,N}}{T}\right\}. \quad (2.106)$$

Just as in the case of the Gibbs distribution, the constant Z_G (most often called the *grand statistical sum*, but sometimes the 'grand partition function') should be determined from the probability normalization condition, now with the summation of probabilities $W_{m,N}$ over all possible values of both m and N:

$$Z_G = \sum_{m,N} \exp\left\{\frac{\mu N - E_{m,N}}{T}\right\}. \quad (2.107)$$

Now, using the general Eq. (2.29) to calculate the entropy for the distribution (2.106) (exactly as we did for the canonical ensemble), we get the following expression,

$$S = -\sum_{m,N} W_{m,N} \ln W_{m,N} = \ln Z_G + \frac{E}{T} - \frac{\mu\langle N\rangle}{T}, \quad (2.108)$$

which is evidently a generalization of Eq. (2.62).[58] We see that now the grand thermodynamic potential Ω (rather than the free energy F) may be expressed directly via the normalization coefficient Z_G:

$$\begin{aligned}\Omega &\equiv F - \mu\langle N\rangle = E - TS - \mu\langle N\rangle \\ &= T\ln\frac{1}{Z_G} = -T\ln\sum_{m,N}\exp\left\{\frac{\mu N - E_{m,N}}{T}\right\}.\end{aligned} \tag{2.109}$$

Finally, solving the last equality for Z_G, and plugging the result back into Eq. (2.106), we can rewrite the grand canonical distribution in the form

$$W_{m,N} = \exp\left\{\frac{\Omega + \mu N - E_{m,N}}{T}\right\}, \tag{2.110}$$

similar to Eq. (2.65) for the Gibbs distribution. Indeed, in the particular case when the number N of particles is fixed, $N = \langle N\rangle$, so that $\Omega + \mu N = \Omega + \mu\langle N\rangle \equiv F$, Eq. (2.110) is reduced right to Eq. (2.65).

2.8 Systems of independent particles

Now let us apply the general statistical distributions discussed above to a simple but very important case when the system we are considering consists of many similar particles whose explicit ('direct') interaction is negligible. As a result, each particular energy value $E_{m,N}$ of such a system may be represented as a sum of energies ε_k of the particles, where the index k numbers single-particle energy levels—rather than those of the whole system, as the index m does.

Let us start with the classical limit. In classical mechanics, the quantization effects are negligible, i.e. there is a formally infinite number of states k within each finite energy interval. However, it is convenient to keep, for the time being, the discrete-state language, with understanding that the average number $\langle N_k\rangle$ of particles in each of these states, usually called the *state occupancy*, is very small. In this case, we may apply the Gibbs distribution to the canonical ensemble of *single particles*, and hence use it with the substitution $E_{m,N} \to \varepsilon_k$, so that Eq. (2.58) becomes

$$\langle N_k\rangle = c\exp\left\{-\frac{\varepsilon_k}{T}\right\} \ll 1, \tag{2.111}$$

where the constant c should be found from the normalization condition:

$$\sum_k \langle N_k\rangle = 1. \tag{2.112}$$

[58] The average number of particles $\langle N\rangle$ is exactly what was called N in thermodynamics (see chapter 1), but I keep this explicit notation here to make a clear distinction between this average value of the variable, and its particular values participating in Eqs. (2.102)–(2.110).

This is the famous *Boltzmann distribution*[59]. Despite its formal similarity to the Gibbs distribution (2.58), let me emphasize the conceptual difference between these two important formulas. The Gibbs distribution describes the probability to find the *whole system* on the energy level E_m, and it is *always* valid—more exactly, for a canonical ensemble of systems in thermodynamic equilibrium. On the other hand, the Boltzmann distribution describes the occupancy of an energy level of a *single particle*, and, as we will see in a minute, for quantum particles may be valid *only in the classical limit* $\langle N_k \rangle \ll 1$, even if they do not interact directly.

The last fact may be surprising, because it may seem that as soon as particles of the system are independent, nothing prevents us from using the Gibbs distribution to derive Eq. (2.111), regardless of the value of $\langle N_k \rangle$. This is indeed true if the particles are *distinguishable*, i.e. may be distinguished from each other—say by their fixed spatial positions, or by the states of certain internal degrees of freedom (say, spin), or any other 'pencil mark'. However, it is an experimental fact that elementary particles of each particular type (say, electrons) are *identical* to each other, i.e. cannot be 'pencil-marked'[60]. For such particles we have to be more careful: even if they do not interact *explicitly*, there is still some *implicit* dependence in their behavior, which is especially evident for the so-called *fermions* (elementary particles with semi-integer spin): they obey the *Pauli exclusion principle* that forbids two identical particles to be in the same quantum state, even if they do not interact explicitly[61].

Note that here the term 'the same quantum state' carries a heavy meaning load here. For example, if two particles are confined to stay in different spatial positions (say, reliably locked in different boxes), they are *distinguishable* even if they are internally *identical*. Thus the Pauli principle, as well as other particle identity effects such as Bose–Einstein condensation, to be discussed in the next chapter, are important only when identical particles may move in the same spatial region. In order to emphasize this fact, it is common to use, instead of 'identical', a more precise (though grammatically rather unpleasant) adjective *indistinguishable*.

In order to take these effects into account, let us examine statistical properties of a system of many non-interacting but indistinguishable particles (at the first stage of calculation, either fermions or bosons) in equilibrium, applying the grand canonical distribution (2.109) to a very unusual particular grand canonical ensemble: a subset of particles in the same quantum state k (figure 2.14).

In this ensemble, the role of environment is played by the particles in all other states $k' \neq k$, because due to infinitesimal interactions, the particles may change their

[59] The distribution was first suggested in 1877 by the founding father of statistical physics, L Boltzmann. For the particular case when ε is the kinetic energy of a free classical particle (and hence has a continuous spectrum), it is reduced to the *Maxwell distribution* (see section 3.1 below), which was derived earlier—in 1860.

[60] This invites a natural question: what particles are 'elementary enough' for their identity? For example, protons and neutrons have an internal structure, in some sense consisting of quarks and gluons; can they be considered elementary? Next, if protons and neutrons are elementary, are atoms? molecules? What about *really* large molecules (such as proteins)? viruses? The general answer to these questions, given by quantum mechanics (or rather experiment :-), is that any particles/systems, no matter how large and complex they are, are identical if they have not only exactly the same internal structure, but also are exactly in the same internal quantum state—for example, in the ground state of all their internal degrees of freedom.

[61] For a more detailed discussion of this issue, see, e.g. *Part QM* section 8.1.

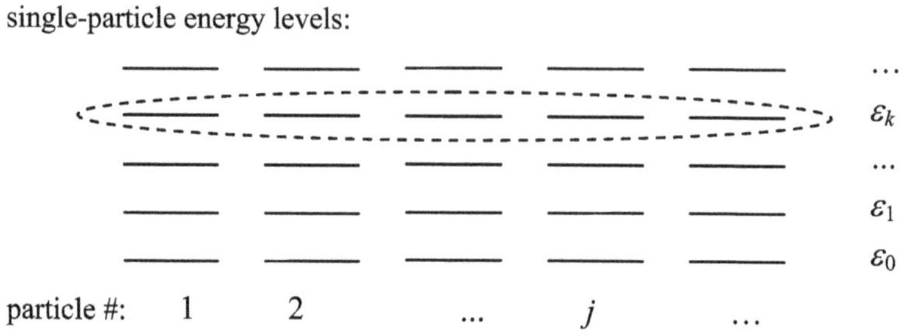

Figure 2.14. Grand canonical ensemble of particles in the same quantum state (with eigenenergy ε_k).

states. In the resulting equilibrium, the chemical potential μ and temperature T of the system should not depend on the state number k, though the grand thermodynamic potential Ω_k of the chosen particle subset may. Replacing N with N_k—the particular (not average!) number of particles in the selected kth state, and the particular energy value $E_{m,N}$ with $\varepsilon_k N_k$, we reduce the final form of Eq. (2.109) to

$$\Omega_k = -T \ln\left(\sum_{N_k} \exp\left\{\frac{\mu N_k - \varepsilon_k N_k}{T}\right\}\right) \equiv -T \ln\left[\sum_{N_k}\left(\exp\left\{\frac{\mu - \varepsilon_k}{T}\right\}\right)^{N_k}\right], \qquad (2.113)$$

where the summation should be carried out over all possible values of N_k. For the final calculation of this sum, the elementary particle type is essential.

On one hand, for *fermions*, obeying the Pauli principle, the numbers N_k in Eq. (2.113) may take only two values, either 0 (the state k is unoccupied) or 1 (the state is occupied), and the summation gives

$$\Omega_k = -T \ln\left[\sum_{N_k = 0,\,1}\left(\exp\left\{\frac{\mu - \varepsilon_k}{T}\right\}\right)^{N_k}\right] \equiv -T \ln\left(1 + \exp\left\{\frac{\mu - \varepsilon_k}{T}\right\}\right). \qquad (2.114)$$

Now the state occupancy may be calculated from the last of Eqs. (1.62)—in this case, with the (average) N replaced with $\langle N_k \rangle$:

$$\langle N_k \rangle = -\left(\frac{\partial \Omega_k}{\partial \mu}\right)_{T,\,V} = \frac{1}{e^{(\varepsilon_k - \mu)/T} + 1}. \qquad (2.115)$$

This is the famous *Fermi–Dirac distribution*, derived in 1926 independently by E Fermi and P Dirac.

On the other hand, *bosons* do not obey the Pauli principle, and for them the numbers N_k can take any non-negative integer values. In this case, Eq. (2.113) turns into the following equality:

$$\Omega_k = -T \ln \sum_{N_k=0}^{\infty}\left(\exp\left\{\frac{\mu - \varepsilon_k}{T}\right\}\right)^{N_k} \equiv -T \ln \sum_{N_k=0}^{\infty} \lambda^{N_k}, \quad \text{with } \lambda \equiv \exp\left\{\frac{\mu - \varepsilon_k}{T}\right\}. \qquad (2.116)$$

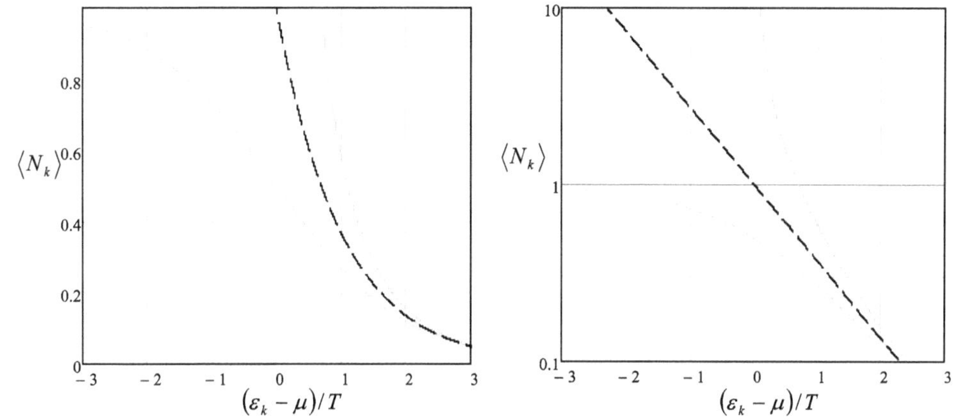

Figure 2.15. The Fermi–Dirac (blue line), Bose–Einstein (red line), and Boltzmann (dashed line) distributions for indistinguishable quantum particles. (The last distribution is valid only asymptotically, at $\langle N_k \rangle \ll 1$.)

This sum is just the usual geometric progression, which converges if $\lambda < 1$, giving

$$\Omega_k = -T \ln \frac{1}{1-\lambda} \equiv T \ln\left(1 - \exp\left\{\frac{\mu - \varepsilon_k}{T}\right\}\right), \quad \text{for } \mu < \varepsilon_k. \tag{2.117}$$

In this case the average occupancy, again calculated using Eq. (1.62) with N replaced with $\langle N_k \rangle$, obeys the *Bose–Einstein distribution*,

$$\langle N_k \rangle = -\left(\frac{\partial \Omega_k}{\partial \mu}\right)_{T,\,V} = \frac{1}{e^{(\varepsilon_k - \mu)/T} - 1}, \quad \text{for } \mu < \varepsilon_k, \tag{2.118}$$

which was derived in 1924 by S Bose (for the particular case $\mu = 0$) and generalized in 1925 by A Einstein for an arbitrary chemical potential. In particular, comparing Eq. (2.118) with Eq. (2.72), we see that harmonic oscillator *excitations*[62], each with energy $\hbar\omega$, may be considered as bosons, with the chemical potential equal to zero. We have already obtained this equality ($\mu = 0$) in a different way—see Eq. (2.93). Its physical interpretation is that the oscillator excitations may be created inside the system, so that there is no energy cost μ of moving them into the system under consideration from its environment.

The simple form of Eqs. (2.115) and (2.118), and their similarity (besides 'only' the difference of the signs before unity in their denominators), is one of the most beautiful results of physics. This similarity should not disguise the fact that the energy dependences of the occupancies $\langle N_k \rangle$, given by these two formulas, are rather different—see their linear and semi-log plots in figure 2.15.

---

[62] As the reader certainly knows, for the electromagnetic field oscillators, such excitations are called *photons*; for mechanical oscillation modes, *phonons*. It is important, however, not to confuse these mode *excitations* with the oscillators as such, and be very careful in prescribing to them certain spatial locations—see, e.g. *Part QM* section 9.1.

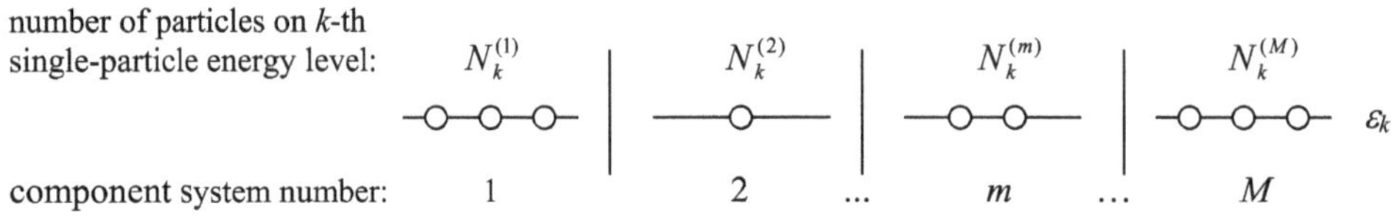

Figure 2.16. Composite system with a certain distribution of $N_k^{(\Sigma)}$ particles in kth state, between M component systems.

In the Fermi–Dirac statistics, the level occupancy is finite (and below 1) at any energy, while in the Bose–Einstein it may be above 1, and even diverges at $\varepsilon_k \to \mu$. However, for any of these distributions, as temperature is increased, it eventually becomes much larger than the difference $(\varepsilon_k - \mu)$ for all k. In this limit, $\langle N_k \rangle \ll 1$, both quantum distributions coincide with each other, as well as with the classical Boltzmann distribution (2.111) with $c = \exp\{\mu/T\}$:

$$\langle N_k \rangle \to \exp\left\{\frac{\mu - \varepsilon_k}{T}\right\}, \quad \text{for } \langle N_k \rangle \to 0. \tag{2.119}$$

This distribution (also shown in figure 2.15) may be, therefore, understood as the high-temperature limit for indistinguishable particles of both sorts.

A natural question now is how to find the chemical potential μ participating in Eqs. (2.115), (2.118), and (2.119). In the grand canonical ensemble as such (figure 2.13), with the number of particles variable, the value of μ is imposed by the system's environment. However, both the Fermi–Dirac and Bose–Einstein distributions are also approximately applicable (in thermal equilibrium) to systems with a *fixed but very large* number N of particles. In these conditions, the role of the environment for some subset of $N' \ll N$ particles is essentially played by the remaining $N - N'$ particles. In this case, μ may be found by the calculation of $\langle N \rangle$ from the corresponding distribution, and then requiring it to be equal to the genuine number of particles in the system. In the next section, we will perform such calculations for several particular systems.

For that and other applications, it will be convenient for us to have ready formulas for the entropy S of a general (i.e. not necessarily equilibrium) state of systems of independent Fermi or Bose particles, expressed not as a function of W_m of the whole system, as Eq. (2.29) does, but via the occupancy numbers $\langle N_k \rangle$. For that, let us consider an ensemble of *composite systems*, each consisting of $M \gg 1$ similar but distinct *component systems*, numbered by index $m = 1, 2, \ldots M$, with independent (i.e. not *explicitly* interacting) particles. We will assume that though in each of the M component systems, the number $N_k^{(m)}$ of particles in its kth quantum state may be different (figure 2.16), their total number $N_k^{(\Sigma)}$ in the composite system is fixed. As a result, the total energy of the composite system is fixed as well,

$$\sum_{m=1}^{M} N_k^{(m)} = N_k^{(\Sigma)} = \text{const}, \qquad E_k = \sum_{m=1}^{M} N_k^{(m)} \varepsilon_k = N_k^{(\Sigma)} \varepsilon_k = \text{const}, \tag{2.120}$$

so that an ensemble of many such composite systems (with the same k), in equilibrium, is microcanonical.

According to Eq. (2.24*a*), the average entropy S_k per component system may be calculated as

$$S_k = \lim_{M\to\infty}\frac{\ln M_k}{M}, \tag{2.121}$$

where M_k is the number of possible different ways such composite system (with fixed $N_k^{(\Sigma)}$) may be implemented. Let us start the calculation of M_k for Fermi particles—for which the Pauli principle is valid. Here the level occupancies $N_k^{(m)}$ may be only equal 0 or 1, so that the distribution problem is solvable only if $N_k^{(\Sigma)} \leqslant M$, and evidently equivalent to the choice of $N_k^{(\Sigma)}$ balls (in *arbitrary* order) from the total number of M *distinct* balls. Comparing this formulation with the definition of the binomial coefficient[63], we immediately get

$$M_k = {}^{M}C_{N_k^{(\Sigma)}} = \frac{M!}{(M - N_k^{(\Sigma)})!N_k^{(\Sigma)}!}. \tag{2.122}$$

From here, using the Stirling formula (again, in its simplest form (2.27)), we get

$$S_k = -\langle N_k\rangle\ \ln\langle N_k\rangle - (1 - \langle N_k\rangle)\ \ln(1 - \langle N_k\rangle), \tag{2.123}$$

where

$$\langle N_k\rangle \equiv \lim_{M\to\infty}\frac{N_k^{(\Sigma)}}{M} \tag{2.124}$$

is exactly the average occupancy of the kth single-particle level in each system that was discussed earlier in this section. Since for a Fermi system, $\langle N_k\rangle$ is always somewhere between 0 and 1; as a result, its entropy (2.123) is always positive.

In the Bose case, where the Pauli limitation is not valid, the number $N_k^{(m)}$ of particles on the kth energy level in each of the systems is an arbitrary (non-negative) integer. Let us consider $N_k^{(\Sigma)}$ particles and $(M - 1)$ partitions (shown by vertical lines in figure 2.16) between M systems as $(M - 1 + N_k^{(\Sigma)})$ similar mathematical objects ordered along one axis. (Each specific location of the partitions evidently fixes all $N_k^{(m)}$.) Hence M_k may be calculated as the number of possible ways to distribute the $(M - 1)$ indistinguishable partitions among these $(M - 1 + N_k^{(\Sigma)})$ ordered objects, i.e. as the following binomial coefficient[64]:

$$M_k = {}^{M+N_k-1}C_{M-1} = \frac{\left(M - 1 + N_k^{(\Sigma)}\right)!}{(M - 1)!N_k^{(\Sigma)}!}. \tag{2.125}$$

Applying the Stirling formula (2.27) again, we get the following result,

$$S_k = -\langle N_k\rangle\ \ln\langle N_k\rangle + (1 + \langle N_k\rangle)\ \ln(1 + \langle N_k\rangle), \tag{2.126}$$

[63] See, e.g. Eq. (A.5).

[64] See also Eq. (A.7).

which again differs from the Fermi case (2.123) 'only' by the signs in the second term, and is valid for any positive $\langle N_k \rangle$.

Expressions (2.123) and (2.126) are valid for an arbitrary (possibly non-equilibrium) case; they may be also used for an alternative derivation of the Fermi–Dirac (2.115) and Bose–Einstein (2.118) distributions, which are valid only in equilibrium. For that, we may use the method of Lagrange multipliers, requiring (just as was done in section 2.2) the total entropy of a system of N independent, similar particles,

$$S = \sum_k S_k, \tag{2.127}$$

considered as a function of state occupancies $\langle N_k \rangle$, to attain its maximum, under the conditions of fixed total number of particles N and total energy E:

$$\sum_k \langle N_k \rangle = N = \text{const}, \qquad \sum_k \langle N_k \rangle \varepsilon_k = E = \text{const}. \tag{2.128}$$

The completion of this calculation is left for the reader's exercise.

In the classical limit, when the average occupancies $\langle N_k \rangle$ of all states are small, both the Fermi and Bose expressions for S_k tend to the same limit

$$S_k = -\langle N_k \rangle \ln \langle N_k \rangle, \qquad \text{for } \langle N_k \rangle \ll 1. \tag{2.129}$$

This expression, frequently referred to as the *Boltzmann* (or 'classical') *entropy*, might be also obtained, for arbitrary $\langle N_k \rangle$, directly from the functionally similar Eq. (2.29), by considering an ensemble of systems, each consisting of just one classical particle, so that $E_m \rightarrow \varepsilon_k$ and $W_m \rightarrow \langle N_k \rangle$. Let me emphasize again that for indistinguishable particles, such identification is generally (i.e. at $\langle N_k \rangle \sim 1$) illegitimate even if the particles do not interact explicitly. As we will see in the next chapter, the indistinguishability may affect statistical properties of similar particles even in the classical limit.

2.9 Problems

Problem 2.1. A famous example of the macroscopic irreversibility was suggested in 1907 by P Ehrenfest. Two dogs share $2N \gg 1$ fleas. Each flea may jump onto another dog, and the rate Γ of such events (i.e. the probability of jumping per unit time) does not depend either on time, or on the location of other fleas. Find the time evolution of the average number of fleas on a dog, and of the flea-related part of the total dogs' entropy (at arbitrary initial conditions), and prove that the entropy can only grow[65].

Problem 2.2. Use the microcanonical distribution to calculate thermodynamic properties (including the entropy, all relevant thermodynamic potentials, and the heat capacity), of a two-level system in thermodynamic equilibrium with its environment, at temperature T that is comparable with the energy gap Δ. For

[65] This is essentially a simpler (and funnier) version of the particle scattering model used by L Boltzmann to prove his famous *H-theorem* (1872). Besides the historic significance of that theorem, the model used in it (see section 6.2 below) is as cartoonish, and hence not more general.

each variable, sketch its temperature dependence, and find asymptotic values (or trends) in the low-temperature and high-temperature limits.

Hint: The two-level system is defined as any system with just two different stationary states, whose energies (say, E_0 and E_1) are separated by a gap $\Delta \equiv E_1 - E_0$. Its most popular (but by no means the only!) example is a spin-½ particle, e.g. an electron, in an external magnetic field[66].

Problem 2.3. Solve the previous problem using the Gibbs distribution. Also, calculate the probabilities of the energy level occupation, and give physical interpretations of your results, in both temperature limits.

Problem 2.4. Calculate the low-field magnetic susceptibility χ of a quantum spin-½ particle with gyromagnetic ratio γ, in thermal equilibrium with environment at temperature T, neglecting its orbital motion. Compare the result with that for a classical spontaneous magnetic dipole $\mathscr{m}$ of a fixed magnitude $\mathscr{m}_0$, free to change its direction in space.

Hint: The low-field magnetic susceptibility of a single particle is defined[67] as

$$\chi = \frac{\partial\langle \mathscr{m}_z\rangle}{\partial\mathscr{H}}\Big|_{\mathscr{H}\to 0},$$

where the z-axis is aligned with the direction of the external magnetic field $\mathscr{H}$.

Problem 2.5. Calculate the low-field magnetic susceptibility of a particle with an arbitrary spin s, neglecting its orbital motion. Compare the result with the solution of the previous problem.

Hint: Quantum mechanics[68] tells us that the Cartesian component $\mathscr{m}_z$ of the magnetic moment of such a particle, in the direction of the applied field, has $(2s + 1)$ stationary values:

$$\mathscr{m}_z = \gamma\hbar s_m, \quad \text{with} \quad s_m = -s, \quad -s+1, \ldots, s-1, \quad s,$$

where γ is the gyromagnetic ratio of the particle, and $\hbar$ is the Planck's constant.

Problem 2.6.* Analyze the possibility of using a system of non-interacting spin-½ particles, placed into a strong, controllable external magnetic field, for refrigeration.

Problem 2.7. The rudimentary 'zipper' model of DNA replication is a chain of N links that may be either open or closed—see the figure below. Opening a link increases the system's energy by $\Delta > 0$; a link may change its state (either open or

---

[66] See, e.g. *Part QM* sections 4.6 and 5.1, for example Eq. (4.167).

[67] This 'atomic' (or 'molecular') susceptibility should be distinguished from the 'volumic' susceptibility $\chi_m \equiv \partial\mathscr{M}_z/\partial\mathscr{H}$, where $\mathscr{M}$ is the magnetization, i.e. the magnetic moment of a unit volume of a system—see, e.g. *Part EM* Eq. (5.111). For a uniform medium with $n \equiv N/V$ non-interacting dipoles per unit volume, $\chi_m = n\chi$.

[68] See, e.g. *Part QM* section 5.7, in particular Eq. (5.169).

close) only if all links to the left of it are already open, while those on the right of it, are already closed. Calculate the average number of open links at thermal equilibrium, and analyze its temperature dependence, especially for the case $N \gg 1$.

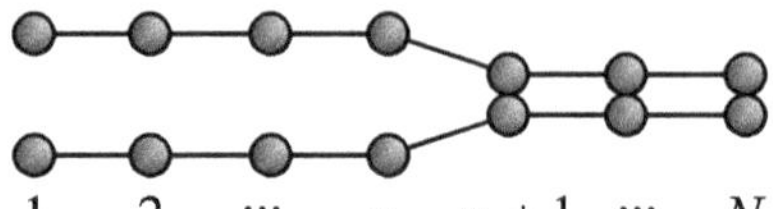

Problem 2.8. Use the microcanonical distribution to calculate the average entropy, energy, and pressure of a classical particle of mass m, with no internal degrees of freedom, free to move in volume V, at temperature T.

Hint: Try to make a more accurate calculation than has been done in section 2.2 for the system of N harmonic oscillators. For that you will need to know the volume V_d of an d-dimensional hypersphere of the unit radius. To avoid being too cruel, I am giving it to you:

$$V_d = \pi^{d/2}/\Gamma\left(\frac{d}{2} + 1\right),$$

where $\Gamma(\xi)$ is the gamma-function[69].

Problem 2.9. Solve the previous problem starting from the Gibbs distribution.

Problem 2.10. Calculate the average energy, entropy, free energy, and the equation of state of a classical 2D particle (without internal degrees of freedom), free to move within area A, at temperature T, starting from:

(i) the microcanonical distribution, and
(ii) the Gibbs distribution.

Hint: For the equation of state, make the appropriate modification of the notion of pressure.

Problem 2.11. A quantum particle of mass m is confined to free motion along a 1D segment of length a. Using any approach you like, calculate the average force the particle exerts on the 'walls' (ends) of such '1D potential well' in thermal equilibrium, and analyze its temperature dependence, focusing on the low-temperature and high-temperature limits.

Hint: You may consider the series $\Theta(\xi) \equiv \sum_{n=1}^{\infty} \exp\{-\xi n^2\}$ a known function of ξ.[70]

[69] For its definition and main properties, see, e.g. Eqs. (A.33)–(A.36).

[70] It may be reduced to the so-called elliptic theta-function $\theta_3(z, \tau)$ for a particular case $z = 0$—see, e.g. section 16.27 in the Abramowitz–Stegun handbook cited in section A.16(ii). However, you do not need that (or any other) handbook to solve this problem.

Problem 2.12.* Rotational properties of diatomic molecules (such as N_2, CO, etc) may be reasonably well described by the 'dumbbell' model: two point particles, of masses m_1 and m_2, with a fixed distance d between them. Ignoring the translational motion of the molecule as the whole, use this model to calculate its heat capacity, and spell out the result in the limits of low and high temperatures. Discuss whether your solution is valid for the so-called *homonuclear* molecules, consisting of two similar atoms, such as H_2, O_2, N_2, etc.

Problem 2.13. Calculate the heat capacity of a heteronuclear diatomic molecule, using the simple model described in the previous problem, but now assuming that the rotation is confined to one plane[71].

Problem 2.14. A classical, rigid, strongly elongated body (such as a thin needle), is free to rotate about its center of mass, and is in thermal equilibrium with its environment. Are the angular velocity vector $\boldsymbol{\omega}$ and the angular momentum vector $\mathbf{L}$, on the average, directed along the elongation axis of the body, or normal to it?

Problem 2.15. Two similar classical electric dipoles, of a fixed magnitude d, are separated by a fixed distance r. Assuming that each dipole moment $\mathbf{d}$ may take any special direction, and that the system is in thermal equilibrium, write the general expressions for its statistical sum Z, average interaction energy E, heat capacity C, and entropy S, and calculate them explicitly in the high-temperature limit.

Problem 2.16. A classical 1D particle of mass m, residing in the potential well

$$U(x) = \alpha \, |x|^{\gamma}, \quad \text{with } \gamma > 0,$$

is in thermal equilibrium with its environment, at temperature T. Calculate the average values of its potential energy U and the full energy E using two approaches:

(i) directly from the Gibbs distribution, and
(ii) using the virial theorem of classical mechanics[72].

Problem 2.17. For a thermally-equilibrium ensemble of slightly anharmonic classical 1D oscillators, with mass m and potential energy

$$U(q) = \frac{\kappa}{2}x^2 + \alpha x^3,$$

with small coefficient α, calculate $\langle x \rangle$ in the first approximation in low temperature T.

Problem 2.18.* A small conductor (in this context, usually called the *single-electron island*) is placed between two conducting electrodes, with voltage V applied between them. The gap between one of the electrodes and the island is so narrow that electrons may tunnel quantum-mechanically through this gap (the 'weak tunnel

[71] This is a reasonable model of the constraints imposed on small atomic groups (e.g. ligands) by their atomic environment inside some large molecules.
[72] See, e.g. *Part CM* problem 1.12.

junction')—see the figure below. Calculate the average charge of the island as a function of V.

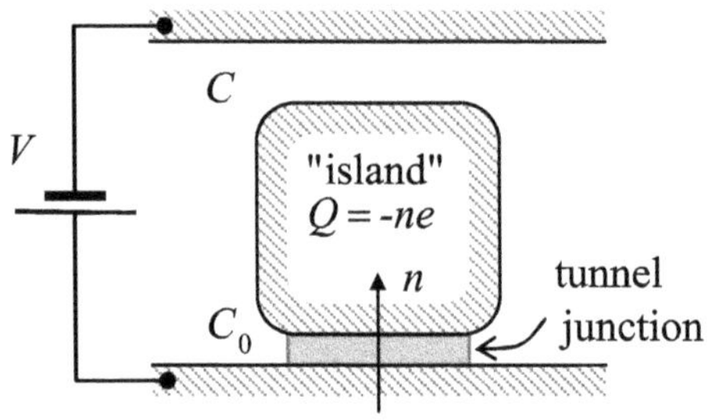

Hint: The quantum-mechanical tunneling of an electron through a weak junction[73] between macroscopic conductors, and its subsequent energy relaxation inside the conductor, may be considered as a single inelastic (energy-dissipating) event, so that the only energy relevant for the thermal equilibrium of the system is its electrostatic potential energy.

Problem 2.19. An LC circuit (see the figure below) is in thermodynamic equilibrium with its environment. Calculate the rms fluctuation $\delta V \equiv \langle V^2 \rangle^{1/2}$ of the voltage across it, for an arbitrary ratio $T/\hbar\omega$, where $\omega = (LC)^{-1/2}$ is the resonance frequency of this 'tank circuit'.

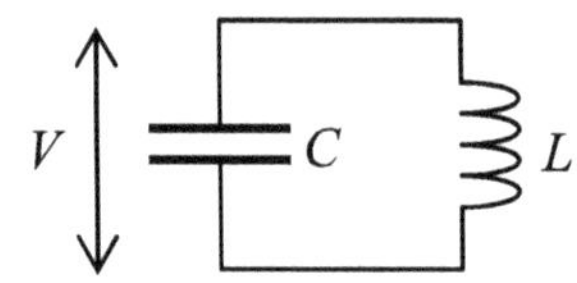

Problem 2.20. Derive Eq. (2.92) from simplistic arguments, representing the blackbody radiation as an ideal gas of photons, treated as ultra-relativistic particles. What do similar arguments give for an ideal gas of classical, non-relativistic particles?

Problem 2.21. Calculate the enthalpy, the entropy, and the Gibbs energy of blackbody electromagnetic radiation with temperature T, and then use these results to find the law of temperature and pressure drop at an adiabatic expansion.

Problem 2.22. As was mentioned in section 2.6(i), the relation between the temperatures $T_\oplus$ of the visible Sun's surface and that (T_o) of the Earth's surface follows from the balance of the thermal radiation they emit. Prove that this relation indeed follows, with a good precision, from a simple model in which the surfaces radiate as perfect black bodies with constant, average temperatures.

[73] In this particular context, the adjective 'weak' denotes a junction with the tunneling transparency so low that the tunneling electron's wavefunction loses its quantum-mechanical coherence before the electron has a chance to tunnel back. In a typical junction of a macroscopic area this condition is fulfilled if its effective resistance is much higher than the quantum unit of resistance (see, e.g. *Part QM* section 3.2), $R_Q \equiv \pi\hbar/2e^2 \approx 6.5$ kΩ.

Hint: You may pick up the experimental values you need from any (reliable) source.

Problem 2.23. If a surface is not perfectly radiation-absorbing ('black'), the electromagnetic power of its thermal radiation differs from the Stefan law (2.89*a*) by a factor $\varepsilon < 1$, called the *emissivity*:

$$\frac{\mathscr{P}}{A} = \varepsilon\sigma T^4.$$

Prove that such a surface reflects the $(1 - \varepsilon)$ part of incident radiation.

Problem 2.24. If two black surfaces, facing each other, have different temperatures (see the figure below), then according to the Stefan radiation law (2.89), there is a net flow of thermal radiation, from a warmer surface to the colder one:

$$\frac{\mathscr{P}_{\text{net}}}{A} = \sigma\left(T_1^4 - T_2^4\right).$$

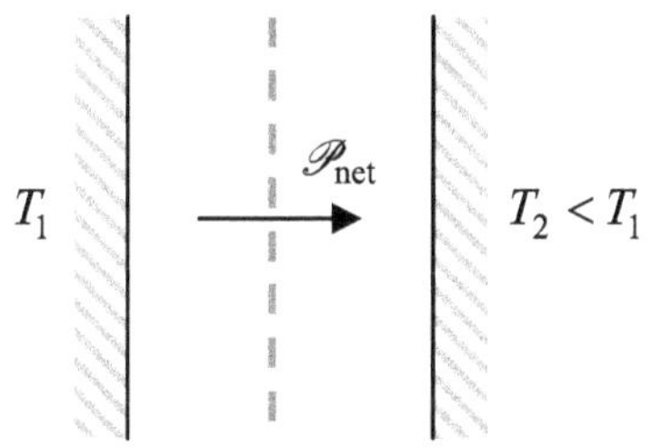

For many applications, notably including most low temperature experiments, this flow is detrimental. One way to suppress it is to reduce the emissivity ε (for its definition, see the previous problem) of both surfaces—say by covering them with shiny metallic films. An alternative way toward the same goal is to place, between the surfaces, a thin layer (usually called the *thermal shield*), with a low emissivity of both surfaces—see the dashed line in the figure above. Assuming that the emissivity is the same in both cases, find out which way is more efficient.

Problem 2.25. Two parallel, well conducting plates of area A are separated by a free-space gap of a constant thickness $t \ll A^{1/2}$. Calculate the energy of the thermally-induced electromagnetic field inside the gap at thermal equilibrium with temperature T in the range

$$\frac{\hbar c}{A^{1/2}} \ll T \ll \frac{\hbar c}{t}.$$

Does the field push the plates apart?

Problem 2.26. Use the Debye theory to estimate the specific heat of aluminum at room temperature (say, 300 K), and express the result in the following popular units:

(i) eV/K per atom,
(ii) J/K per mole, and
(iii) J/K per gram.

Compare the last number with the experimental value (from a reliable book or online source).

Problem 2.27. Low-temperature specific heat of some solids has a considerable contribution from thermal excitation of spin waves, whose dispersion law scales as $\omega \propto k^2$ at $\omega \to 0$.[74] Neglecting anisotropy, calculate the temperature dependence of this contribution to C_V at low temperatures and discuss conditions of its experimental observation.

Hint: Just as the photons and phonons discussed in section 2.6, the quantum excitations of spin waves (called *magnons*) may be considered as non-interacting bosonic quasiparticles with zero chemical potential, whose statistics obeys Eq. (2.72).

Problem 2.28. Derive a general expression for the specific heat of a very long, straight chain of similar particles of mass m, confined to move only in the direction of the chain, and elastically connected with effective spring constants κ—see the figure below. Spell out the result in the limits of very low and very high temperatures.

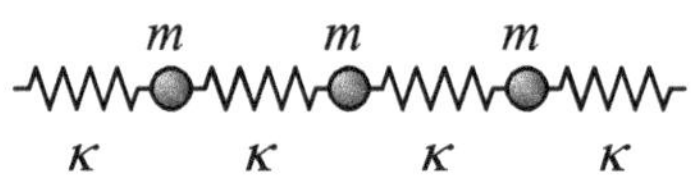

Hint: You may like to use the following integral[75]:

$$\int_0^{+\infty} \frac{\xi^2 d\xi}{\sinh^2 \xi} = \frac{\pi^2}{6}.$$

Problem 2.29. Calculate the rms thermal fluctuation of the middle point of a uniform guitar string of length l, stretched by force $\mathscr{T}$, at temperature T. Evaluate your result for $l = 0.7$ m, $\mathscr{T} = 10^3$N, and room temperature.

Hint: You may like to use the following series:

$$1 + \frac{1}{3^2} + \frac{1}{5^2} + \ldots \equiv \sum_{m=0}^{\infty} \frac{1}{(2m+1)^2} = \frac{\pi^2}{8}.$$

Problem 2.30. Use the general Eq. (2.123) to re-derive the Fermi–Dirac distribution (2.115) for a system in equilibrium.

Problem 2.31. Each of two similar particles, not interacting directly, may be in any of two quantum states, with single-particle energies ε equal to 0 and Δ. Write down

[74] Note that by the same dispersion law is typical for bending waves in thin elastic rods—see, e.g. *Part CM* section 7.8.

[75] It may be reduced, via integration by parts, to the table integral Eq. (A.35*d*) with $n = 1$.

the statistical sum Z of the system, and use it to calculate its average total energy E at temperature T, for the cases when the particles are:

(i) distinguishable;
(ii) indistinguishable fermions;
(iii) indistinguishable bosons.

Analyze and interpret the temperature dependence of $\langle E \rangle$ for each case, assuming that $\Delta > 0$.

Problem 2.32. Calculate the chemical potential of a system of $N \gg 1$ independent fermions, kept at a fixed temperature T, if each particle has two non-degenerate energy levels, separated by gap Δ.

References

[1] Arnold V and Avez A 1989 *Ergodic Problems of Classical Mechanics* (Addison-Wesley)
[2] Pierce J 1980 *An Introduction to Information Theory* (Dover)
[3] Schuster H and Just W 2005 *Deterministic Chaos* 4th edn (Wiley-VHS)
[4] Bérut A *et al* 2012 *Nature* **483** 187
[5] Koski J *et al* 2014 *PNAS USA* **111** 13786
[6] Jun Y *et al* 2014 *Phys. Rev. Lett.* **113** 190601
[7] Peterson J *et al* 2016 *Proc. Roy. Soc. A* **472** 20150813
[8] Bennett C 1973 *IBM J. Res. Devel.* **17** 525
[9] Bennett C 1982 *Int. J. Theor. Phys.* **21** 905
[10] Likharev K 1982 *Int. J. Theor. Phys.* **21** 311
[11] Feynman R 1998 *Statistical Mechanics* (CRC Press)

Chapter 3

Ideal and not-so-ideal gases

In this chapter, the general principles of thermodynamics and statistics, discussed in the previous two chapters, are applied to examine the basic physical properties of gases, i.e. collections of identical particles (for example, atoms or molecules) that are free to move inside a certain volume, either not interacting or weakly interacting with each other. We will see that due to the quantum statistics, properties of even the simplest, so-called ideal gases, with negligible direct interactions between particles, may be highly nontrivial.

3.1 Ideal classical gas

Direct interactions of typical atoms and molecules are well localized, i.e. rapidly decreasing with distance r between them, becoming negligible at a certain distance r_0. In a gas of N particles inside volume V, the average distance $\langle r \rangle$ between the particles is of the order of $(V/N)^{1/3}$. As a result, if the gas density $n \equiv N/V \sim \langle r \rangle^{-3}$ is much lower than r_0^{-3}, i.e. if $nr_0^3 \ll 1$, the chance for its particles to approach each other and interact is rather small. The model in which such direct interactions are completely ignored is called the *ideal gas*.

Let us start with a *classical* ideal gas, which may be defined as the gas in whose behavior the quantum effects are negligible. As we saw in section 2.8, the condition of that is to have the average occupancy of each quantum state low:

$$\langle N_k \rangle \ll 1. \tag{3.1}$$

It may seem that we have already found properties of such a system, in particular the equilibrium occupancy of its states—see Eq. (2.111):

$$\langle N_k \rangle = \text{const} \times \exp\left\{-\frac{\varepsilon_k}{T}\right\}. \tag{3.2}$$

doi:10.1088/2053-2563/aaf503ch3

In some sense this is true, but we still need, first, to see what exactly Eq. (3.2) means for the gas, i.e. a system with an essentially continuous energy spectrum, and, second, to show that, rather surprisingly, the particles' indistinguishability affects some properties of even classical gases.

The first of these tasks is evidently easiest for a gas out of external fields, and with no internal degrees of freedom[1]. In this case ε_k is just the kinetic energy of the particle, which is isotropic and parabolic:

$$\varepsilon_k = \frac{p^2}{2m} = \frac{p_x^2 + p_y^2 + p_z^2}{2m}. \tag{3.3}$$

Now we have to use two facts from other fields of physics, hopefully well known to the reader. First, in quantum mechanics, the linear momentum $\mathbf{p}$ is associated with the wavevector $\mathbf{k}$ of the de Broglie wave, $\mathbf{p} = \hbar\mathbf{k}$. Second, the eigenvalues of $\mathbf{k}$ for *any* waves (including de Broglie waves) in free space are uniformly distributed in the momentum space, with a constant density of states, given by Eq. (2.82):

$$\frac{dN_{\text{states}}}{d^3k} = \frac{gV}{(2\pi)^3}, \quad \text{i.e.} \quad \frac{dN_{\text{states}}}{d^3p} = \frac{gV}{(2\pi\hbar)^3}, \tag{3.4}$$

where g is the degeneracy of particle's internal states (say, for electrons, the spin degeneracy $g = 2$). Even regardless of the exact proportionality coefficient between dN_{states} and d^3p, the very fact that this coefficient is a constant means that the probability dW to find the particle in a small region $d^3p = dp_1dp_2dp_3$ of the momentum space is proportional to the right-hand side of Eq. (3.2), with ε_k given by Eq. (3.3):

$$dW = C\exp\left\{-\frac{p^2}{2mT}\right\}d^3p \equiv C\exp\left\{-\frac{p_1^2 + p_2^2 + p_3^2}{2mT}\right\}dp_1dp_2dp_3. \tag{3.5}$$

This is the famous *Maxwell distribution*[2]. The normalization constant C may be readily found from the last form of Eq. (3.5), by requiring the integral of dW over all the momentum space to equal 1. Indeed, the integral is evidently a product of three similar 1D integrals over each Cartesian component p_j of the momentum ($j = 1, 2, 3$),

[1] In more realistic cases when particles do have internal degrees of freedom, but they are all in a certain (say, ground) quantum state, Eq. (3.3) is valid as well, with ε_k referring to the internal ground-state energy. The effect of thermal excitation of the internal degrees of freedom will be briefly discussed at the end of this section.

[2] This formula was suggested by J C Maxwell as early as in 1860, i.e. well before the Boltzmann and Gibbs distributions. Note also that the term 'Maxwell distribution' is often associated with the distribution of particle's momentum (or velocity) *magnitude*,

$$dW = 4\pi Cp^2\exp\left\{-\frac{p^2}{2mT}\right\}dp = 4\pi Cm^3v^2\exp\left\{-\frac{mv^2}{2T}\right\}dv, \quad \text{with } 0 \leqslant p, v < \infty,$$

which immediately follows from the first form of Eq. (3.5), combined with the expression $d^3p = 4\pi p^2dp$ due to the spherical symmetry of the distribution in the momentum/velocity space.

which may be readily reduced to the well-known dimensionless Gaussian integral[3], so that we get

$$C = \left[\int_{-\infty}^{+\infty} \exp\left\{-\frac{p_j^2}{2mT}\right\} dp_j\right]^{-3} \equiv \left[(2mT)^{1/2}\int_{-\infty}^{+\infty} e^{-\xi^2} d\xi\right]^{-3} = (2\pi mT)^{-3/2}. \quad (3.6)$$

As a sanity check, let us use the Maxwell distribution to calculate the average energy corresponding to each half-degree of freedom:

$$\begin{aligned}\left\langle \frac{p_j^2}{2m} \right\rangle &= \int \frac{p_j^2}{2m} dW \\ &= \left[C^{1/3}\int_{-\infty}^{+\infty} \frac{p_j^2}{2m} \exp\left\{-\frac{p_j^2}{2mT}\right\} dp_j\right] \times \left[C^{1/3}\int_{-\infty}^{+\infty} \exp\left\{-\frac{p_{j'}^2}{2mT}\right\} dp_{j'}\right]^2 \\ &= \frac{T}{\pi^{1/2}}\int_{-\infty}^{+\infty} \xi^2 e^{-\xi^2} d\xi.\end{aligned} \quad (3.7)$$

The last, dimensionless integral equals $\sqrt{\pi}/2$,[4] so that, finally,

$$\left\langle \frac{p_j^2}{2m} \right\rangle \equiv \left\langle \frac{mv_j^2}{2} \right\rangle = \frac{T}{2}. \quad (3.8)$$

This result is (fortunately :-) in agreement with the equipartition theorem (2.48). It also means that the rms velocity of the particles is

$$\delta v \equiv \langle v^2 \rangle^{1/2} = \left\langle \sum_{j=1}^{3} v_j^2 \right\rangle^{1/2} = \langle 3v_j^2 \rangle^{1/2} = \left(3\frac{T}{m}\right)^{1/2}. \quad (3.9)$$

For a typical gas (say, for N_2, the air's main component), with $m \approx 28m_p \approx 4.7 \times 10^{-26}$ kg at room temperature ($T = k_B T_K \approx k_B \times 300$ K $\approx 4.1 \times 10^{-21}$ J), this velocity is about 500 m s^{-1}, comparable with the sound velocity in the same gas—and with the muzzle velocity of a typical handgun bullet. Still, it is measurable using even the simple table-top equipment (say, a set of two concentric, rapidly rotating cylinders with a thin slit collimating an atomic beam emitted at the axis) that was already available at the end of the 19th century. Experiments using such equipment gave convincing early confirmations of Maxwell's theory.

This is all very simple (isn't it?), but actually the thermodynamic properties of a classical gas, especially its entropy, are more intricate. To show that, let us apply the

[3] See, e.g. Eq. (A.36*b*).
[4] See, e.g. Eq. (A.36*c*).

Gibbs distribution to a gas portion consisting of N particles, rather than just one of them. If the particles are exactly similar, the eigenenergy spectrum $\{\varepsilon_k\}$ of each of them is also exactly the same, and each value E_m of the total energy is just the sum of particular energies $\varepsilon_{k(l)}$ of the particles, where $k(l)$, with $l = 1, 2, \ldots N$, is the number of the energy level on which the lth particle resides. Moreover, since the gas is classical, $\langle N_k \rangle \ll 1$, the probability of having two or more particles in any state may be ignored. As a result, we can use Eq. (2.59) to write

$$\begin{aligned} Z &\equiv \sum_m \exp\left\{-\frac{E_m}{T}\right\} = \sum_{k(l)} \exp\left\{-\frac{1}{T}\sum_l \varepsilon_{k(l)}\right\} \\ &= \sum_{k(1)}\sum_{k(2)} \ldots \sum_{k(N)} \prod_l \exp\left\{-\frac{\varepsilon_{k(l)}}{T}\right\}, \end{aligned} \tag{3.10}$$

where the summation has to be carried over all possible states of each particle. Since the summation over each set $\{k(l)\}$ concerns only one of the operands of the product of exponents under the sum, it is tempting to complete the calculation as follows:

$$\begin{aligned} Z \rightarrow Z_{\text{dist}} &= \sum_{k(1)} \exp\left\{-\frac{\varepsilon_{k(1)}}{T}\right\} \cdot \sum_{k(2)} \exp\left\{-\frac{\varepsilon_{k(2)}}{T}\right\} \ldots \cdot \sum_{k(N)} \exp\left\{-\frac{\varepsilon_{k(N)}}{T}\right\} \\ &= \left(\sum_k \exp\left\{-\frac{\varepsilon_k}{T}\right\}\right)^N, \end{aligned} \tag{3.11}$$

where the final summation is over all states of one particle. This formula is indeed valid for distinguishable particles[5]. However, if the particles are indistinguishable (again, meaning that they are identical *and* free to move within the same spatial region), Eq. (3.11) has to be modified by what is called the *correct Boltzmann counting*:

$$Z = \frac{1}{N!}\left(\sum_k \exp\left\{-\frac{\varepsilon_k}{T}\right\}\right)^N, \tag{3.12}$$

that considers all quantum states, differing only by particle permutations in the gas portion, as one.

This expression is valid for any set $\{\varepsilon_k\}$ of eigenenergies. Now let us use it for the translational 3D motion of free particles, taking into account that the fundamental relation (3.4), which implies the following rule for the replacement of a sum over

[5] Since, by our initial assumption, each particle belongs to the same portion of gas, i.e. cannot be distinguished from others by its spatial position, this requires some internal 'pencil mark', for example a specific structure or a specific quantum state of its internal degrees of freedom.

quantum states of such motion with an integral in the classical limit—whose exact conditions are still to be specified[6]:

$$\sum_k(\ldots) \to \int(\ldots)\, dN_{\text{states}} = \frac{gV}{(2\pi)^3}\int(\ldots)d^3k = \frac{gV}{(2\pi\hbar)^3}\int(\ldots)\, d^3p. \tag{3.13}$$

In application to Eq. (3.12), this rule yields

$$Z = \frac{1}{N!}\left(\frac{gV}{(2\pi\hbar)^3}\left[\int_{-\infty}^{+\infty}\exp\left\{-\frac{p_j^2}{2mT}\right\}dp_j\right]^3\right)^N. \tag{3.14}$$

The integral in the square brackets is the same one as in Eq. (3.6), i.e. equal to $(2\pi mT)^{1/2}$, so that finally

$$Z = \frac{1}{N!}\left(\frac{gV}{(2\pi\hbar)^3}(2\pi mT)^{3/2}\right)^N \equiv \frac{1}{N!}\left[gV\left(\frac{mT}{2\pi\hbar^2}\right)^{3/2}\right]^N. \tag{3.15}$$

Now, assuming that $N \gg 1$,[7] and applying the Stirling formula, we can calculate gas' free energy,

$$F = T\ln\frac{1}{Z} = -NT\ln\frac{V}{N} + Nf(T), \tag{3.16a}$$

with

$$f(T) \equiv -T\left\{\ln\left[g\left(\frac{mT}{2\pi\hbar^2}\right)^{3/2}\right] + 1\right\}. \tag{3.16b}$$

The first of these relations is exactly Eq. (1.45), which was derived in section 1.4 from the equation of state $PV = NT$, using thermodynamic identities. At that stage, this equation of state was just postulated, but now we can finally *derive* it by calculating the pressure from the second of Eqs. (1.35), and Eq. (3.16*a*):

$$P = -\left(\frac{\partial F}{\partial V}\right)_T = \frac{NT}{V}. \tag{3.17}$$

So, the equation of state of the ideal classical gas, with density $n \equiv N/V$, is indeed given by Eq. (1.44):

$$P = \frac{NT}{V} \equiv nT. \tag{3.18}$$

[6] As a reminder, we have already used this rule (twice) in section 2.6, with particular values of g.

[7] For the opposite limit when $N = g = 1$, Eq. (3.15) yields the results obtained, by two alternative methods, in the solutions of problems 2.8 and 2.9. Indeed, for $N = 1$, the 'correct Boltzmann counting' factor $N!$ equals 1, so that the particle distinguishability effects vanish—naturally.

Hence we may use Eqs. (1.46)–(1.51), derived from this equation of state, to calculate all other thermodynamic variables of the gas. For example, using Eq. (1.47) with $f(T)$ given by Eq. (3.16*b*), for the internal energy and the specific heat of the gas we immediately get

$$E = N\left[f(T) - T\frac{df(T)}{dT}\right] = \frac{3}{2}NT, \quad c_V \equiv \frac{C_V}{N} = \frac{1}{N}\left(\frac{\partial E}{\partial T}\right)_V = \frac{3}{2}, \tag{3.19}$$

in full agreement with Eq. (3.8) and hence with the equipartition theorem. Much less trivial is the result for entropy, which may be obtained by combining Eqs. (1.46) and (3.16*a*):

$$S = -\left(\frac{\partial F}{\partial T}\right)_V = N\left[\ln\frac{V}{N} - \frac{df(T)}{dT}\right]. \tag{3.20}$$

This formula[8], in particular, provides the means to resolve the following *gas mixing paradox* (sometimes called the 'Gibbs paradox'). Consider two volumes, V_1 and V_2, separated by a partition, each filled with the same gas, with the same density n, at the same temperature T, and hence with the same pressure P. Now let us remove the partition and let the gas portions mix; would the total entropy change? According to Eq. (3.20), it would not, because the ratio $V/N = n$, and hence the expression in square brackets is the same in the initial and the final state, so that the entropy is additive, as any extensive variable should be. This makes full sense if the gas particles in both parts of the volume are truly identical, i.e. the partition's removal does not change our information about the system. However, let us assume that all particles are distinguishable; then the entropy should clearly increase, because the mixing would certainly decrease our information about the system, i.e. increase its disorder. A quantitative description of this effect may be obtained using Eq. (3.11). Repeating for Z_{dist} the calculations made above for Z, we readily get a different formula for entropy:

$$S_{\text{dist}} = N\left[\ln V - \frac{df_{\text{dist}}(T)}{dT}\right], \quad f_{\text{dist}}(T) \equiv -T\ln\left[g\left(\frac{mT}{2\pi\hbar^2}\right)^{3/2}\right]. \tag{3.21}$$

Please notice that in contrast to the S given by Eq. (3.20), due to the term ln V, the entropy S_{dist} is *not* proportional to N (at fixed temperature T and density N/V). While for distinguishable particles this fact does not present any conceptual problem, for indistinguishable particles it would mean that entropy were not an extensive variable, i.e. it would contradict the basic assumptions of thermodynamics. This fact emphasizes again the necessity of the correct Boltzmann counting in the latter case.

[8] The result represented by Eq. (3.20), with the function f given by Eq. (3.16*b*), was obtained independently by O Sackur and H Tetrode as early as 1911, i.e. well before the final formulation of quantum mechanics in the late 1920s.

Using Eq. (3.21), we can calculate the change of entropy due to mixing two gas portions, with N_1 and N_2 distinguishable particles, at a fixed temperature T (and hence at unchanged function f_{dist}):

$$\begin{aligned}\Delta S_{\text{dist}} &= (N_1 + N_2)\ln(V_1 + V_2) - (N_1 \ln V_1 + N_2 \ln V_2)\\ &= N_1 \ln \frac{V_1 + V_2}{V_1} + N_2 \ln \frac{V_1 + V_2}{V_2} > 0.\end{aligned} \tag{3.22}$$

Note that for a particular case, $V_1 = V_2 = V/2$, Eq. (3.22) reduces to the simple result $\Delta S_{\text{dist}} = (N_1 + N_2)\ln 2$, which may be readily understood from the point of view of the information theory. Indeed, allowing each particle of the total number $N = N_1 + N_2$ to spread to twice larger volume, we loose one bit of information per particle, i.e. $\Delta I = (N_1 + N_2)$ bits for the whole system. Let me leave it for the reader to show that Eq. (3.22) is also valid if particles in each sub-volume are indistinguishable from each other, but different from those in another sub-volume, i.e. for mixing of two different gases[9]. However, it is certainly *not* applicable to the system where *all* particles are identical, stressing again that the correct Boltzmann counting (3.12) does indeed affect the gas entropy, even though it may be not be as consequential as the Maxwell distribution (3.5), the equation of state (3.18), and the average energy (3.19).

In this context, one may wonder whether the change (3.22) (called the *mixing entropy*) is experimentally observable. The answer is yes. For example, after free mixing of two different gases, one can use a thin movable membrane that is *semipermeable*, i.e. penetrable by particles of one type only, to separate them again, thus reducing the entropy back to the initial value, and measure either the necessary mechanical work $\Delta\mathscr{W} = T\Delta S_{\text{dist}}$ or the corresponding heat discharge into the heat bath. Practically, measurements of this type are easier in *weak solutions*[10]—systems with a small concentration $c \ll 1$ of particles of one sort (*solute*) within much more abundant particles of another sort (*solvent*). The mixing entropy also affects thermodynamics of chemical reactions in gases and liquids[11]. It is curious that besides purely thermal measurements, the mixing entropy in some conducting solutions (*electrolytes*) is also measurable by a purely electrical method, called *cyclic voltammetry*, in which a low-frequency ac voltage, applied between solid-state electrodes embedded in the solution, is used to periodically separate different ions, and then mix them again[12].

---

[9] By the way, if an ideal classical gas consists of particles of several different sorts, its full pressure is a sum of independent *partial pressures* exerted by each component—the so-called *Dalton law*. While this fact was an important experimental discovery in the early 1800s, from the point of view of statistical physics this is just a straightforward corollary of Eq. (3.18), because in an ideal gas, the component particles do not interact.

[10] It is interesting that statistical mechanics of weak solutions is very similar to that of ideal gases, with Eq. (3.18) recast into the following formula (derived in 1885 by J van't Hoff), $PV = cNT$, for the partial pressure of the solute. One of its corollaries is that the net force (called the *osmotic pressure*) exerted on a semipermeable membrane is proportional to the difference of solute concentrations it is supporting.

[11] Unfortunately, I do not have time for even a brief introduction into this important field, and have to refer the interested reader to specialized textbooks—for example, [1], or [2], or [3].

[12] See, e.g. either chapter 6 in [4] (which is a good introduction to electrochemistry as the whole), or section II.8.3.1 in [5].

Now let us briefly discuss two generalizations of our results for ideal classical gases. First, let us consider such a gas in an external field of potential forces. It may be described by replacing Eq. (3.3) with

$$\varepsilon_k = \frac{p_k^2}{2m} + U(\mathbf{r}_k), \tag{3.23}$$

where $\mathbf{r}_k$ is the position of the particular particle, and $U(\mathbf{r})$ is the potential energy per particle. If the potential $U(\mathbf{r})$ is changing in space sufficiently slowly[13], Eq. (3.4) is still applicable, but only to small volumes, $V \to dV = d^3r$ whose linear size is much smaller than the spatial scale of substantial variations of the function $U(\mathbf{r})$. Hence, instead of Eq. (3.5), we may only write the probability dW of finding the particle in a small volume d^3rd^3p of the six-dimensional phase space:

$$dW = w(\mathbf{r}, \mathbf{p})d^3rd^3p, \quad w(\mathbf{r}, \mathbf{p}) = \text{const} \times \exp\left\{-\frac{p^2}{2mT} - \frac{U(\mathbf{r})}{T}\right\}. \tag{3.24}$$

Hence, the Maxwell distribution of particle velocities is still valid at each point $\mathbf{r}$, so that the equation of state (3.18) is valid also locally. A more interesting issue here is the spatial distribution of the total density,

$$n(\mathbf{r}) \equiv N\int w(\mathbf{r}, \mathbf{p})d^3p, \tag{3.25}$$

of all gas particles, regardless of their momentum/velocity. For this variable, Eq. (3.24) yields[14]

$$n(\mathbf{r}) = n(0)\exp\left\{-\frac{U(\mathbf{r})}{T}\right\}, \tag{3.26}$$

where the potential energy reference is at the origin, and the local gas pressure may be still calculated from Eq. (3.18):

$$P(\mathbf{r}) = n(\mathbf{r})T = P(0)\exp\left\{-\frac{U(\mathbf{r})}{T}\right\}. \tag{3.27}$$

An important example of application of Eq. (3.27) is an approximate description of the Earth atmosphere. At all heights $h \ll R_E \sim 6 \times 10^6$ m above the Earth's surface (say, the sea level), we may describe the Earth gravity effect by the potential $U = mgh$, and Eq. (3.27) yields the so-called *barometric formula*

$$P(h) = P(0)\exp\left\{-\frac{h}{h_0}\right\}, \quad \text{with } h_0 \equiv \frac{T}{mg} = \frac{k_B T_K}{mg}. \tag{3.28}$$

---

[13] Qualitatively, the effective distance of substantial variations of the potential, $T/|\nabla U(\mathbf{r})|$, has to be much larger than the *mean free path* l of the gas particles, i.e. the average distance a particle passes its successive collisions with its counterparts. (For more on this notion, see chapter 6 below.)

[14] In some textbooks, Eq. (3.26) is also called the *Boltzmann distribution*, though it certainly should be distinguished from Eq. (2.111).

For the same N_2, the main component of the atmosphere, at $T_K = 300$ K, $h_0 \approx 7$ km. This gives the right order of magnitude of the Earth atmosphere's thickness, though the exact law of the pressure change differs somewhat from Eq. (3.28) because of a certain drop of the absolute temperature T with height, with the so-called *lapse rate* of about 2% per km.

The second generalization I need to discuss is to particles with internal degrees of freedom. Now ignoring the potential energy $U(\mathbf{r})$, we may describe them by replacing Eq. (3.3) with

$$\varepsilon_k = \frac{p^2}{2m} + \varepsilon_k{}', \tag{3.29}$$

where $\varepsilon_k{}'$ describes the internal energy spectrum of the kth particle. If the particles are similar, we may repeat all the above calculations, and see that all the results (including the Maxwell distribution, and the equation of state) are still valid, with the only exception of Eq. (3.16), which now becomes

$$f(T) = -T\left\{\ln\left[g\left(\frac{mT}{2\pi\hbar^2}\right)^{3/2}\right] + 1 + \ln\left[\sum_{\varepsilon_k{}'}\exp\left\{-\frac{\varepsilon_k{}'}{T}\right\}\right]\right\}. \tag{3.30}$$

As we already know from Eqs. (1.50) and (1.51), this change may affect both specific heats of the ideal gas—though not their difference, $c_V - c_P = 1$. They may be readily calculated for usual atoms and molecules, at not very high temperatures (say room temperature of ~25 meV), because in these conditions, $\varepsilon_k{}' \gg T$ for most of their internal degrees of freedom, including the electronic and vibrational ones. (The typical energy of the lowest electronic excitations is of the order of a few eV, and that of the lowest vibrational excitations is only an order of magnitude lower.) As a result, these degrees of freedom are *frozen out*: their contributions $\exp\{-\varepsilon_k{}'/T\}$ into the sum in Eq. (3.30), and hence to the heat capacity, are negligible. In monoatomic gases, this is true for all degrees of freedom besides those of the translational motion, already taken into account by the first term in Eq. (3.30), i.e. by Eq. (3.16*b*), so that their specific heat is typically well described by Eq. (3.19).

The most important exception is the rotational degrees of freedom of diatomic and polyatomic molecules. As quantum mechanics shows[15], the excitation energy of these degrees of freedom scales as $\hbar^2/2I$, where I is the molecule's relevant moment of inertia. In most important molecules, this energy is rather low (e.g. for N_2, it is close to 0.25 meV, i.e. ~1% of the room temperature), so that at usual conditions they are well excited and, moreover, behave virtually as classical degrees of freedom, each giving a quadratic contribution to the molecule's energy, and hence obeying the equipartition theorem, i.e. giving an extra contribution of $T/2$ into the energy, and of ½ to the specific heat[16]. In polyatomic molecules, there are three such classical

[15] See, e.g. the model solution of problem 2.12, and references therein.

[16] This result may be readily obtained again from the last term of Eq. (3.30) by treating it exactly as the first one was obtained, and then applying the general Eq. (1.50).

degrees of freedom (corresponding to its rotations about three principal axes[17]), but in diatomic molecules, only two[18]. To describe these contributions, Eq. (3.19) is usually generalized as

$$c_V = \begin{cases} 3/2, & \text{for monoatomic gases,} \\ 5/2, & \text{for gases of diatomic molecules,} \\ 3, & \text{for gases of polyatomic molecules.} \end{cases} \tag{3.31}$$

Please keep in mind, however, that as the above discussion shows, this approximation is invalid at very low and very high temperatures; the most notable reason for that is the thermal activation of vibrational degrees of freedom for many important molecules at temperatures of a few thousand K.

3.2 Calculating μ

Now let us discuss properties of ideal gases of free, indistinguishable particles in more detail, paying special attention to the chemical potential μ—which, for some readers, may still be a somewhat mysterious aspect of the Fermi and Bose distributions. Note that the particle indistinguishability *requires* the absence of thermal excitations of their internal degrees of freedom, so that in the balance of this chapter such excitations will be ignored, and the particle's energy ε_k will be associated with its 'external' energy alone—for an ideal gas of free particles, with its kinetic energy (3.3).

Let us start from the classical gas, and recall the conclusion of thermodynamics that μ is the Gibbs potential per unit particle—see Eq. (1.56). Hence we can calculate $\mu = G/N$ from Eqs. (1.49) and (3.16*b*). The result,

$$\mu = -T\ln\frac{V}{N} + f(T) + T = T\ln\left[\frac{N}{gV}\left(\frac{2\pi\hbar^2}{mT}\right)^{3/2}\right], \tag{3.32a}$$

which may be rewritten as

$$\exp\left\{\frac{\mu}{T}\right\} = \frac{N}{gV}\left(\frac{2\pi\hbar^2}{mT}\right)^{3/2}. \tag{3.32b}$$

This formula is very important, because it gives us some information about μ not only for a classical gas, but for quantum (Fermi and Bose) gases as well. Indeed, we already know that for indistinguishable particles, the Boltzmann distribution (2.111) is valid only if $\langle N_k\rangle \ll 1$. Comparing this condition with the quantum statistics (2.115) and (2.118), we see again that the condition of the gas' classicity may be expressed as

---

[17] See, e.g. *Part CM* section 4.1.

[18] This conclusion of the quantum theory may be interpreted as the indistinguishability of rotations about the molecule's symmetry axis.

$$\exp\left\{\frac{\mu - \varepsilon_k}{T}\right\} \ll 1 \tag{3.33}$$

for all ε_k. Since the lowest value of ε_k given by Eq. (3.3) is zero, Eq. (3.33) may be satisfied only if $\exp\{\mu/T\} \ll 1$. This means that the chemical potential of the classical has to be not just negative, but also 'strongly negative' in the sense

$$-\mu \gg T. \tag{3.34a}$$

According to Eq. (3.32), this important condition may be represented as

$$T \gg T_0, \tag{3.34b}$$

with T_0 defined as

$$T_0 \equiv \frac{\hbar^2}{m}\left(\frac{N}{gV}\right)^{2/3} \equiv \frac{\hbar^2}{m}\left(\frac{n}{g}\right)^{2/3} \equiv \frac{\hbar^2}{g^{2/3} m r_{\text{ave}}^2}, \tag{3.35}$$

where r_{ave} is the average distance between the particles:

$$r_{\text{ave}} \equiv \frac{1}{n^{1/3}} = \left(\frac{V}{N}\right)^{1/3}. \tag{3.36}$$

In this form, the condition (3.34) is very transparent physically: disregarding the factor $g^{2/3}$ (which is typically of the order of 1), it means that the average thermal energy of a particle, which is always of the order of T, has to be much larger than the energy of quantization of particle's motion at the length r_{ave}. An alternative form of this condition is[19]

$$r_{\text{ave}} \gg g^{-1/3} r_{\text{c}}, \quad \text{with} \quad r_{\text{c}} \equiv \frac{\hbar}{(mT)^{1/2}}. \tag{3.37}$$

For a typical gas (say, N_2, with $m \approx 14m_p \approx 2.3 \times 10^{-26}$ kg) at the standard room temperature ($T = k_B \times 300\text{ K} \approx 4.1 \times 10^{-21}$ J), the correlation length r_c is close to 10^{-11} m, i.e. is significantly smaller than the physical size $a \sim 3 \times 10^{-10}$ m of the molecule. This estimate shows that at room temperature, as soon as any practical gas is rare enough to be ideal ($r_{\text{ave}} \gg a$), it is classical, i.e. the only way to observe the quantum effects in the translation motion of molecules is a very deep refrigeration. According to Eq. (3.37), for the same nitrogen molecule, taking $r_{\text{ave}} \sim 10^3 a \sim 10^{-7}$ m (to ensure that direct interaction effects are negligible), temperature should be well below 1 μK.

In order to analyze quantitatively what happens with gases when T is reduced to such low values, we need to calculate μ for an arbitrary ideal gas of indistinguishable particles. Let us use the lucky fact that the Fermi–Dirac and the Bose–Einstein statistics may be represented with one formula:

[19] In quantum mechanics, the parameter r_c so defined is frequently called the *correlation length*—see, e.g. *Part QM* section 7.2 and in particular Eq. (7.37).

$$\langle N(\varepsilon)\rangle = \frac{1}{e^{(\varepsilon-\mu)/T} \pm 1}, \tag{3.38}$$

where (and everywhere in the balance of this section) the top sign stands for fermions and the lower one for bosons, to discuss the fermionic and bosonic gases in one shot.

If we deal with a member of the *grand canonical* ensemble (figure 2.13), in which μ is externally fixed, we may use Eq. (3.38) to calculate the *average* number N of particles in volume V. If the volume is so large that $N \gg 1$, we may use the general state counting rule (3.13) to get

$$\begin{aligned} N &= \frac{gV}{(2\pi)^3}\int \langle N(\varepsilon)\rangle\, d^3k = \frac{gV}{(2\pi\hbar)^3}\int \frac{d^3p}{e^{[\varepsilon(p)-\mu]/T} \pm 1} \\ &= \frac{gV}{(2\pi\hbar)^3}\int_0^\infty \frac{4\pi p^2\, dp}{e^{[\varepsilon(p)-\mu]/T} \pm 1}. \end{aligned} \tag{3.39}$$

In most practical cases, however, the number N of gas particles is fixed by particle confinement (i.e. the gas portion under study is a member of a *canonical* ensemble—see figure 2.6), and hence μ rather than N should be calculated. Let us use the trick already mentioned in section 2.8: if N is very large, the relative fluctuation of the particle number, at fixed μ, is negligibly small ($\delta N/N \sim 1/\sqrt{N} \ll 1$), and the relation between the *average* values of N and μ should not depend on which of these variables is exactly fixed. Hence, Eq. (3.39), with μ having the sense of the *average* chemical potential, should be valid even if N is exactly fixed, so that small fluctuations of N are replaced with (equally small) fluctuations of μ. Physically, in this case the role of the μ-fixing environment for any gas' sub-portion is played by the rest of the gas, and Eq. (3.39) expresses the condition of self-consistency of such chemical equilibrium.

So, at $N \gg 1$, Eq. (3.39) may be used for calculating the average μ as a function of two independent parameters: N (i.e. the gas density $n = N/V$) and temperature T. For carrying out this calculation, it is convenient to convert the right-hand side of Eq. (3.39) to an integral over the particle's energy $\varepsilon(p) = p^2/2m$, so that $p = (2m\varepsilon)^{1/2}$, and $dp = (m/2\varepsilon)^{1/2}d\varepsilon$, obtaining

$$N = \frac{gVm^{3/2}}{\sqrt{2}\pi^2\hbar^3}\int_0^\infty \frac{\varepsilon^{1/2}d\varepsilon}{e^{(\varepsilon-\mu)/T} \pm 1}. \tag{3.40}$$

This key result may be represented in two other, more convenient forms. First, Eq. (3.40), derived for our current (3D, isotropic and parabolic-dispersion) approximation (3.3), is just a particular case of the following general relation

$$N = \int_0^\infty g(\varepsilon)\langle N(\varepsilon)\rangle d\varepsilon, \tag{3.41}$$

where

$$g(\varepsilon) \equiv \frac{dN_{\text{states}}}{d\varepsilon} \tag{3.42}$$

is the temperature-independent density of all quantum states of a particle—regardless of whether they are occupied or not. Indeed, according to the general Eq. (3.4), for our simple model (3.3),

$$\begin{aligned} g(\varepsilon) = g_3(\varepsilon) &\equiv \frac{dN_{\text{states}}}{d\varepsilon} = \frac{d}{d\varepsilon}\left(\frac{gV}{(2\pi\hbar)^3}\frac{4\pi}{3}p^3\right) \\ &= \frac{4\pi gV}{3(2\pi\hbar)^3}\frac{d(p^3)}{d\varepsilon} = \frac{gVm^{3/2}}{\sqrt{2}\pi^2\hbar^3}\varepsilon^{1/2}, \end{aligned} \tag{3.43}$$

so that we return to Eq. (3.39).

On the other hand, for some calculations, it is convenient to introduce a dimensionless energy variable $\xi \equiv \varepsilon/T$ to express Eq. (3.40) via a dimensionless integral:

$$N = \frac{gV(mT)^{3/2}}{\sqrt{2}\pi^2\hbar^3}\int_0^\infty \frac{\xi^{1/2}d\xi}{e^{\xi-\mu/T} \pm 1}. \tag{3.44}$$

As a sanity check, in the classical limit (3.34), the exponent in the denominator of the fraction under the integral is much larger than 1, and Eq. (3.44) reduces to

$$\begin{aligned} N &= \frac{gV(mT)^{3/2}}{\sqrt{2}\pi^2\hbar^3}\int_0^\infty \frac{\xi^{1/2}d\xi}{e^{\xi-\mu/T}} \\ &\approx \frac{gV(mT)^{3/2}}{\sqrt{2}\pi^2\hbar^3}\exp\left\{\frac{\mu}{T}\right\}\int_0^\infty \xi^{1/2}e^{-\xi}d\xi, \quad \text{at} \quad -\mu \gg T. \end{aligned} \tag{3.45}$$

By the definition of the gamma-function $\Gamma(\xi)$,[20] this dimensionless integral is just $\Gamma(3/2) = \sqrt{\pi}/2$, and we get

$$\exp\left\{\frac{\mu}{T}\right\} = N\frac{\sqrt{2}\pi^2\hbar^3}{gV(mT)^{3/2}}\frac{2}{\sqrt{\pi}} = \left(2\pi\frac{T_0}{T}\right)^{3/2}, \tag{3.46}$$

which is exactly the same result as given by Eq. (3.32), which has been obtained in a rather different way—from the Boltzmann distribution and thermodynamic identities.

Unfortunately, in the general case of arbitrary μ the integral in Eq. (3.44) cannot be worked out analytically[21]. The best we can do is to use T_0, defined by Eq. (3.35), to rewrite Eq. (3.44) in the following convenient, fully dimensionless form:

$$\frac{T}{T_0} = \left[\frac{1}{\sqrt{2}\pi^2}\int_0^\infty \frac{\xi^{1/2}d\xi}{e^{\xi-\mu/T} \pm 1}\right]^{-2/3}, \tag{3.47}$$

---

[20] See, e.g. Eq. (A.34*a*).

[21] For the reader's reference only: for the upper sign, the integral in Eq. (3.40) is a particular form (for $s = ½$) of a special function called the *complete Fermi–Dirac integral* F_s, while for the lower sign, it is a particular case (for $s = 3/2$) of another special function called the *polylogarithm* Li_s. (In what follows, I will not use these notations.)

and use this relation to calculate the ratio T/T_0, and then the ratio $\mu/T_0 \equiv (\mu/T) \times (T/T_0)$, as functions of μ/T numerically. After that, we may plot the results versus each other, now considering the first ratio as the argument. Figure 3.1 below shows the resulting plots, for both particle types. They show that at high temperatures, $T \gg T_0$, the chemical potential is negative and approaches the classical behavior given by Eq. (3.46) for both fermions and bosons—just as we could expect. However, at temperatures $T \sim T_0$ the type of statistics becomes crucial. For fermions, the reduction of temperature leads to μ changing its sign from negative to positive, and then approaching a constant positive value called the *Fermi energy*, $\varepsilon_F \approx 7.595\ T_0$ at $T \to 0$. Conversely, the chemical potential of a gas of bosons stays negative, and then turns into zero at certain *critical temperature* $T_c \approx 3.313\ T_0$. Both these limits, which are very important for applications, may (and will be) explored analytically, separately for each statistics.

Before carrying out such studies (in the next two sections), let me show that, rather surprisingly, for any non-relativistic, ideal quantum gas, the relation between the product PV and the energy,

$$PV = \frac{2}{3}E, \tag{3.48}$$

is the same as follows from Eqs. (3.18) and (3.19) for the classical gas, and hence does *not* depend on the particle statistics. In order to prove this, it is sufficient to use

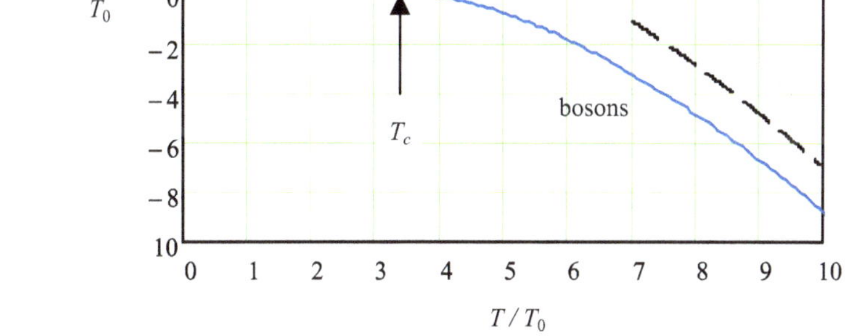

Figure 3.1. The chemical potential of an ideal gas of $N \gg 1$ indistinguishable quantum particles, as a function of temperature at a fixed gas density $n \equiv N/V$ (i.e. fixed parameter $T_0 \propto n^{3/2}$), for two different particle types. The dashed line shows the classical approximation (3.46), valid only at $T \gg T_0$.

Eqs. (2.114) and (2.117) for the grand thermodynamic potential of each quantum state, which may be conveniently represented by a single formula,

$$\Omega_k = \mp T \ln\left(1 \pm e^{(\mu-\varepsilon_k)/T}\right), \tag{3.49}$$

and sum them over all states k, using the general summation formula (3.13). The result for the total grand potential of a 3D gas with the dispersion law (3.3) is

$$\begin{aligned}\Omega &= \mp T\frac{gV}{(2\pi\hbar)^3}\int_0^\infty \ln\left(1 \pm e^{(\mu-p^2/2m)/T}\right)4\pi p^2\,dp\\ &= \mp T\frac{gVm^{3/2}}{\sqrt{2}\,\pi^2\hbar^3}\int_0^\infty \ln\left(1 \pm e^{(\mu-\varepsilon)/T}\right)\varepsilon^{1/2}\,d\varepsilon.\end{aligned} \tag{3.50}$$

Working out this integral by parts, exactly as we did it with the one in Eq. (2.90), we get

$$\Omega = -\frac{2}{3}\frac{gVm^{3/2}}{\sqrt{2}\,\pi^2\hbar^3}\int_0^\infty \frac{\varepsilon^{3/2}d\varepsilon}{e^{(\varepsilon-\mu)/T}\pm 1} = -\frac{2}{3}\int_0^\infty \varepsilon g_3(\varepsilon)\langle N(\varepsilon)\rangle d\varepsilon. \tag{3.51}$$

But the last integral is just the total energy E of the gas:

$$\begin{aligned}E &= \frac{gV}{(2\pi\hbar)^3}\int_0^\infty \frac{p^2}{2m}\frac{4\pi p^2\,dp}{e^{[\varepsilon(p)-\mu]/T}\pm 1}\\ &= \frac{gVm^{3/2}}{\sqrt{2}\,\pi^2\hbar^3}\int_0^\infty \frac{\varepsilon^{3/2}d\varepsilon}{e^{(\varepsilon-\mu)/T}\pm 1} = \int_0^\infty \varepsilon g_3(\varepsilon)\langle N(\varepsilon)\rangle d\varepsilon,\end{aligned} \tag{3.52}$$

so that for any temperature and any particle type, $\Omega = -(2/3)E$. But since, from thermodynamics, $\Omega = -PV$, we have Eq. (3.48) proved. This universal relation[22] will be repeatedly used below.

3.3 Degenerate Fermi gas

Analysis of low-temperature properties of a *Fermi gas* is very simple in the limit $T = 0$. Indeed, in this limit, the Fermi–Dirac distribution (2.115) is just the step function:

$$\langle N(\varepsilon)\rangle = \begin{cases}1, & \text{for } \varepsilon < \mu,\\ 0, & \text{for } \mu < \varepsilon,\end{cases} \tag{3.53}$$

—see by the bold line in figure 3.2a. Since $\varepsilon = p^2/2m$ is isotropic in the momentum space, in that space the particles, at $T = 0$, fully occupy all possible quantum states inside a sphere (frequently called either the *Fermi sphere* or the *Fermi sea*) with some radius p_F (figure 3.2b), while all states above the sea surface are empty. Such *degenerate Fermi gas* is a striking manifestation of the Pauli principle: though in thermodynamic equilibrium at $T = 0$ all particles try to lower their energies as much

[22] For gases of diatomic and polyatomic molecules at relatively high temperatures, when some of their internal degrees of freedom are thermally excited, Eq. (3.48) is valid only for the translational-motion energy.

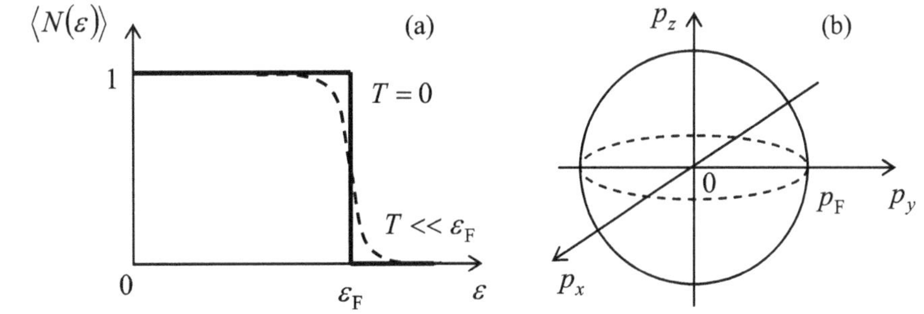

Figure 3.2. Representation of the Fermi sea: (a) on the energy axis, and (b) in the momentum space.

as possible, only g of them may occupy each translational ('orbital') quantum state. As a result, the sphere's volume is proportional to the particle number N, or rather to their density $n = N/V$.

Indeed, the radius p_F may be readily related to the number of particles N using Eq. (3.39), with the upper sign, whose integral in this limit is just the Fermi sphere's volume:

$$N = \frac{gV}{(2\pi\hbar)^3}\int_0^{p_F} 4\pi p^2 dp = \frac{gV}{(2\pi\hbar)^3}\frac{4\pi}{3}p_F^3. \tag{3.54}$$

Now we can use Eq. (3.3) to express via N the chemical potential μ (again, in the limit $T = 0$, it bears the special name of the *Fermi energy* ε_F)[23]:

$$\varepsilon_F \equiv \mu\,|_{T=0} = \frac{p_F^2}{2m} = \frac{\hbar^2}{2m}\left(6\pi^2\frac{N}{gV}\right)^{2/3} \equiv \left(\frac{9\pi^4}{2}\right)^{1/3} T_0 \approx 7.595\, T_0, \tag{3.55a}$$

where T_0 is the quantum temperature scale defined by Eq. (3.35). This formula quantifies the low-temperature trend of the function $\mu(T)$, clearly visible in figure 3.1, and in particular explains the ratio ε_F/T_0 mentioned in section 3.2. Note also a simple and very useful relation,

$$\varepsilon_F = \frac{3}{2}\frac{N}{g_3(\varepsilon_F)}, \quad \text{i.e. } g_3(\varepsilon_F) = \frac{3}{2}\frac{N}{\varepsilon_F}, \tag{3.55b}$$

which may be obtained immediately from the comparison of Eqs. (3.43) and (3.54).

The total energy of the degenerate Fermi gas may be (equally easily) calculated from Eq. (3.52):

$$E = \frac{gV}{(2\pi\hbar)^3}\int_0^{p_F}\frac{p^2}{2m}4\pi p^2 dp = \frac{gV}{(2\pi\hbar)^3}\frac{4\pi}{2m}\frac{p_F^5}{5} = \frac{3}{5}\varepsilon_F N, \tag{3.56}$$

showing that the average energy, $\langle\varepsilon\rangle \equiv E/N$, of a particle inside the Fermi sea is equal to 3/5 = 60% of that (ε_F) of particles in the most energetic occupied states, on

[23] Note that in the electronic engineering literature, μ is usually called the *Fermi level*, at any temperature.

the Fermi surface. Since, according to the formulas of chapter 1, at zero temperature $H = G = N\mu$, and $F = E$, the only thermodynamic variable still to be calculated is the gas pressure P. For it, we could use any of thermodynamic relations $P = (H - E)/V$ or $P = -(\partial F/\partial V)_T$, but it is even easier to use our recent result (3.48). Together with Eq. (3.56), it yields

$$P = \frac{2}{3}\frac{E}{V} = \frac{2}{5}\varepsilon_F\frac{N}{V} = \left(\frac{36\pi^4}{125}\right)^{1/3} P_0 \approx 3.035 P_0,\\ \text{where } P_0 \equiv nT_0 = \frac{\hbar^2 n^{5/3}}{mg^{2/3}}. \tag{3.57}$$

From here, it is straightforward to calculate the *bulk modulus* (reciprocal *compressibility*)[24],

$$K \equiv -V\left(\frac{\partial P}{\partial V}\right)_T = \frac{2}{3}\varepsilon_F\frac{N}{V}, \tag{3.58}$$

which is simpler to measure experimentally than P.

Perhaps the most important example[25] of the degenerate Fermi gas is the *conduction electrons* in metals—the electrons that belong to outer shells of the isolated atoms but become shared in solid metals, and as a result can move through the crystal lattice almost freely. Though the electrons (which are fermions with spin $s = ½$ and hence with the spin degeneracy $g = 2s + 1 = 2$) are negatively charged, the Coulomb interaction of conduction electrons with each other is substantially compensated by the positively charged ions of the atomic lattice, so that they follow the simple model discussed above, in which the interaction is disregarded, reasonably well. This is especially true for alkali metals (forming Group 1 of the periodic table of elements), whose experimentally measured Fermi surfaces are spherical within 1%—even within 0.1% for Na.

Table 3.1 lists, in particular, the experimental values of the bulk modulus for such metals, together with the values given by Eq. (3.58) using the ε_F calculated from Eq. (3.55*a*) with the experimental density of the conduction electrons. The agreement is pretty impressive, taking into account that the simple theory described above completely ignores the Coulomb and exchange interactions of the electrons. This agreement implies that, surprisingly, the experimentally observed rigidity of solids (or at least metals) is predominantly due to the kinetic energy (3.3) of the conduction electrons, rather than any electrostatic interactions—though, to be fair, these interactions are the crucial factor defining the equilibrium value of n. Numerical

[24] For a general discussion of this notion, see, e.g. *Part CM* Eqs. (7.32) and (7.36).

[25] Recently, degenerate gases (with $\varepsilon_F \sim 5T$) have been formed of weakly interacting Fermi atoms as well—see, e.g. [6] and references therein. Another interesting example of the system that may be approximately treated as a degenerate Fermi gas is the set of $Z \gg 1$ electrons in heavy atoms. However in this system the account of electron interaction via the electrostatic field they create is important. Since for this *Thomas–Fermi model* of atoms, the thermal effects are negligible, it was discussed already in the quantum-mechanical part of this series (see *Part QM* chapter 8), and in this course its analysis is left for the reader's exercise.

Table 3.1. Experimental and theoretical parameters of electrons' Fermi sea in some alkali metals[28].

Metal	ε_F (eV) Eq. (3.55)	K (GPa) Eq. (3.58)	K (GPa) experiment	γ (mcal mole^{-1}·K^{-2}) Eq. (3.69)	γ (mcal mole^{-1}·K^{-2}) experiment
Na	3.24	923	642	0.26	0.35
K	2.12	319	281	0.40	0.47
Rb	1.85	230	192	0.46	0.58
Cs	1.59	154	143	0.53	0.77

calculations using more accurate approximations (e.g. the density functional theory[26]), which agree with experiment with a few percent accuracy, confirm this conclusion[27].

Now looking at the values of ε_F listed in the table, note that room temperatures ($T_K \sim 300$ K) correspond to $T \sim 25$ meV. As a result, virtually all experiments with metals, at least in their solid or liquid form, are performed in the limit $T \ll \varepsilon_F$. According to Eq. (3.39), at such temperatures the occupancy step described by the Fermi–Dirac distribution has a non-zero but relatively small width of the order of T—see the dashed line in figure 3.2a. Calculations in this case are much facilitated by the so-called *Sommerfeld expansion formula*[29] for the integrals like those in Eqs. (3.41) and (3.52):

$$I(T) \equiv \int_0^\infty \varphi(\varepsilon)\langle N(\varepsilon)\rangle d\varepsilon \approx \int_0^\mu \varphi(\varepsilon)d\varepsilon + \frac{\pi^2}{6}T^2\frac{d\varphi(\mu)}{d\mu}, \quad \text{at } T \ll \mu, \tag{3.59}$$

where $\varphi(\varepsilon)$ is an arbitrary function that is sufficiently smooth at $\varepsilon = \mu$ and integrable at $\varepsilon = 0$. In order to prove this formula, let us introduce another function,

$$f(\varepsilon) \equiv \int_0^\varepsilon \varphi(\varepsilon')d\varepsilon', \quad \text{so that} \quad \varphi(\varepsilon) = \frac{df(\varepsilon)}{d\varepsilon}, \tag{3.60}$$

and work out the integral $I(T)$ by parts:

$$\begin{aligned} I(T) &\equiv \int_0^\infty \frac{df(\varepsilon)}{d\varepsilon}\langle N(\varepsilon)\rangle d\varepsilon = \int_{\varepsilon=0}^{\varepsilon=\infty} \langle N(\varepsilon)\rangle df \\ &= [\langle N(\varepsilon)\rangle f]_{\varepsilon=0}^{\varepsilon=\infty} - \int_{\varepsilon=0}^{\varepsilon=\infty} f(\varepsilon)d\langle N(\varepsilon)\rangle \\ &= \int_0^\infty f(\varepsilon)\left[-\frac{\partial\langle N(\varepsilon)\rangle}{\partial\varepsilon}\right]d\varepsilon. \end{aligned} \tag{3.61}$$

[26] See, e.g. *Part QM* section 8.4.

[27] Note also a huge difference between the very high bulk modulus of metals ($K \sim 10^{11}$ Pa) and its very low values in usual, atomic gases (for them, at ambient conditions, $K \sim 10^5$ Pa). About four orders of magnitude of this difference in due to that in the particle density N/V, but the balance is due to the electron gas' degeneracy. Indeed, in an ideal classical gas, $K = P = T(N/V)$, so that the factor (2/3) ε_F in Eq. (3.58), of the order of a few eV in metals, should be compared with the factor $T \approx 25$ meV in the classical gas at room temperature.

[28] Data from [7].

[29] Named after A Sommerfeld, who was the first (in 1927) to apply quantum mechanics to degenerate Fermi gases, in particular to electrons in metals, and may be credited for most of the results discussed in this section.

As evident from Eq. (2.115) and/or figure 3.2a, at $T \ll \mu$ the function $(-\partial\langle N(\varepsilon)\rangle/\partial\varepsilon)$ approaches zero for all energies, besides a narrow peak, of the unit area, at $\varepsilon \approx \mu$. Hence, if we expand the function $f(\varepsilon)$ in the Taylor series near this point, just a few leading terms of the expansion should give us a good approximation:

$$\begin{aligned} I(T) &\approx \int_0^\infty \left[f(\mu) + \frac{df}{d\varepsilon}\bigg|_{\varepsilon=\mu}(\varepsilon-\mu) + \frac{1}{2}\frac{d^2f}{d\varepsilon^2}\bigg|_{\varepsilon=\mu}(\varepsilon-\mu)^2 \right]\left[-\frac{\partial\langle N(\varepsilon)\rangle}{\partial\varepsilon}\right]d\varepsilon \\ &= \int_0^\mu \varphi(\varepsilon')d\varepsilon' \int_0^\infty \left(-\frac{\partial\langle N(\varepsilon)\rangle}{\partial\varepsilon}\right)d\varepsilon + \varphi(\mu)\int_0^\infty (\varepsilon-\mu)\left[-\frac{\partial\langle N(\varepsilon)\rangle}{\partial\varepsilon}\right]d\varepsilon \\ &+ \frac{1}{2}\frac{d\varphi(\mu)}{d\mu}\int_0^\infty (\varepsilon-\mu)^2\left[-\frac{\partial\langle N(\varepsilon)\rangle}{\partial\varepsilon}\right]d\varepsilon. \end{aligned} \tag{3.62}$$

In the last form of this relation, the first integral over ε equals $\langle N(\varepsilon=0)\rangle - \langle N(\varepsilon=\infty)\rangle = 1$, the second one vanishes (because the function under it is asymmetric with respect to the point $\varepsilon = \mu$), and only the last one needs to be dealt with explicitly, by working it out by parts and then using a table integral[30]:

$$\begin{aligned} \int_0^\infty (\varepsilon-\mu)^2\left[-\frac{\partial\langle N(\varepsilon)\rangle}{\partial\varepsilon}\right]d\varepsilon &\approx T^2\int_{-\infty}^{+\infty} \xi^2\frac{d}{d\xi}\left(-\frac{1}{e^\xi+1}\right)d\xi \\ &= 4T^2\int_0^{+\infty}\frac{\xi d\xi}{e^\xi+1} = 4T^2\frac{\pi^2}{12}. \end{aligned} \tag{3.63}$$

Being plugged into Eq. (3.62), this result proves the Sommerfeld formula (3.59). The last preparatory step we need to make is to account for a possible small difference (as we will see below, also proportional to T^2) between the temperature-dependent chemical potential $\mu(T)$ and the Fermi energy defined as $\varepsilon_F \equiv \mu(0)$, in the largest (first) term on the right-hand side of Eq. (3.59), to write

$$\begin{aligned} I(T) &\approx \int_0^{\varepsilon_F} \varphi(\varepsilon)d\varepsilon + (\mu - \varepsilon_F)\varphi(\mu) + \frac{\pi^2}{6}T^2\frac{d\varphi(\mu)}{d\mu} \\ &= I(0) + (\mu - \varepsilon_F)\varphi(\mu) + \frac{\pi^2}{6}T^2\frac{d\varphi(\mu)}{d\mu}. \end{aligned} \tag{3.64}$$

Now, applying this formula to Eq. (3.41) and the last form of Eq. (3.52), we get the following results (which are valid for any dispersion law $\varepsilon(\mathbf{p})$ and even any dimensionality of the gas):

$$N(T) = N(0) + (\mu - \varepsilon_F)g(\mu) + \frac{\pi^2}{6}T^2\frac{dg(\mu)}{d\mu}, \tag{3.65}$$

$$E(T) = E(0) + (\mu - \varepsilon_F)\mu g(\mu) + \frac{\pi^2}{6}T^2\frac{d}{d\mu}[\mu g(\mu)]. \tag{3.66}$$

[30] See, e.g. Eqs. (A.35*c*) and (A.15*b*), with $n = 1$.

If the number of particles does not change with temperature, $N(T) = N(0)$, as in most experiments, Eq. (3.65) gives the following formula for finding the temperature-induced change of μ:

$$\mu - \varepsilon_F = -\frac{\pi^2}{6}T^2\frac{1}{g(\mu)}\frac{dg(\mu)}{d\mu}. \tag{3.67}$$

Note that the change is quadratic in T and negative, in agreement with the numerical results shown with the red line in figure 3.1. Plugging this expression (which is only valid when the magnitude of the change is much smaller than ε_F) into Eq. (3.66), we get the finite-temperature correction to the energy:

$$E(T) - E(0) = \frac{\pi^2}{6}g(\mu)T^2, \tag{3.68}$$

where within the accuracy of our approximation, μ may be replaced with ε_F. (Due to the universal relation (3.48), this result also gives the temperature correction to the pressure.)

Now we may use Eq. (3.68) to calculate the heat capacity of the degenerate Fermi gas:

$$C_V \equiv \left(\frac{\partial E}{\partial T}\right)_V = \gamma T, \quad \text{with} \quad \gamma = \frac{\pi^2}{3}g(\varepsilon_F). \tag{3.69}$$

According to Eq. (3.55*b*), in the particular case of a 3D gas with the isotropic and parabolic dispersion law (3.3), Eq. (3.69) reduces to

$$\gamma = \frac{\pi^2}{2}\frac{N}{\varepsilon_F}, \quad \text{i.e.} \; c_V \equiv \frac{C_V}{N} = \frac{\pi^2}{2}\frac{T}{\varepsilon_F} \ll 1. \tag{3.70}$$

This important result deserves a discussion. First, note that within the range of validity of the Sommerfeld approximation ($T \ll \varepsilon_F$), the specific heat of the degenerate gas is much smaller than that of the classical gas, even without internal degrees of freedom, $c_V = 3/2$—see Eq. (3.19). The physical reason for such a low heat capacity is that the particles deep inside the Fermi sea cannot pick up thermal excitations with available energies of the order of $T \ll \varepsilon_F$, because all states above them are already occupied. The only particles (or rather quantum states, due to the particle indistinguishability) that may be excited with such small energies are those at the very Fermi surface, more exactly within a surface layer of thickness $\Delta\varepsilon \sim T \ll \varepsilon_F$, and Eq. (3.70) presents a very vivid expression of this fact.

The second important feature of Eqs. (3.69) and (3.70) is the linear dependence of the heat capacity on temperature, which decreases with a reduction of T much slower than that of crystal vibrations—see Eq. (2.99). This means that in metals the specific heat at temperatures $T \ll T_D$ is dominated by the conduction electrons. Indeed, experiments confirm not only the linear dependence (3.70) of the specific heat[31], but also the values of the proportionality coefficient $\gamma \equiv C_V/T$ for cases when

[31] Solids, with their low thermal expansion coefficients, provide a virtually fixed-volume confinement for the electron gas, so that the specific heat measured at ambient conditions may be legitimately compared with the calculated c_V.

ε_F can be calculated independently, for example for alkali metals—see the two rightmost columns of table 3.1 above. More typically, Eq. (3.69) is used for the experimental measurement of the density of states on the Fermi surface, $g(\varepsilon_F)$—the factor which participates in many theoretical results, in particular in transport properties of degenerate Fermi gases (see chapter 6 below).

3.4 Bose–Einstein condensation

Now let us explore what happens at cooling of an ideal gas of bosons. Figure 3.3a shows on a more appropriate, log–log scale, the same plot as figure 3.1, i.e. the result of a numerical solution of Eq. (3.47) with the appropriate (lower) sign in the denominator. One can see that that the chemical potential μ indeed tends to zero at some finite 'critical temperature' T_c. This temperature may be found by taking $\mu = 0$ in Eq. (3.47), which is then reduced to a table integral[32]:

$$T_c = T_0\left[\frac{1}{\sqrt{2}\pi^2}\int_0^\infty \frac{\xi^{1/2}d\xi}{e^\xi - 1}\right]^{-2/3} = T_0\left[\frac{1}{\sqrt{2}\pi^2}\Gamma\left(\frac{3}{2}\right)\zeta\left(\frac{3}{2}\right)\right]^{-2/3} \approx 3.313\ T_0, \quad (3.71)$$

the result explaining the T_c/T_0 ratio mentioned in section 3.2.

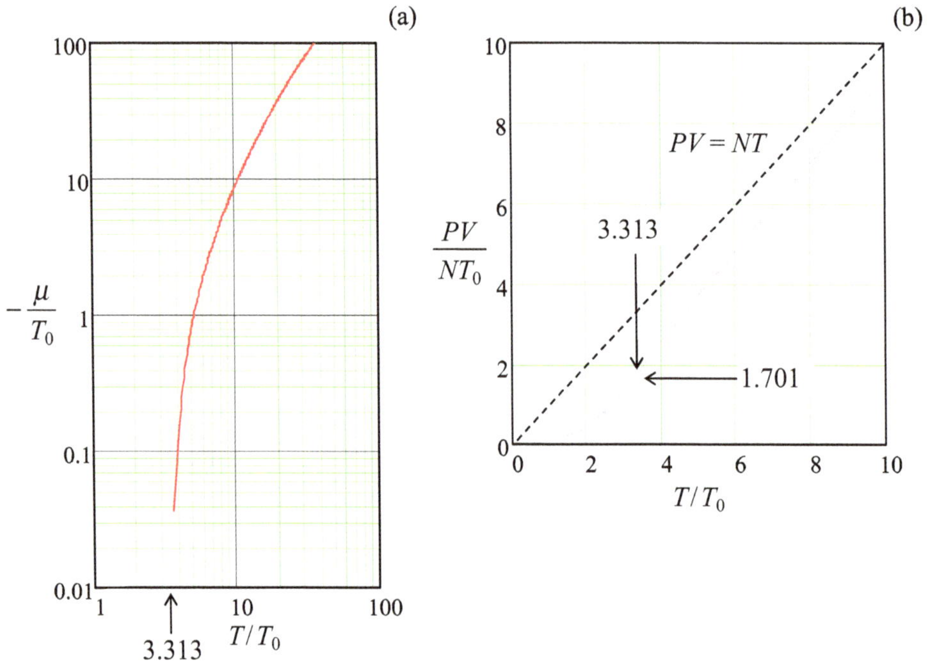

Figure 3.3. The Bose–Einstein condensation: (a) the chemical potential of the gas and (b) its pressure, as functions of temperature. The dashed line corresponds to the classical-gas approximation.

[32] See, e.g. Eqs. (A.35*b*), (A.10*b*) and (A.34*e*) with $s = 3/2$.

Hence we should have a good look at the temperature interval $0 < T < T_c$, which cannot be directly described by Eq. (3.40) (with the appropriate negative sign in the denominator), and hence may look rather mysterious. Indeed, within this range, the chemical potential μ cannot be either negative or zero, because according to Eq. (3.71), in this case Eq. (3.40) would give a value of N smaller than the number of particles we actually have. On the other hand, μ cannot be positive either, because the integral (3.40) would diverge at $\varepsilon \to \mu$ due to the divergence of $\langle N(\varepsilon)\rangle$—see, e.g. figure 2.15. The only possible resolution of the paradox, suggested by A Einstein in 1925, is as follows: at $T < T_c$, the chemical potential of each particle of the system still equals exactly zero, but a certain number (N_0 of N) of them are in the ground state (with $\varepsilon \equiv p^2/2m = 0$), forming the so-called *Bose–Einstein condensate*, very frequently referred to as the BEC. Since the condensate particles do not contribute to Eq. (3.40) (because of the factor $\varepsilon^{1/2} = 0$), their number N_0 may be calculated by using that formula (or, equivalently, Eq. (3.44)) with $\mu = 0$, to find the number $(N - N_0)$ of particles still remaining in the gas, i.e. having energy $\varepsilon > 0$:

$$N - N_0 = \frac{gV(mT)^{3/2}}{\sqrt{2}\pi^2\hbar^3}\int_0^\infty \frac{\xi^{1/2}d\xi}{e^\xi - 1}. \tag{3.72}$$

This result is even simpler than it may look. Indeed, let us write it for the case $T = T_c$, when $N_0 = 0$:[33]

$$N = \frac{gV(mT_c)^{3/2}}{\sqrt{2}\pi^2\hbar^3}\int_0^\infty \frac{\xi^{1/2}d\xi}{e^\xi - 1}. \tag{3.73}$$

Since the dimensionless integral in both relations is the same, we may just eliminate it, getting an extremely simple and elegant result:

$$\frac{N - N_0}{N} = \left(\frac{T}{T_c}\right)^{3/2}, \quad \text{so that} \quad N_0 = N\left[1 - \left(\frac{T}{T_c}\right)^{3/2}\right], \quad \text{at } T \leqslant T_c. \tag{3.74a}$$

Please note that this result is only valid for the particles whose motion, within the volume V, is free—in other words, for a spatially-uniform system of particles confined in a rigid-wall box of volume V. In most experiments with the Bose–Einstein condensation of diluted gases of neutral (and hence very weakly interacting) atoms, they are held not in such a box, but at the bottom of a 'soft' potential well, which may be well approximated by a 3D quadratic parabola: $U(\mathbf{r}) = m\omega^2r^2/2$. It is straightforward (and hence left for the reader's exercise) to show that in this case the temperature dependence of N_0 is somewhat different:

$$N_0 = N\left[1 - \left(\frac{T}{T_c^*}\right)^3\right], \quad \text{at } T \leqslant T_c^*, \tag{3.74b}$$

[33] This is, of course, just another form of Eq. (3.71).

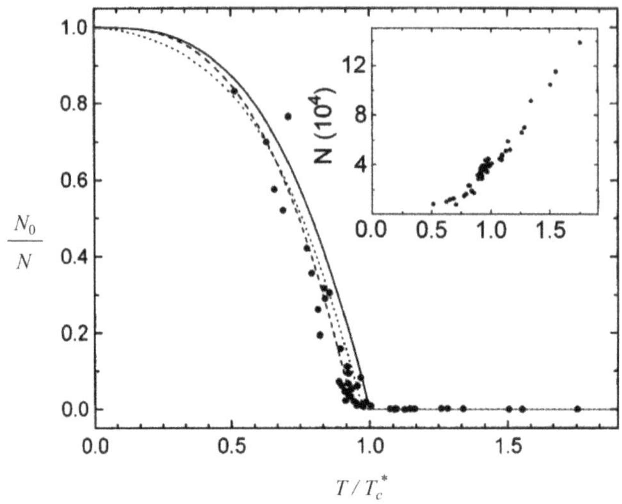

Figure 3.4. The total number N of trapped ^{87}Rb atoms (inset) and their ground-state fraction N_0/N, as functions of the ratio T/T_c, as measured in one of the pioneering experiments by J Ensher *et al* in 1996. In this experiment, T_c^* was as low as 0.28×10^{-6} K. The solid line shows the simple theoretical dependence $N(T)$, given by Eq. (3.74*b*), while other lines correspond to more complex theories taking into account the finite number N of trapped atoms. Reprinted with permission from [8]. Copyright 1996 by the American Physical Society.

where T_c^* is a different critical temperature, which now depends on $\hbar\omega$, i.e. the confining potential's 'steepness'. (In this case V is not exactly fixed; however, the effective volume occupied by the particles at $T = T_c^*$ is related to this temperature by a formula close to Eq. (3.71), so that all estimates given above are still valid.) Figure 3.4 shows one of the first sets of experimental data for the Bose–Einstein condensation of dilute gases of neutral atoms. Taking into account the finite number of particles in the experiment, the agreement with the simple theory is surprisingly good.

Now returning to the spatially-uniform gas, let us explore what happens below the critical temperature with other gas parameters. Eq. (3.52) with the appropriate (lower) sign shows that approaching T_c from higher temperatures, the gas energy and hence its pressure do not vanish—see the red line in figure 3.3b. Indeed, at $T = T_c$ (where $\mu = 0$), that formula yields[34]

$$\begin{aligned} E(T_c) &= gV\frac{m^{3/2}T_c^{5/2}}{\sqrt{2}\pi^2\hbar^3}\int_0^\infty \frac{\xi^{3/2}d\xi}{e^\xi - 1} \\ &= gV\frac{m^{3/2}T_c^{5/2}}{\sqrt{2}\pi^2\hbar^3}\Gamma\left(\frac{5}{2}\right)\zeta\left(\frac{5}{2}\right) \approx 0.7701\ NT_c, \end{aligned} \tag{3.75}$$

so that using the universal relation (3.48), we get the pressure value,

[34] For the involved dimensionless integral see, e.g. Eqs. (A.35*b*), (A.10*b*) and (A.34*e*) with $s = 5/2$.

$$P(T_c) = \frac{2}{3}\frac{E(T_c)}{V} = \frac{\zeta(5/2)}{\zeta(3/2)}\frac{N}{V}T_c \approx 0.5134\frac{N}{V}T_c \approx 1.701P_0, \tag{3.76}$$

which is somewhat lower than, but comparable to $P(0)$ for the fermions—cf. Eq. (3.57).

Now we can use the same Eq. (3.52), also with $\mu = 0$, to calculate the energy of the gas at $T < T_c$,

$$E(T) = gV\frac{m^{3/2}T^{5/2}}{\sqrt{2}\pi^2\hbar^3}\int_0^\infty \frac{\xi^{3/2}d\xi}{e^\xi - 1}. \tag{3.77}$$

Comparing this relation with the first form of Eq. (3.75), which features the same integral, we immediately get one more simple temperature dependence:

$$E(T) = E(T_c)\left(\frac{T}{T_c}\right)^{5/2}, \quad \text{at } T \leqslant T_c. \tag{3.78}$$

From the universal relation (3.48), we immediately see that the gas pressure follows the same dependence:

$$P(T) = P(T_c)\left(\frac{T}{T_c}\right)^{5/2}, \quad \text{at } T \leqslant T_c. \tag{3.79}$$

This temperature dependence of pressure is shown with the blue line in figure 3.3b. The plot shows that for all temperatures (both below and above T_c) the pressure is below that of the classical gas of the same density. Now note also that since, according to Eqs. (3.57) and (3.76), $P(T_c) \propto P_0 \propto V^{-5/3}$, while, according to Eqs. (3.35) and (3.71), $T_c \propto T_0 \propto V^{-2/3}$, the pressure (3.79) is proportional to $V^{-5/3}/(V^{-2/3})^{5/2} = V^0$, i.e. does not depend on the volume at all! The physics of this result (which is valid at $T < T_c$ only) is that as we decrease the volume at a fixed total number N of particles, more and more of them go to the condensate, decreasing the number $(N - N_0)$ of particles in the gas phase, but not changing its pressure. Such behavior is very typical for the coexistence of two phases—see, in particular, the next chapter.

The last thermodynamic variable of major interest is the heat capacity, because it may be readily measured. For temperatures $T \leqslant T_c$, it may be easily calculated from Eq. (3.78):

$$C_V(T) \equiv \left(\frac{\partial E}{\partial T}\right)_{N,V} = E(T_c)\frac{5}{2}\frac{T^{3/2}}{T_c^{5/2}}, \tag{3.80}$$

so that below T_c, the capacity *increases* with temperature, at the critical temperature reaching the value

$$C_V(T_c) = \frac{5}{2}\frac{E(T_c)}{T_c} \approx 1.925\,N, \tag{3.81}$$

which is approximately 28% above that $(3N/2)$ of the classical gas. (As a reminder, in both cases we ignore the contributions from the internal degrees of freedom.) The analysis for $T \geqslant T_c$ is a little bit more cumbersome, because differentiating E over temperature—say, using Eq. (3.52)—one should also take into account the temperature dependence of μ that follows from Eq. (3.40)—see also figure 3.1. However, the most important feature of the result may be predicted without the calculation—which is being left for the reader's exercise. Namely, since at $T \gg T_c$ the heat capacity has to approach the classical value 1.5 N, starting from the value (3.81), it must *decrease* with temperature at $T > T_c$, thus forming a sharp maximum (a 'cusp') at the critical point $T = T_c$—see figure 3.5.

Such a cusp is good indication of the Bose–Einstein condensation in virtually any experimental system, especially because inter-particle interactions (unaccounted for in our simple discussion) typically make this feature even more substantial, turning it into a weak (logarithmic) singularity. Historically, such a singularity was the first noticed, though not immediately understood sign of the Bose–Einstein condensation, observed in 1931 by W Keesom and K Clusius in liquid ^{4}He at the *λ-point* (so called exactly because of the characteristic shape of the $C_V(T)$ dependence) $T = T_c \approx 2.17$ K. Other milestones of the Bose–Einstein condensation studies include:

- the experimental discovery of superconductivity, which was later explained as the result of the Bose–Einstein condensation of electron pairs, by H Kamerlingh-Onnes in 1911;
- the development of the Bose–Einstein statistics, and predicting the condensation, by S Bose and A Einstein, in 1924–25;
- the discovery of superfluidity in liquid ^{4}He by P Kapitza and (independently) by J Allen and D Misener in 1937, and its explanation as a result of the Bose–

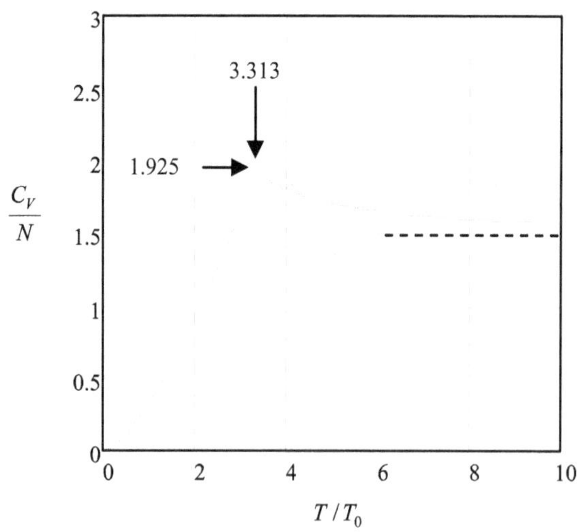

Figure 3.5. Temperature dependences of the heat capacity of an ideal Bose–Einstein gas, numerically calculated from Eqs. (3.52) and (3.40) for $T \geqslant T_c$, and given by Eq. (3.80). for $T \leqslant T_c$.

Einstein condensation by F and H Londons and L Titza, with further elaborations by L Landau—all in 1938;
- the explanation of superconductivity as a result of electron binding to *Cooper pairs*, with simultaneous condensation of the resulting bosons, by J Bardeen, L Cooper, and J Schrieffer in 1957;
- the discovery of superfluidity of two different phases of ^{3}He, due to the similar Bose–Einstein condensation of pairs of its fermion atoms, by D Lee, D Osheroff, and R Richardson in 1972;
- the first observation of the Bose–Einstein condensation in dilute gases (^{87}Ru by E Cornell, C Wieman *et al* and ^{23}Na by W Ketterle *et al*) in 1995.

The importance of the last achievement, and of the continuing intensive research work in this direction, stems from the fact that in contrast to other Bose–Einstein condensates, in dilute gases (with the typical density n as low as $\sim 10^{14}$ cm^{-3}) the particles interact very weakly, and hence many experimental results are very close to the simple theory described above and its straightforward elaborations—see, e.g. figure 3.4.[35] On the other hand, the importance of other implementations of the Bose–Einstein condensates, which involve more complex and challenging physics, should not be underestimated—as it sometimes is.

Perhaps the most important feature of any Bose–Einstein condensate is that all N_0 condensed particles are in the same quantum state, and hence are described by exactly the same wavefunction. This wavefunction is substantially less 'feeble' than that of a single particle—in the following sense. In the second quantization language[36], the well-known Heisenberg's uncertainty relation may be rewritten for the creation/annihilation operators; in particular, for bosons,

$$|\delta\hat{a}\,\delta\hat{a}^{\dagger}| \geqslant 1. \tag{3.82}$$

Since $\hat{a}$ and $\hat{a}^{\dagger}$ are the quantum-mechanical operators of the complex amplitude $a = A\exp\{i\varphi\}$ and its complex conjugate $a^* = A\exp\{-i\varphi\}$, where A and φ are real amplitude and phase of the wavefunction, Eq. (3.82) yields the following approximate uncertainty relation (strict in the limit $\delta\varphi \ll 1$) between the number of particles $N = AA^*$ and the phase φ

$$\delta N \delta\varphi \geqslant 1/2. \tag{3.83}$$

This means that a condensate of $N \gg 1$ bosons may be in a state with both phase and amplitude of the wavefunction behaving virtually as c-numbers, with very small relative uncertainties: $\delta N \ll N$, $\delta\varphi \ll 1$. Moreover, such states are much less susceptible to perturbations by experimental instruments. For example, the electric current carried along a superconducting wire by a coherent Bose–Einstein

[35] Such controllability of theoretical description has motivated the use of dilute-gas BECs for modeling of renowned problem of many-body physics—see, e.g. the review [9]. These efforts are assisted by the development of better techniques for reaching the necessary sub-μK temperatures—see, e.g. the recent work [10]. For a more general, detailed discussion see, e.g. [11].

[36] See, e.g. *Part QM* section 8.3.

condensate of Cooper pairs may be as high as hundreds of amperes. As a result, the 'strange' behaviors predicted by the quantum mechanics are not averaged out as in the usual particle ensembles (see, e.g. the discussion of the density matrix in section 2.1), but may be directly revealed in macroscopic, measurable dynamics of the condensate.

For example, the density $\mathbf{j}$ of the electric current may be described by the same formula as the well-known usual probability current density of a single quantum particle[37], just multiplied by the electric charge $q = -2e$ of a single pair and the Cooper pair density n:

$$\mathbf{j} = qn\frac{\hbar}{m}\left(\nabla\varphi - \frac{q}{\hbar}\mathbf{A}\right), \tag{3.84}$$

where $\mathbf{A}$ is the vector-potential of the (electro)magnetic field. If a superconducting wire is not extremely thin, the supercurrent does not penetrate into its interior[38]. As a result, the integral of Eq. (3.84), taken along a closed superconducting loop, inside its interior (where $\mathbf{j} = 0$), yields

$$\frac{q}{\hbar}\oint_C \mathbf{A} \cdot d\mathbf{r} = \Delta\varphi = 2\pi M, \tag{3.85}$$

where M is an integer. But, according to the basic electrodynamics, the integral on the left-hand side of this relation is nothing more than the flux Φ of the magnetic field $\mathscr{B}$ piercing the wire loop area A. Thus we immediately arrive at the famous *magnetic flux quantization* effect

$$\Phi \equiv \int_A \mathscr{B}_n d^2r = M\Phi_0, \quad \text{where } \Phi_0 \equiv \frac{2\pi\hbar}{|q|} \approx 2.07 \times 10^{-15}\,\text{Wb}, \tag{3.86}$$

which was theoretically predicted in 1950 and experimentally observed in 1961. Amazingly, this effect holds even 'over miles of dirty lead wire', citing H Casimir's famous expression, sustained by the coherence of the Bose–Einstein condensate of Cooper pairs. Other prominent examples of such *macroscopic quantum effects* in Bose–Einstein condensates include not only the superfluidity and superconductivity as such, but also the Josephson effect, quantized Abrikosov vortices, etc. Some of these effects are briefly discussed in other parts of this series[39].

3.5 Gases of weakly interacting particles

Now let us discuss the effects of weak particle interaction effects on properties of their gas. (Unfortunately, I will have time to do that only for classical gases[40].) In

---

[37] See, e.g. *Part QM* Eq. (3.28).

[38] This is the *Meissner–Ochsenfeld* (or just 'Meissner') *effect* which may be also readily explained using Eq. (3.84), combined with the Maxwell equations—see, e.g. *Part EM* section 6.4.

[39] See *Part QM* sections 1.6 and 3.1, and *Part EM* sections 6.4–6.5.

[40] A concise discussion of effects of weak interactions on properties of *quantum* gases may be found, for example, in chapter 10 of the textbook by K Huang [12].

most cases of interest, particle interaction may be described by a certain potential energy U, so that the total energy is

$$E = \sum_{k=1}^{N} \frac{p_k^2}{2m} + U(\mathbf{r}_1, \ldots, \mathbf{r}_j, \ldots, \mathbf{r}_N), \tag{3.87}$$

where $\mathbf{r}_k$ is the position of the kth particle's center. Let us see how far the statistical physics would allow us to proceed for an arbitrary potential U. For $N \gg 1$, at the calculation of the Gibbs statistical sum (2.59), we may perform the usual transfer from the summation over all quantum states of the system to integration over the $6N$-dimensional space, with the correct Boltzmann counting:

$$\begin{aligned} Z = \sum_m e^{-E_m/T} &\rightarrow \frac{1}{N!} \frac{g^N}{(2\pi\hbar)^{3N}} \int \exp\left\{-\sum_{k=1}^{N} \frac{p_j^2}{2mT}\right\} d^3p_1 \ldots d^3p_N \\ &\times \int \exp\left\{-\frac{U(\mathbf{r}_1, \ldots \mathbf{r}_N)}{T}\right\} d^3r_1 \ldots d^3r_N \\ &\equiv \left(\frac{1}{N!} \frac{g^N V^N}{(2\pi\hbar)^{3N}} \int \exp\left\{-\sum_{k=1}^{N} \frac{p_j^2}{2mT}\right\} d^3p_1 \ldots d^3p_N \right) \\ &\times \left(\frac{1}{V^N} \int \exp\left\{-\frac{U(\mathbf{r}_1, \ldots \mathbf{r}_N)}{T}\right\} d^3r_1 \ldots d^3r_N \right). \end{aligned} \tag{3.88}$$

But according to Eq. (3.14), the first operand in the last product is just the statistical sum of an ideal gas (with the same g, N, V, and T), so that we may use Eq. (2.63) to write

$$\begin{aligned} F &= F_{\text{ideal}} - T \ln\left[\frac{1}{V^N} \int d^3r_1 \ldots d^3r_N e^{-U/T}\right] \\ &\equiv F_{\text{ideal}} - T \ln\left[1 + \frac{1}{V^N} \int d^3r_1 \ldots d^3r_N (e^{-U/T} - 1)\right], \end{aligned} \tag{3.89}$$

where F_{ideal} is the free energy of the ideal gas (i.e. the same gas but with $U = 0$), given by Eq. (3.16).

I believe that Eq. (3.89) is a very convincing demonstration of the enormous power of the statistical physics methods. Instead of trying to solve an impossibly complex problem of classical dynamics of $N \gg 1$ (think of $N \sim 10^{23}$) interacting particles, and calculating appropriate ensemble averages later on, the Gibbs approach reduces finding the free energy (and then, from thermodynamic relations, all other thermodynamic variables) to the calculation of just one integral on its right-hand side of Eq. (3.89). Still, this integral is $3N$-dimensional and may be worked out analytically only if the particle interaction is weak in some sense. Indeed, the last form of Eq. (3.89) makes its especially evident that if $U \rightarrow 0$ everywhere, the term in the parentheses under the integral vanishes, and so does the integral itself, and hence the addition to F_{ideal}.

Now let us see what this integral would yield for the simplest, *short-range* interactions, in which the potential U is substantial only when the mutual distance $\mathbf{r}_{kk'} \equiv \mathbf{r}_k - \mathbf{r}_{k'}$ between the centers of two particles is smaller than certain value $2r_0$, where r_0 may be interpreted as the particle radius. If the gas is sufficiently dilute, so that the radius r_0 is much smaller than the average distance r_{ave} between the particles, the integral in the last form of Eq. (3.89) is of the order of $(2r_0)^{3N}$, i.e. much smaller than $r_{\text{ave}}^{3N} \sim V^N$. Then we may expand the logarithm in that form into the Taylor series with respect to the small second term in the square brackets, and keep only its first nonvanishing term:

$$F \approx F_{\text{ideal}} - \frac{T}{V^N}\int d^3r_1 \ldots d^3r_N \left(e^{-U/T} - 1\right). \tag{3.90}$$

Moreover, if the gas density is so low, the chances for three or more particles to come close to each other and interact (collide) are typically very small, so that *pair collisions* are the most important. In this case, we may recast the integral in Eq. (3.90) as a sum of $N(N-1)/2 \approx N^2/2$ similar terms describing such pair interactions, each of the type

$$V^{N-2}\int \left(e^{-U(\mathbf{r}_{kk'})/T} - 1\right) d^3r_k d^3r_{k'}. \tag{3.91}$$

It is convenient to think about $\mathbf{r}_{kk'}$ as the radius-vector of the particle number k in the reference frame with the origin placed at the center of the particle number k'—see figure 3.6a.

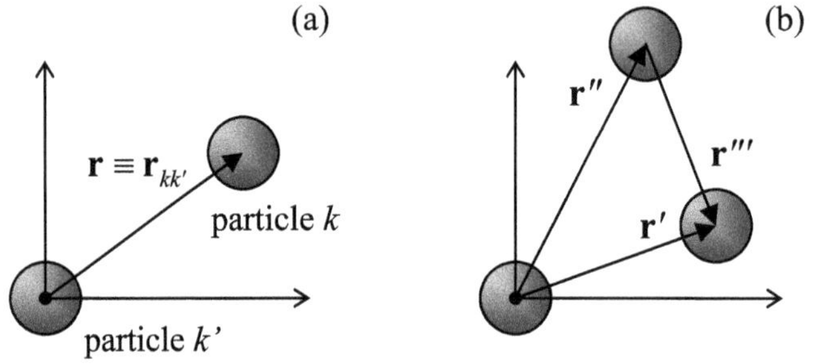

Figure 3.6. The definition of the interparticle distance vectors at their (a) pair and (b) triple interactions.

Then it is clear that in Eq. (3.91), we may first calculate the integral over $\mathbf{r}_{k'}$, while keeping the distance vector $\mathbf{r}_{kk'}$, and hence $U(\mathbf{r}_{kk'})$, constant, getting one more factor V. Moreover, since all particle pairs are similar, in the remaining integral over $\mathbf{r}_{kk'}$ we may drop the radius-vector index, so that Eq. (3.90) becomes

$$F = F_{\text{ideal}} - \frac{T}{V^N}\frac{N^2}{2}V^{N-1}\int \left(e^{-U(\mathbf{r})/T} - 1\right) d^3r = F_{\text{ideal}} + \frac{T}{V}N^2 B(T), \tag{3.92}$$

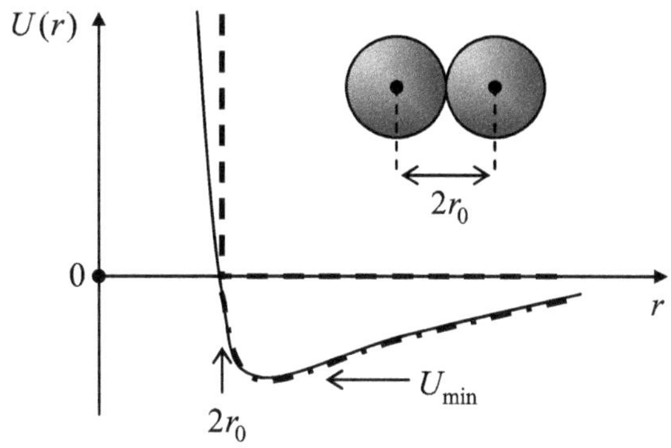

Figure 3.7. Pair interactions of particles. Solid line: a typical interaction potential; dashed line: its hardball model (3.95); dash-dotted line: the improved model (3.97)—all schematically. The inset illustrates the hardball model's physics.

where the function $B(T)$, called the *second virial coefficient*[41], has an especially simple form for spherically-symmetric interactions:

$$B(T) \equiv \frac{1}{2}\int \left(1 - e^{-U(\mathbf{r})/T}\right) d^3r \rightarrow \frac{1}{2}\int_0^{\infty} 4\pi r^2 dr \left(1 - e^{-U(r)/T}\right). \tag{3.93}$$

From Eq. (3.92), and the second of the thermodynamic relations (1.35), we already know something particular about the equation of state $P(V, T)$:

$$P = -\left(\frac{\partial F}{\partial V}\right)_{T,N} = P_{\text{ideal}} + \frac{N^2 T}{V^2}B(T) = T\left[\frac{N}{V} + B(T)\frac{N^2}{V^2}\right]. \tag{3.94}$$

We see that at a fixed gas density $n = N/V$, the pair interaction creates additional pressure, proportional to $(N/V)^2 = n^2$ and a function of temperature, $B(T)T$.

Let us calculate $B(T)$ for a couple of simple models of particle interactions. The solid curve in figure 3.7 shows (schematically) a typical form of the interaction potential between electrically neutral atoms/molecules. At large distances the interaction of particles that do not have their own permanent electrical dipole moment $\mathbf{p}$, is dominated by the attraction (the so-called *London dispersion force*) between correlated components of the spontaneously induced dipole moments, giving $U(r) \rightarrow r^{-6}$ at $r \rightarrow \infty$.[42] At closer distances the potential is repulsive, growing very fast at $r \rightarrow 0$, but its

[41] The term 'virial', from Latin *viris* (meaning 'force'), was introduced to molecular physics by R Clausius. The motivation for the adjective 'second' for $B(T)$ is evident from the last form of Eq. (3.94), with the 'first virial coefficient', standing before the N/V ratio and sometimes denoted $A(T)$, equal to 1—see also Eq. (3.100) below.

[42] Indeed, the independent fluctuation-induced components $\mathbf{p}(t)$ and $\mathbf{p}'(t)$ of dipole moments of two particles have random mutual orientation, so that the time average of their interaction energy, proportional to r^{-3}, vanishes. However, the electric field $\mathscr{E}$ of each dipole $\mathbf{p}$, proportional to r^{-3}, induces a correlated component of $\mathbf{p}'$, also proportional to r^{-3}, giving an interaction energy proportional to $\mathbf{p}' \cdot \mathscr{E} \propto r^{-6}$, with a nonvanishing statistical average. Quantitative discussions of this effect, within several models, may be found in chapters 3, 5 and 6 of *Part QM* of this series.

quantitative form is specific for each particular pair of particles[43]. The crudest description of such repulsion is given by the so-called *hardball model*:

$$U(r) = \begin{cases} +\infty, & \text{for } 0 < r < 2r_0, \\ 0, & \text{for } 2r_0 < r < \infty, \end{cases} \tag{3.95}$$

—see the dashed line and the inset in figure 3.7. As Eq. (3.93) shows, in this model the second virial coefficient is temperature-independent:

$$B(T) = b \equiv \frac{1}{2}\int_0^{2r_0} 4\pi r^2 dr = \frac{2\pi}{3}(2r_0)^3, \tag{3.96}$$

(and is four times larger than the hardball's volume $V_0 = (4\pi/3)r_0^{\,3}$), so that the equation of state (3.94) still gives a linear dependence of pressure on temperature.

A correction to this result may be obtained by the following approximate account of the long-range attraction (see the dash-dotted line in figure 3.7)[44]:

$$U(r) = \begin{cases} +\infty, & \text{for } 0 < r < 2r_0, \\ U(r), \ \text{with } |U| \ll T, & \text{for } 2r_0 < r < \infty. \end{cases} \tag{3.97}$$

For this improved model, Eq. (3.93) yields:

$$B(T) = b + \frac{1}{2}\int_{2r_0}^{\infty} 4\pi r^2 dr \frac{U(r)}{T} = b - \frac{a}{T}, \quad \text{with} \quad a \equiv 2\pi \int_{2r_0}^{\infty} r^2 dr\, |U(r)|. \tag{3.98}$$

In this model, the equation of state (3.94) acquires a temperature-independent term:

$$P = T\left[\frac{N}{V} + \left(\frac{N}{V}\right)^2\left(b - \frac{a}{T}\right)\right] \equiv T\left[\frac{N}{V} + b\left(\frac{N}{V}\right)^2\right] - a\left(\frac{N}{V}\right)^2. \tag{3.99}$$

Still, the correction to the ideal-gas pressure is proportional to $(N/V)^2$, and has to be relatively small for Eq. (3.99) to be valid, so that its right-hand side may be considered as the sum of two leading terms in the general expansion of P into the Taylor series in the density $n = N/V$ of the gas:

$$P = T\left[\frac{N}{V} + B(T)\left(\frac{N}{V}\right)^2 + C(T)\left(\frac{N}{V}\right)^3 + \dots\right], \tag{3.100}$$

[43] Note that the particular form of the first term in the approximation $U(r) = a/r^{12} - b/r^6$ (called the *Lennard-Jones potential* or the '12-6 potential'), suggested in 1924, lacks physical justification, and in professional physics was soon replaced with other approximations, including the so-called *exp-6 model*, which fits most experimental data much better. However, the Lennard-Jones potential keeps creeping from one undergraduate textbook to another one, apparently just for enabling a simple calculation of the equilibrium distance between the particles at $T = 0$.

[44] The strong inequality between U and T in this model is necessary not only to make calculations simpler. A deeper reason is that if $(-U_{\min})$ becomes comparable with, or larger than T, particles may become trapped in the potential well formed by this potential, forming a different phase—a liquid or a solid. In such phases, the probability to find more than two particles interacting simultaneously is high, so that approximation (3.92), on which all our further results are based, becomes invalid.

where $C(T)$ is called the *third virial coefficient.* It is natural to ask how can we calculate $C(T)$ and the higher virial coefficients.

Generally, this may be done just by a careful analysis of Eq. (3.90),[45] but I would like to use this occasion to demonstrate a different, very interesting and counter-intuitive approach, called the *cluster expansion method*[46], which allows one to streamline such calculations. Let us apply to our system, with the energy (3.87), the grand canonical distribution. (Just as in section 3.2, we may argue that if the average number $\langle N\rangle$ of particles in a member of a grand canonical ensemble, with fixed μ and T, is much larger than 1, the relative fluctuations of N are small, so that all its thermodynamic properties should be similar to those when N is exactly fixed.) For our current case, Eq. (2.109) takes the form

$$\Omega = -T\ln\sum_{N=0}^{\infty} Z_N, \quad \text{with} \quad Z_N \equiv e^{\mu N/T}\sum_m e^{-E_{m,N}/T},$$
$$E_{m,N} = \sum_{k=1}^{N}\frac{p_k^2}{2m} + U(r_1, \ldots, r_N). \tag{3.101}$$

(Notice that here, as in all discussions of the grand canonical distribution, N means a particular rather than the average number of particles.) Now, let us try to forget for a second that in real systems of interest the number of particles is extremely large, and start to calculate, one by one, the first terms Z_N.

In the term with $N = 0$, both contributions to $E_{m,N}$ vanish, and so does the factor $\mu N/T$, so that $Z_0 = 1$. In the next term, with $N = 1$, the interaction term vanishes, so that $E_{m,1}$ is reduced to the kinetic energy of one particle, giving

$$Z_1 = e^{\mu/T}\sum_k \exp\left\{-\frac{p_k^2}{2mT}\right\}. \tag{3.102}$$

Making the usual transition from the summation to integration, we may write

$$Z_1 = ZI_1, \quad \text{where } Z \equiv e^{\mu/T}\frac{gV}{(2\pi\hbar)^3}\int \exp\left\{-\frac{p^2}{2mT}\right\}d^3p, \quad \text{and } I_1 \equiv 1. \tag{3.103}$$

This is the same simple (Gaussian) integral as in Eq. (3.6), giving

$$Z = e^{\mu/T}\frac{gV}{(2\pi\hbar)^3}(2\pi mT)^{3/2} = e^{\mu/T}gV\left(\frac{mT}{2\pi\hbar^2}\right)^{3/2}. \tag{3.104}$$

[45] L Boltzmann has used that way to calculate the 3rd and 4th virial coefficients for the hardball model—as much as can be done analytically.

[46] This method was developed in 1937–38 by J Mayer and collaborators for a classical gas, and generalized to quantum systems in 1938 by B Kahn and G Uhlenbeck.

Now let us explore the next term, with $N = 2$, which describes, in particular, pair interactions $U = U(\mathbf{r})$, with $\mathbf{r} = \mathbf{r} - \mathbf{r}'$. Due to the assumed particle indistinguishability, this term needs the 'correct Boltzmann counting' factor 1/2!—cf. Eqs. (3.12) and (3.88):

$$Z_2 = e^{2\mu/T}\frac{1}{2!}\sum_{k,k'}\left[\exp\left\{-\frac{p_k^2}{2mT} - \frac{p_{k'}^2}{2mT}\right\} e^{-U(\mathbf{r})/T}\right]. \tag{3.105}$$

Since U is coordinate-dependent, here the transfer from the summation to integration should be done more carefully than in the first term—cf. Eqs. (3.24) and (3.88):

$$\begin{aligned} Z_2 = e^{2\mu/T}\frac{1}{2!}\frac{(gV)^2}{(2\pi\hbar)^6}\int \exp\left\{-\frac{p^2}{2mT}\right\}d^3p \\ \times \int \exp\left\{-\frac{p'^2}{2mT}\right\}d^3p' \times \frac{1}{V}\int e^{-U(\mathbf{r})/T}d^3r. \end{aligned} \tag{3.106}$$

Comparing this expression with the definition (3.104) of the parameter Z, we get

$$Z_2 = \frac{Z^2}{2!}I_2, \quad \text{where } I_2 \equiv \frac{1}{V}\int e^{-U(\mathbf{r})/T}d^3r. \tag{3.107}$$

Acting absolutely similarly, for the third term of the grand canonical sum we may get

$$Z_3 = \frac{Z^3}{3!}I_3, \quad \text{where } I_3 \equiv \frac{1}{V^2}\int e^{-U(\mathbf{r}',\mathbf{r}'')/T}d^3r'd^3r'', \tag{3.108}$$

where $\mathbf{r}'$ and $\mathbf{r}''$ are the vectors characterizing the mutual positions of three particles —see figure 3.6b.

These results may be extended by induction to an arbitrary N. Plugging the expression for Z_N into Eq. (3.101) and recalling that $\Omega = -PV$, we get the equation of state of the gas in the form

$$P = \frac{T}{V}\ln\left(1 + ZI_1 + \frac{Z^2}{2!}I_2 + \frac{Z^3}{3!}I_3 + \ldots\right). \tag{3.109}$$

As a sanity check, at $U = 0$, all integrals I_N are equal to 1, and the expression under the logarithm in just the Taylor expansion of the function e^Z, giving $P = TZ/V$, and $\Omega = -PV = -TZ$. In this case, according to the last of Eqs. (1.62), the *average* number of particles of particles in the system is $\langle N\rangle = -(\partial\Omega/\partial\mu)_{T,V} = Z$, because since $Z \propto \exp\{\mu/T\}$, $\partial Z/\partial\mu = Z/T$.[47] Thus, in this limit we have happily recovered the equation of state of the ideal gas.

[47] Actually, the fact that in that case $Z = \langle N\rangle$, could have been noted earlier—just by comparing Eq. (3.104) with Eq. (3.32).

Returning to the general case of nonvanishing interactions, let us assume that the logarithm in Eq. (3.109) may be represented as a Taylor expansion in Z:

$$P = \frac{T}{V}\sum_{l=1}^{\infty}\frac{J_l}{l!}Z^l. \tag{3.110}$$

(The lower limit of the sum reflects the fact that according to Eq. (3.109), at $Z = 0$, $P = (T/V)\ln 1 = 0$, so that the coefficient J_0 in Eq. (3.110) has to be equal 0 as well.) According to Eq. (1.60), this expansion corresponds to the grand potential

$$\Omega = -PV = -T\sum_{l=1}^{\infty}\frac{J_l}{l!}Z^l. \tag{3.111}$$

Again using the last of Eqs. (1.62), we get

$$\langle N\rangle = \sum_{l=1}^{\infty}\frac{J_l}{(l-1)!}Z^l. \tag{3.112}$$

This equation may be used for finding Z for the given $\langle N\rangle$, and hence for the calculation of the equation of state from Eq. (3.110). The only remaining conceptual action item is to express the coefficients J_l via the integrals I_N participating in the expansion (3.109). This may be done using the well-known Taylor expansion of the logarithm function[48],

$$\ln(1+\xi) = \sum_{l=1}^{\infty}(-1)^{l+1}\frac{\xi^l}{n}. \tag{3.113}$$

Using it together with Eq. (3.109), we get a Taylor series in Z, starting as

$$P = \frac{T}{V}\left[Z + \frac{Z^2}{2!}(I_2 - 1) + \frac{Z^3}{3!}[(I_3 - 1) - 3(I_2 - 1)] + \ldots\right]. \tag{3.114}$$

Comparing this expression with Eq. (3.110), we see that

$$\begin{aligned}
J_1 &= 1,\\
J_2 &= I_2 - 1 = \frac{1}{V}\int\left(e^{-U(\mathbf{r})/T} - 1\right)d^3r,\\
J_3 &= (I_3 - 1) - 3(I_2 - 1)\\
&= \frac{1}{V^2}\int\left(e^{-U(\mathbf{r}',\,\mathbf{r}'')/T} - e^{-U(\mathbf{r}')/T}\right.\\
&\quad\left. - e^{-U(\mathbf{r}'')/T} - e^{-U(\mathbf{r}''')/T} + 2\right)d^3r'd^3r'',\ \ldots
\end{aligned} \tag{3.115}$$

[48] Looking at Eq. (3.109), one may think that since $\xi = Z + Z^2I_2/2 + \ldots$ is of the order of at least $Z \sim \langle N\rangle \gg 1$, the expansion (3.113), which converges only if $|\xi| < 1$, is illegitimate. However, the expansion is justified by its result (3.114), in which the nth term is of the order of $\langle N\rangle^n(V_0/V)^{n-1}/n!$, so that the series does converge if the gas density is sufficiently low: $\langle N\rangle/V \ll 1/V_0$, i.e. $r_{\text{ave}} \gg r_0$. This is the very beauty of the cluster expansion, whose few first terms, rather unexpectedly, give a good approximation even for a gas with $\langle N\rangle \gg 1$ particles.

where $\mathbf{r}''' \equiv \mathbf{r}' - \mathbf{r}''$—see figure 3.6b. The expression of J_2, describing the pair interactions of particles, is (besides a different numerical factor) equal to the second virial coefficient $B(T)$—see Eq. (3.93). As a reminder, the subtraction of 1 from the integral I_2 in the second of Eqs. (3.115) makes the contribution of each elementary 3D volume d^3r into the integral J_2 nonvanishing only if at this $\mathbf{r}$ two particles interact ($U \neq 0$). Very similarly, in the last of Eqs. (3.115), the subtraction of three pair-interaction terms from $(I_3 - 1)$ makes the contribution from an elementary 6D volume $d^3r'd^3r''$ into the integral J_3 nonvanishing only if at that mutual location of particles *all three* of them interact simultaneously, etc.

In order to illustrate the cluster expansion method at work, let us eliminate the factor Z from the system of equations (3.110) and (3.112), keeping (for the sake of simplicity) only the terms up to $O(Z^3)$—as was done in Eq. (3.114). Spelling out Eq. (3.110) and (3.112),

$$\frac{PV}{T} = J_1 Z + \frac{J_2}{2} Z^2 + \frac{J_3}{6} Z^3 + \ldots, \tag{3.116}$$

$$\langle N \rangle = J_1 Z + J_2 Z^2 + \frac{J_3}{2} Z^3 + \ldots, \tag{3.117}$$

and dividing these two expressions, we get a result,

$$\begin{aligned} \frac{PV}{\langle N \rangle T} &\approx \frac{1 + (J_2/2J_1)Z + (J_3/6J_1)Z^2}{1 + (J_2/J_1)Z + (J_3/2J_1)Z^2} \\ &= 1 - \frac{J_2}{2J_1} Z + \left(\frac{J_2^2}{2J_1^2} - \frac{J_3}{3J_1} \right) Z^2 + \ldots, \end{aligned} \tag{3.118}$$

which is accurate to terms $O(Z^2)$. In this approximation, we may use Eq. (3.117), solved for Z with the same accuracy:

$$Z \approx \langle N \rangle - \frac{J_2}{J_1} \langle N \rangle^2. \tag{3.119}$$

Plugging this expression into Eq. (3.118), we get the virial expansion (3.100) with

$$B(T) = -\frac{J_2}{2J_1} V, \quad C(T) = \left(\frac{J_2^2}{J_1^2} - \frac{J_3}{3J_1} \right) V^2. \tag{3.120}$$

The first of these relations, combined with the first two of Eqs. (3.115), yields for the 2nd virial coefficient the same Eq. (3.93) that was obtained from the Gibbs distribution, while the second one allows us to calculate the 3rd virial coefficient $C(T)$. (Let me leave the calculation of J_3 and $C(T)$, for the hardball model, for the reader's exercise.) Evidently, a more accurate expansion of Eqs. (3.110), (3.112), and (3.114) may be used to calculate an arbitrary virial coefficient, though starting from the 5th coefficient, such calculations may be completed only numerically even in the simplest hardball model.

3.6 Problems

Problem 3.1. Use the Maxwell distribution for an alternative (statistical) calculation of the mechanical work performed by the Szilard engine discussed in section 2.3.

Hint: You may assume the simplest geometry of the engine—see figure 2.4.

Problem 3.2. Use the Maxwell distribution to calculate the *drag coefficient* $\eta \equiv -\partial\langle\mathscr{F}\rangle/\partial u$, where $\mathscr{F}$ is the force exerted by an ideal classical gas on a piston moving with a low velocity u, in the simplest geometry shown in the figure below, assuming that collisions of gas particles with the piston are elastic.

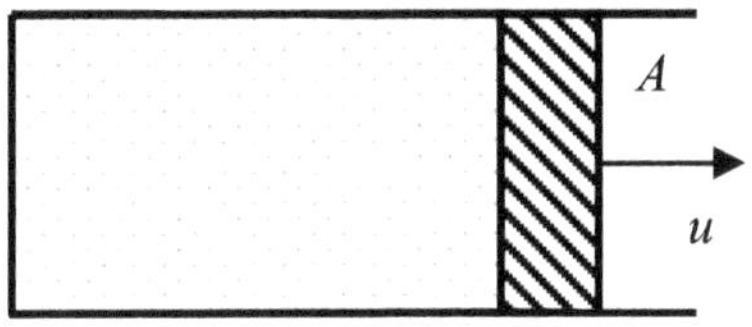

Problem 3.3. Derive the equation of state of the ideal classical gas from the grand canonical distribution.

Problem 3.4. Prove that Eq. (3.22),

$$\Delta S = N_1 \ln \frac{V_1 + V_2}{V_1} + N_2 \ln \frac{V_1 + V_2}{V_2},$$

derived for the change of entropy at mixing of two ideal classical gases of completely distinguishable particles (that initially had equal densities N/V and temperatures T), is also valid if particles in each of the initial volumes are identical to each other, but different from those in the counterpart volume. Assume that masses and internal degeneracy factors g of all the particles are equal.

Problem 3.5. A round cylinder of radius R and length L, containing an ideal classical gas of $N \gg 1$ particles of mass m each, is rotated about its symmetry axis with angular velocity ω. Assuming that the gas as the whole rotates with the cylinder, and is in thermal equilibrium at temperature T,

(i) calculate the gas pressure distribution along its radius, and analyze it temperature dependence, and
(ii) neglecting the internal degrees of freedom of the particles, calculate the total energy of the gas and its heat capacity in the high- and low-temperature limits.

Problem 3.6. $N \gg 1$ classical, non-interacting, indistinguishable particles of mass m are confined in a parabolic, spherically-symmetric 3D potential well $U(\mathbf{r}) = \kappa r^2/2$. Use two different approaches to calculate all major thermodynamic characteristics of the system, in thermal equilibrium at temperature T, including its heat capacity. What of the results should be changed if the particles are distinguishable, and how?

Hint: Suggest a replacement of the notions of volume and pressure, appropriate for this system.

Problem 3.7. In the simplest model of thermodynamic equilibrium between the liquid and gas phases of the same molecules, temperature and pressure do not affect the molecule's condensation energy Δ. Calculate the concentration and pressure of such *saturated vapor*, assuming that it behaves as an ideal gas of classical particles.

Problem 3.8. An ideal classical gas of $N \gg 1$ particles is placed into a container of volume V and wall surface area A. The particles may condense on container walls, releasing energy Δ per particle, and forming an ideal 2D gas. Calculate the equilibrium number of condensed particles and the gas pressure, and discuss their temperature dependences.

Problem 3.9. The inner surfaces of the walls of a closed container of volume V, filled with $N \gg 1$ particles, have $N_S \gg 1$ similar traps (small potential wells). Each trap can hold only one particle, at potential energy $-\Delta < 0$. Assuming that the gas of the particles in the volume is ideal and classical, derive the equation for the chemical potential μ of the system in equilibrium, and use it to calculate the potential and the gas pressure in the limits of small and large values of the ratio N/N_S.

Problem 3.10. Calculate the magnetic response (the *Pauli paramagnetism*) of a degenerate ideal gas of spin-½ particles to a weak external magnetic field, due to a partial spin alignment with the field.

Problem 3.11. Calculate the magnetic response (the *Landau diamagnetism*) of a degenerate ideal gas of electrically charged fermions to a weak external magnetic field, due to their orbital motion.

Problem 3.12.* Explore the *Thomas–Fermi model* of a heavy atom, with nuclear charge $Q = Ze \gg e$, in which the electrons are treated as a degenerate Fermi gas, interacting with each other only via their contribution to the common electrostatic potential $\phi(\mathbf{r})$. In particular, derive the ordinary differential equation obeyed by the radial distribution of the potential, and use it to estimate the effective radius of the atom[49].

Problem 3.13.* Use the Thomas–Fermi model, explored in the previous problem, to calculate the total binding energy of a heavy atom. Compare the result with that for the simpler model, in which the Coulomb electron–electron interaction of electrons is completely ignored.

[49] Since this problem, and the next one, are important for atomic physics, and that at their solution the thermal effects may be ignored, they were given in chapter 8 of *Part QM* of the series as well, for the benefit of readers who would not take this *SM* course. Note, however, that the solution of these two problems is streamlined by using the notion of the chemical potential μ, which was introduced only in this course.

Problem 3.14. Calculate the characteristic *Thomas–Fermi length* λ_{TF} of weak electric field's screening by conduction electrons in a metal, modeling their ensemble as an ideal, degenerate, isotropic Fermi gas.

Hint: Assume that λ_{TF} is much larger than the Bohr radius r_B.

Problem 3.15. For a degenerate ideal 3D Fermi gas of N particles, confined in a rigid-wall box of volume V, calculate the temperature dependences of its pressure P and the heat capacity difference $(C_P - C_V)$, in the leading approximation in $T \ll \varepsilon_F$. Compare the results with those for the ideal classical gas.

Hint: You may like to use the solution of problem 1.9.

Problem 3.16. How would the Fermi statistics of an ideal gas affect the barometric formula (3.28)?

Problem 3.17. Derive general expressions for the energy E and the chemical potential μ of a uniform Fermi gas of $N \gg 1$ non-interacting, indistinguishable, ultra-relativistic particles[50]. Calculate E, and also the gas pressure P explicitly in the degenerate gas limit $T \to 0$. In particular, is Eq. (3.48) valid in this case?

Problem 3.18. Use Eq. (3.49) to calculate the pressure of an ideal gas of ultra-relativistic, indistinguishable quantum particles, for an arbitrary temperature, as a function of the total energy E of the gas, and its volume V. Compare the result with the corresponding relations for the electromagnetic blackbody radiation and an ideal gas of non-relativistic particles.

Problem 3.19.* Calculate the speed of sound in an ideal gas of ultra-relativistic fermions of density n at negligible temperature.

Problem 3.20. Calculate basic thermodynamic characteristics, including all relevant thermodynamic potentials, specific heat, and the surface tension for a uniform non-relativistic 2D electron gas with given areal density $n \equiv N/A$:

(i) at $T = 0$, and

(ii) at low but nonvanishing temperatures (in the lowest substantial order in $T/\varepsilon_F \ll 1$), neglecting the Coulomb interaction effects[51].

Problem 3.21. Calculate the effective latent heat $\Lambda_{ef} \equiv -N(\partial Q/\partial N_0)_{N,V}$ of evaporation of the spatially-uniform Bose–Einstein condensate as a function of

[50] This is, for example, an approximate but reasonable model for electrons in white dwarf stars, whose Coulomb interaction is mostly compensated by the charge of nuclei of fully ionized helium atoms.

[51] This condition may be approached reasonably well, for example, in 2D electron gases formed in semiconductor heterostructures (see, e.g. the discussion in *Part QM* section 1.6, and the solution of problem 3.2 of that course), due to the electron field's compensation by background ionized atoms, and its screening by highly doped semiconductor bulk.

temperature T. Here Q is the heat absorbed by the (condensate + gas) system of $N \gg 1$ particles as a whole, while N_0 is the number of particles in the condensate alone.

Problem 3.22.* For an ideal, spatially-uniform Bose gas, calculate the law of the chemical potential's disappearance at $T \to T_c$, and use the result to prove that the heat capacity C_V is a continuous function of temperature at the critical point $T = T_c$.

Problem 3.23. In chapter 1 of these notes, several thermodynamic relations involving entropy have been discussed, including the first of Eqs. (1.39):

$$S = -(\partial G/\partial T)_P.$$

If we combine this expression with Eq. (1.56), $G = \mu N$, it looks like that for the Bose–Einstein condensate, whose chemical potential μ equals zero at temperatures below the critical point T_c, the entropy should vanish as well. On the other hand, dividing both parts of Eq. (1.19) by dT, and assuming that at this temperature change the volume is kept constant, we get

$$C_V = T(\partial S/\partial T)_V.$$

(This equality was also mentioned in chapter 1.) If C_V is known as a function of temperature, the last relation may be integrated over T to calculate S:

$$S = \int_{V=\text{const}} \frac{C_V(T)}{T} dT + \text{const}.$$

According to Eq. (3.80), the specific heat for the Bose–Einstein condensate is proportional to $T^{3/2}$, so that the integration gives a nonvanishing entropy $S \propto T^{3/2}$. Resolve this apparent contradiction, and calculate the genuine entropy at $T = T_c$.

Problem 3.24. The standard analysis of the Bose–Einstein condensation, outlined in section 3.4, may seem to ignore the energy quantization of the particles confined in volume V. Use the particular case of a cubic confining volume $V = a \times a \times a$ with rigid walls to analyze whether the main conclusions of the standard theory, in particular Eq. (3.71) for the critical temperature of the system of $N \gg 1$ particles, are affected by such quantization.

Problem 3.25.* $N \gg 1$ non-interacting bosons are confined in a soft, spherically-symmetric potential well $U(\mathbf{r}) = m\omega^2 r^2/2$. Develop the theory of the Bose–Einstein condensation in this system; in particular, prove Eq. (3.74*b*), and calculate the critical temperature T_c^*. Looking at the solution, what is the most straightforward way to detect the condensation in experiment?

Problem 3.26. Calculate the chemical potential of an ideal, uniform 2D gas of spin-0 Bose particles as a function of its areal density n (the number of particles per unit area), and find out whether such a gas can condense at low temperatures. Review your result for the case of a large ($N \gg 1$) but finite number of particles.

Problem 3.27. Can the Bose–Einstein condensation happen in a 2D system of $N \gg 1$ non-interacting bosons placed into a soft, axially-symmetric potential well, whose potential may be approximated as $U(\mathbf{r}) = m\omega^2\rho^2/2$, where $\rho^2 \equiv x^2 + y^2$, and $\{x, y\}$ are the Cartesian coordinates in the particle confinement plane? If yes, calculate the critical temperature of the condensation.

Problem 3.28. Use Eqs. (3.115) and (3.120) to calculate the third virial coefficient $C(T)$ for the hardball model of particle interactions.

Problem 3.29. Assuming the hardball model, with volume V_0 per molecule, for the liquid phase, describe how the results of problem 3.7 change if the liquid forms spherical drops of radius $R \gg V_0^{1/3}$. Briefly discuss the implications of the result for water cloud formation.

Hint: Surface effects in macroscopic volumes of liquids may be well described by attributing an additional energy γ (called the *surface tension*) to unit surface area[52].

References

[1] Rock P 1983 *Chemical Thermodynamics* (University Science Books)
[2] Atkins P 1994 *Physical Chemistry* 5th ed (Freeman)
[3] Barrow G 1996 *Physical Chemistry* 6th ed (McGraw-Hill)
[4] Bard A and Falkner L 2000 *Electrochemical Methods* 2nd ed (Wiley)
[5] Scholz F (ed) 2010 *Electroanalytical Methods* 2nd ed (Springer)
[6] Aikawa K *et al* 2014 *Phys. Rev. Lett.* **112** 010404
[7] Ashcroft N and Mermin N 1976 *Solid State Physics* (W. B. Sounders)
[8] Ensher J *et al* 1996 *Phys. Rev. Lett.* **77** 4984
[9] Bloch I *et al* 2008 *Rev. Mod. Phys.* **80** 885
[10] Hu J *et al* 2017 *Science* **358** 1078
[11] Pethick C and Smith H 2008 *Bose–Einstein Condensation in Dilute Gases* 2nd ed (Cambridge University Press)
[12] Huang K 2003 *Statistical Mechanics* 2nd ed (Wiley)

[52] See, e.g. *Part CM* section 8.2.

Chapter 4

Phase transitions

This chapter gives a rather brief discussion of coexistence between different states ('phases') of collections of similar particles, and the laws of transitions between these phases. Due to the complexity of these phenomena, which involve particle interactions, quantitative analytical results in this field have been obtained only for a few very simple models, typically giving only a very approximate description of real systems.

4.1 First-order phase transitions

From everyday experience, say with ice, liquid water, and water vapor, we know that one chemical substance (i.e. a set of many similar particles) may exist in several stable states—*phases*. A typical substance may have:

(i) a dense *solid phase*, in which interatomic forces keep all atoms/molecules in virtually fixed relative positions, with just small thermal fluctuations about them;

(ii) a *liquid phase*, of comparable density, in which the relative distances between atoms or molecules are almost constant, but the particles are virtually free to move around each other, and

(iii) the *gas phase*, typically of a much lower density, in which molecules are virtually free to move all around the containing volume[1].

Experience also tells us that at certain conditions, two phases may be in thermal and chemical equilibrium—say, ice floating on water, with temperature at the

[1] The plasma phase, in which atoms are partly or completely ionized, is frequently mentioned on one more phase, on equal footing with the three phases listed above, but one has to remember that in contrast to them, a typical electroneutral plasma consists of particles of two different sorts—ions and electrons.

doi:10.1088/2053-2563/aaf503ch4

freezing point. Actually, in section 3.4 we already discussed a qualitative theory of one such equilibrium, the Bose–Einstein condensate coexistence with the uncondensed 'vapor' of similar particles. However, this is a rather rare case when the phase coexistence is due to the quantum nature of the particles (bosons) that may not interact directly. Much more frequently, the formation of different phases, and transitions between them, is due to particle interactions.

Phase transitions are sometimes classified by their *order*[2]. I will start my discussion with the *first-order phase transitions* that feature nonvanishing *latent heat* Λ—the amount of heat that is necessary to give one phase in order to turn it into another phase completely, even if temperature and pressure are kept constant[3]. Very unfortunately, even the simplest 'microscopic' models of particle interaction, such as those discussed in section 3.5, give rather complex equations of state. (As a reminder, even the simplest hardball model leads to the series (3.100), whose higher virial coefficients defy analytical calculation.) This is why I will follow the tradition to discuss the first-order phase transitions using a simple 'macroscopic' (phenomenological) model suggested in 1873 by J van der Waals.

For its introduction, it is useful to recall that in section 3.5 we have derived Eq. (3.99)—the equation of state for a classical gas of weakly interacting particles, which takes into account (albeit approximately) both interaction components necessary for a realistic description of gas condensation—the long-range attraction of the particles and their short-range repulsion. Let us rewrite that result as follows:

$$P + a\frac{N^2}{V^2} = \frac{NT}{V}\left(1 + \frac{Nb}{V}\right). \tag{4.1}$$

As we saw at the derivation of this formula, the physical meaning of the constant b is the effective volume of space taken by a particle pair collision—see Eq. (3.96). The relation (4.1) is quantitatively valid only if the second term in the parentheses is small, $Nb \ll V$, i.e. if the total volume excluded from particles' free motion because of their collisions is much smaller than the whole volume V. In order to describe the condensed phase (which I will call 'liquid')[4], we need to generalize this relation to the case $Nb \sim V$. Since the effective volume left for particles' motion is $V - Nb$, it is very natural to make the following replacement: $V \rightarrow V - Nb$, in the ideal gas' equation of state. If we still keep on the left-hand side the term aN^2/V^2, which describes the long-range attraction of particles, we get the *van der Waals equation* of state:

$$P + a\frac{N^2}{V^2} = \frac{NT}{V - Nb}. \tag{4.2}$$

---

[2] Such classification schemes, started by P Ehrenfest, have been repeatedly modified, and only the 'first-order phase transition' is still a generally accepted term, but with a definition different from the original one.

[3] For example, for water the latent heat of vaporization at the ambient pressure is as high as $\sim 2.2 \times 10^6$ J kg^{-1}, i.e. ~ 0.4 eV per molecule, making this ubiquitous liquid indispensable for many practical purposes—including effective fire fighting. (The latent heat of water ice's melting is an order of magnitude lower.)

[4] Due to the phenomenological character of the van der Waals model, one cannot say whether the condensed phase it predicts corresponds to a liquid or a solid. However, in most real substances at ambient conditions, gas coexists with liquid, hence the name.

The advantage of this simple model is that in the rare gas limit, $Nb \ll V$, it reduces back to the microscopically-justified Eq. (4.1). (To verify this, it is sufficient to Taylor-expand the right-hand side of Eq. (4.2) in small $Nb/V \ll 1$, and retain only two leading terms.) Let us explore properties of this model.

It is frequently convenient to discuss any equation of state in terms of its isotherms, i.e. the $P(V)$ curves plotted at constant T. As Eq. (4.2) shows, in the van der Waals model such a plot depends on four parameters (a, b, N, and T). For its analysis it is convenient to introduce dimensionless variables: pressure $p \equiv P/P_c$, volume $v \equiv V/V_c$, and temperature $t \equiv T/T_c$, normalized to their so-called *critical values*,

$$P_c \equiv \frac{1}{27}\frac{a}{b^2}, \qquad V_c \equiv 3Nb, \qquad T_c \equiv \frac{8}{27}\frac{a}{b}, \tag{4.3}$$

whose nature will be clear in a minute. In this notation, Eq. (4.2) acquires the following form,

$$\left(p + \frac{3}{v^2}\right) = \frac{8t}{(3v - 1)}, \tag{4.4}$$

so that the normalized isotherms $p(v)$ depend on only one parameter, the normalized temperature t—see figure 4.1. The most important property of these plots is that the isotherms have qualitatively different shapes in two temperature regions. At $t > 1$, i.e. $T > T_c$, pressure increases monotonically at gas compression (qualitatively, as in an ideal classical gas, with $P = NT/V$, to which the van der Waals system evidently tends at $T \gg T_c$), i.e. with $(\partial P/\partial V)_T < 0$ at all points of the isotherm[5]. However, below the

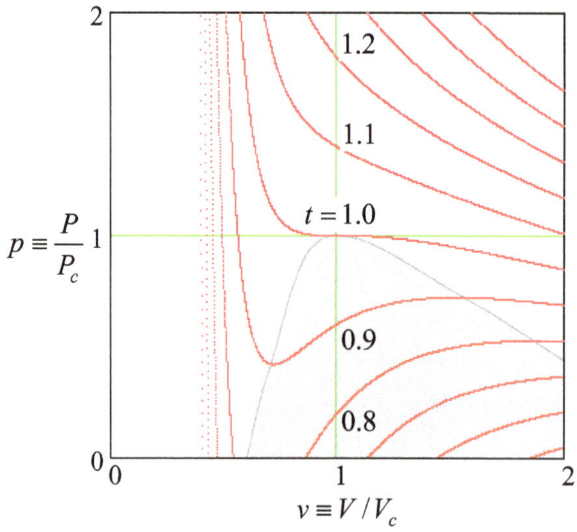

Figure 4.1. The van der Waals equation plotted on the $[p, v]$ plane for several values of the reduced temperature $t \equiv T/T_c$. Shading shows the single-phase instability range in which $(\partial P/\partial V)_T > 0$.

[5] The special choice of numerical coefficients in Eq. (4.3) is motivated by making the border between these two regions to take place exactly at $t = 1$, i.e. at temperature T_c, with the critical point coordinates equal to P_c and V_c.

critical temperature T_c, any isotherm features a segment with $(\partial P/\partial V)_T > 0$. It is easy to understand that, as least in a constant pressure experiment (see, for example, figure 1.5),[6] these segments describe a mechanically unstable equilibrium. Indeed, if due to a random fluctuation, the volume deviated upward from the equilibrium value, the pressure would also increase, forcing the environment (say, the heavy piston in figure 1.5) to allow a further expansion of the system, leading to even higher pressure, etc. A similar deviation of volume downward would lead to a similar avalanche-like decrease of the volume. Such avalanche instability would develop further and further until the system had reached one of the stable branches with a negative slope $(\partial P/\partial V)_T$. In the range where the single-phase equilibrium state is unstable, the system as a whole may be stable only if it consists of the two phases (one with a smaller, and another with a higher density $n = N/V$) that are described by the two stable branches—see figure 4.2.

In order to understand the basic properties of this two-phase system, let us recall the general conditions of the thermodynamic equilibrium of two systems, which have been discussed in chapter 1:

$$T_1 = T_2 \text{ (thermal equilibrium)}, \tag{4.5}$$

$$\mu_1 = \mu_2 \text{ ('chemical' equilibrium)}, \tag{4.6}$$

the latter condition meaning that the average energy of a single ('probe') particle in both systems has to be the same. To those, we should add the evident condition of mechanical equilibrium,

$$P_1 = P_2 \text{ (mechanical equilibrium)}, \tag{4.7}$$

which immediately follows from the balance of normal forces exerted on an inter-phase boundary.

If we discuss isotherms, Eq. (4.5) is fulfilled automatically, while Eq. (4.7) means that the effective isotherm $P(V)$ describing a two-phase system should be a horizontal line—see figure 4.2:

$$P = P_0(T). \tag{4.8}$$

Along this line[7], internal properties of each phase do not change; only the particle distribution does: it evolves gradually from all particles being in the liquid phase at point 1 to all particles being in the gas phase at point 2.[8] In particular, according to

[6] Actually, this assumption is not crucial for our analysis of mechanical stability, because if a fluctuation takes place in a small part of the total volume V, its other parts play the role of pressure-fixing environment.

[7] Frequently, $P_0(T)$ is called the *saturated vapor pressure*.

[8] An important question is: why does the phase-equilibrium line $P = P_0(T)$ stretch all the way from point 1 to point 2? Indeed, the branches 1–1′ and 2–2′ of the single-phase isotherm also have negative derivative $(\partial P/\partial V)_T$ and hence are mechanically stable with respect to small perturbations. The answer is that these branches are actually *metastable*, i.e. have larger Gibbs energy per particle (i.e. μ) than the counterpart phase and are hence unstable to *larger* perturbations—such as foreign microparticles (say, dust), protrusions on the confining wall, etc. In very controlled conditions, these single-phase 'superheated' and 'supercooled' states can survive virtually all the way to the zero-derivative points 1′ and 2′, leading to sudden jumps of the system into the counterpart phase. (For fixed-pressure conditions, such jumps are shown by dashed lines in figure 4.2.) In particular, purified water may be supercooled to almost −50 °C, and superheated to nearly +270 °C—all at the atmospheric pressure. However, at more realistic conditions, perturbations result in the two-phase coexistence formation close to points 1 and 2.

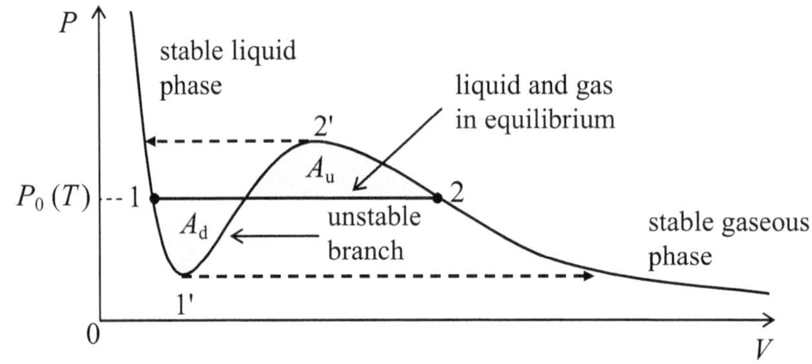

Figure 4.2. Phase equilibrium at $T < T_c$ (schematically).

Eq. (4.6), the chemical potentials μ of the phases should be equal at each point of the horizontal line (4.8). This fact enables us to find the line's position: it has to connect points 1 and 2 in which the chemical potentials of the phases are equal to each other. Let us recast this condition as

$$\int_1^2 d\mu = 0, \quad \text{i.e.} \int_1^2 dG = 0, \tag{4.9}$$

where the integral may be taken along the single-phase isotherm. (For this mathematical calculation, the mechanical instability of states at some part of this curve is not important.) By its construction, along that curve, $N = \text{const}$ and $T = \text{const}$, so that according to Eq. (1.53*c*), $dG = -SdT + VdP + \mu dN$, for a slow (reversible) change, $dG = VdP$. Hence Eq. (4.9) yields

$$\int_1^2 VdP = 0. \tag{4.10}$$

From figure 4.2, it is easy to see that geometrically this equation means that the shaded areas A_d and A_u should be equal, and hence Eq. (4.10) may be rewritten in the form of the so-called *Maxwell's rule*

$$\int_1^2 [P - P_0(T)]\, dV = 0. \tag{4.11}$$

This relation is more convenient for calculations than Eq. (4.10) if the equation of state may be explicitly solved for P—as is the case for the van der Waals equation (4.2). Such calculation (left for the reader's exercise) shows that for that model, the temperature dependence of the saturated vapor pressure at low T is exponential,

$$P_0(T) \propto P_c \exp\left\{-\frac{\Delta}{T}\right\}, \quad \text{with} \quad \Delta \equiv \frac{a}{b} = \frac{27}{8} T_c, \quad \text{for } T \ll T_c, \tag{4.12}$$

corresponding very well to the physical picture of particle activation from potential wells of depth Δ.[9]

The signature parameter of the first-order phase transition, the latent heat of evaporation

$$\Lambda \equiv \int_1^2 dQ, \tag{4.13}$$

may also be found by a similar integration along the single-phase isotherm. Indeed, using Eq. (1.19), $dQ = TdS$, we get

$$\Lambda \equiv \int_1^2 TdS = T(S_2 - S_1). \tag{4.14}$$

Let us express the right-hand side of Eq. (4.14) via the equation of state. For that, let us take the full derivative of both sides of Eq. (4.6) over temperature, considering each value of $G = N\mu$ as a function of P and T, and taking into account that according to Eq. (4.7), $P_1 = P_2 = P_0(T)$:

$$\left(\frac{\partial G_1}{\partial T}\right)_P + \left(\frac{\partial G_1}{\partial P}\right)_T \frac{dP_0}{dT} = \left(\frac{\partial G_2}{\partial T}\right)_P + \left(\frac{\partial G_2}{\partial P}\right)_T \frac{dP_0}{dT}. \tag{4.15}$$

According to the first of Eqs. (1.39), the partial derivative $(\partial G/\partial T)_P$ is just minus the entropy, while according to the second of those equations, $(\partial G/\partial P)_T$ is the volume. Thus Eq. (4.15) becomes

$$-S_1 + V_1\frac{dP_0}{dT} = -S_2 + V_2\frac{dP_0}{dT}. \tag{4.16}$$

Solving this equation for $(S_2 - S_1)$, and plugging the result into Eq. (4.14), we get the *Clapeyron–Clausius formula*

$$\Lambda = T(V_2 - V_1)\frac{dP_0}{dT}. \tag{4.17}$$

For the van der Waals model, this formula may be readily used for the analytical calculation of Λ in two limits: $T \ll T_c$ and $(T_c - T) \ll T_c$—the exercises left for the reader. In the latter limit, $\Lambda \propto (T_c - T)^{1/2}$, naturally vanishing at the critical temperature.

Finally, some important properties of the van der Waals' model may be revealed more easily by looking at the set of its isochores $P = P(T)$ for V = const, rather than at the isotherms. Indeed, as Eq. (4.2) shows, all single-phase isochores are straight lines. However, if we interrupt these lines at the points when the single phase becomes metastable, and complement them with the (very nonlinear!) dependence $P_0(T)$, we get the pattern (called the *phase diagram*) shown schematically in figure 4.3a.

[9] It is fascinating how well is this Arrhenius exponent hidden in the polynomial van der Waals equation (4.2)!

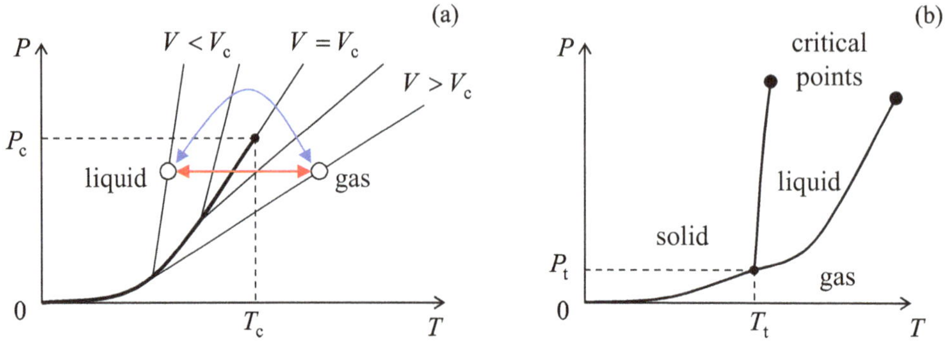

Figure 4.3. (a) Van der Waals model's isochores, the saturated gas pressure diagram and the critical point, and (b) the phase diagram of a typical three-phase system (schematically).

At this plot, one more meaning of the critical point $\{P_c, T_c\}$ becomes very vivid. At fixed pressure $P < P_c$, the liquid and gaseous phases are clearly separated by the saturated pressure line $P_0(T)$, so if we achieve the transition between the phases just by changing temperature (see the red horizontal line in figure 4.3), we have to pass through the phase equilibrium point, either giving to the system or taking out the latent heat. However, if we perform the transition between the same initial and final points by changing both the pressure and temperature, going around the critical point (see the blue line in figure 4.3), no definite point of transition may be observed: the substance stays in a single phase, and it is a subjective judgment of the observer in which region that phase should be called the liquid, and in which region—the gas. For water, the critical point corresponds to the temperature of 647 K (374 °C), and $P_c \approx$ 22.1 MPa (i.e. ~200 bars), so that a lecture demonstration of its critical behavior would require substantial safety precautions. This is why such demonstrations are typically carried out with other gases such as the diethyl ether[10], with much lower T_c (194 °C) and P_c (3.6 MPa), or the now-infamous carbon dioxide CO_2, with even lower T_c (31.1 °C), though higher P_c (7.4 MPa). Though these substances are colorless and clear in both gas and liquid phases, their separation (by gravity) is visible, due to small differences in the optical refraction coefficients, at $P < P_c$, but not above P_c.[11]

Thus, in the van der Waals model, two phases may coexist, though only at certain conditions—in particular, $T < T_c$. Now a natural, more general question is whether the coexistence of more than two phases of the same substance is possible. For example, can the water ice, the liquid water, and the water vapor (steam) all be in thermodynamic equilibrium? The answer is essentially given by Eq. (4.6). From thermodynamics, we know that for a uniform system, i.e. one phase, pressure and temperature completely define the chemical potential $\mu(P, T)$. Hence, dealing with

[10] $(CH_3–CH_2)–O–(CH_2–CH_3)$, historically the first popular general anesthetic.

[11] It is interesting that very close to the critical point the substance suddenly becomes opaque—in the case of ether, whitish. The qualitative explanation of this effect, called the *critical opalescence*, is simple: at this point the difference of the Gibbs energies per particle (i.e. the chemical potentials) of the two phases becomes so small that unavoidable thermal fluctuations lead to spontaneous appearance and disappearance of relatively large (a-few-μm-scale) single-phase regions in all the volume. A large concentration of boundaries of such randomly-shaped regions leads to strong light scattering.

two phases, we had to satisfy just *one* chemical equilibrium condition (4.6) for *two* common arguments P and T. Evidently, this leaves us with one extra degree of freedom, so that the two-phase equilibrium is possible within a certain range of P at fixed T (or vice versa)—see again the horizontal line in figure 2 and the bold line in figure 4.3a. Now, if we want *three* phases to be in equilibrium, we need to satisfy *two* equations for these variables:

$$\mu_1(P, T) = \mu_2(P, T) = \mu_3(P, T). \tag{4.18}$$

Typically, the functions $\mu(P, T)$ are monotonic, so that the two Eqs. (4.18) have just one solution, the so-called *triple point* $\{P_t, T_t\}$. Of course, the triple point $\{P_t, T_t\}$ of equilibrium between *three* phases should not to be confused with the critical points $\{P_c, T_c\}$ of transitions between *two* phase pairs. Figure 4.3b shows, *very* schematically, their relation for a typical three-phase system solid–liquid–gas. For example, water, ice, and water vapor are at equilibrium at a triple point corresponding to $P_t \approx$ 0.612 kPa[12] and T_t = 273.16 K. The practical importance of this particular temperature point is that by an international agreement it has been accepted for the definition of not only the Kelvin temperature scale, but also of the Celsius scale, as 0.01 °C, so that the absolute temperature zero corresponds to exactly −273.15 °C. More generally, triple points of pure substances (such as H_2, N_2, O_2, Ar, Hg, and H_2O) are broadly used for thermometer calibration, defining the so-called *international temperature scales* including the currently accepted scale ITS-90.

This analysis may be readily generalized to multi-component systems consisting of particles of several (say, L) sorts[13]. If such a system is in a single phase, i.e. is macroscopically uniform, its chemical potential may be defined by the natural generalization of Eq. (1.53*c*):

$$dG = -SdT + VdP + \sum_{l=1}^{L} \mu^{(l)} dN^{(l)}. \tag{4.19}$$

Typically, a single phase is not a pure substance, but has certain concentrations of other components, so that $\mu^{(l)}$ may depend not only on P and T, but also on *concentrations* $c^{(l)} \equiv N^{(l)}/N$ of particles of each sort. If the total number N of particles is fixed, the number of independent concentrations is $(L - 1)$. For the chemical equilibrium of R phases, all R values of $\mu_r^{(l)}$ $(r = 1, 2, \ldots, R)$ have to be equal for particles of each sort: $\mu_1^{(l)} = \mu_2^{(l)} = \ldots = \mu_R^{(l)}$, with each $\mu_r^{(l)}$ depending on $(L - 1)$ concentrations $c_r^{(l)}$, and also on P and T. This requirement gives $L(R - 1)$ equations for $(L - 1)R$ concentrations $c_r^{(l)}$, plus two common arguments P and T, i.e. for $[(L - 1)R + 2]$ independent variables. This means that the number of phases has to satisfy the limitation

[12] Please note that P_t for water is several orders of magnitude lower than P_c of the water–vapor transition, so that figure 4.3b is indeed very much schematic!

[13] Perhaps the most practically important example is the air/water system. For its detailed discussion, based on Eq. (4.19), the reader may be referred, e.g. to section 3.9 in [1]. Other important applications include liquid solutions, and metallic *alloys*—solid solutions of metal elements.

$$L(R-1) \leqslant (L-1)R + 2, \quad \text{i.e. } R \leqslant L + 2, \tag{4.20}$$

where the equality sign may be reached in just one point in the whole parameter space. This is the *Gibbs phase rule*. As a sanity check, for a single-component system, $L = 1$, the rule yields $R \leqslant 3$—exactly the result we have already discussed.

4.2 Continuous phase transitions

As figure 4.2 illustrates, if we fix pressure P in a system with a first-order phase transition, and start changing its temperature, then the complete crossing of the transition point, defined by the equation $P_0(T) = P$, requires the insertion (or extraction) of some non-zero latent heat Λ. Eqs. (4.14) and (4.17) show that Λ is directly related to non-zero differences between the entropies and volumes of the two phases (at the same pressure). As we know from chapter 1, both S and V may be represented as first derivatives of appropriate thermodynamic potentials. This is why P Ehrenfest called such transitions, involving jumps of potentials' *first* derivatives, *first-order* phase transitions[14].

On the other hand, there are phase transitions that have no first derivative jumps at the transition temperature T_c, so that the temperature point may be clearly marked, for example, by a jump of a *second* derivative of a thermodynamic potential—for example, the derivative $\partial C/\partial T$ which, according to Eq. (1.24), equals to $\partial^2 E/\partial T^2$. In the initial Ehrenfest's classification, this was an example of a *second-order* phase transition. However, most features of such phase transitions are also pertinent to some systems in which the second derivatives of potentials are continuous as well. For this reason, I will use a more recent terminology (suggested by M Fisher), in which all phase transitions with $\Lambda = 0$ are called *continuous*.

Most (though not all) continuous phase transitions result from particle interactions. Here are some representative examples:

(i) At temperatures above ~120 °C, the crystal lattice of barium titanate ($BaTiO_3$) is cubic, with a Ba ion in the center of each Ti-cornered cube (or vice versa)—see figure 4.4a. However, as temperature is being lowered below that critical

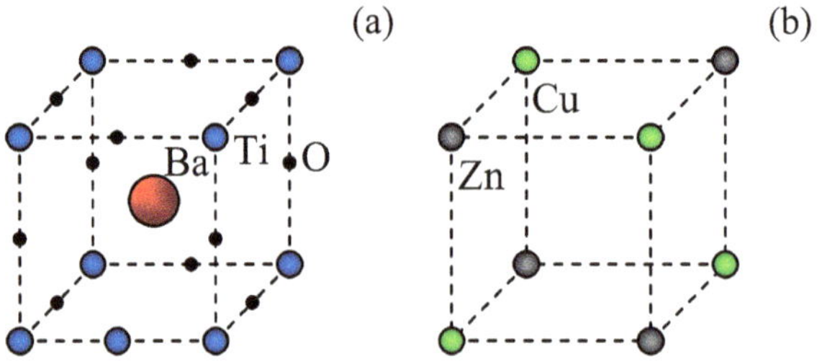

Figure 4.4. Single cells of crystal lattices of (a) $BaTiO_3$ and (b) CuZn.

[14] One of the reasons why his classification is now considered incomplete is that it does not distinguish the cases when these derivatives diverge at the transition point approach.

value, the sublattice of Ba ions starts moving along one of 6 sides of the TiO_3 sublattice, leading to a small deformation of both lattices—which become tetragonal. This is a typical example of a *structural transition*, in this particular case combined with a *ferroelectric transition*, because (due to the positive electric charge of the Ba ions) below the critical temperature the $BaTiO_3$ crystal acquires a spontaneous electric polarization even in the absence of external electric field.

(ii) A different kind of phase transition happens, for example, in Cu_xZn_{1-x} alloys—so-called *brasses*. Their crystal lattice is always cubic, but above a certain critical temperature T_c (which depends on x) any of its nodes may be occupied by either a copper or a zinc atom, at random. At $T < T_c$, a trend toward ordered atom alternation arises, and at low temperatures, the atoms are fully ordered, as shown in figure 4.4b for the stoichiometric case $x = 0.5$. This is a good example of an *order–disorder* transition.

(iii) At *ferromagnetic* transitions (such as the one taking place, for example, in Fe at 1388 K) and *antiferromagnetic* transitions (e.g. in MnO at 116 K), lowering of temperature below the critical value[15] does not change atom positions substantially, but results in a partial ordering of atomic spins, eventually leading to their full ordering (figure 4.5).

Note that, as follows from Eqs. (1.1)–(1.3), for ferroelectric transitions in cylindrical samples, the role of pressure is played by the external electric field $\mathscr{E}$, and for the ferromagnetic transitions, by the external magnetic field $\mathscr{H}$. As we will see very soon, even in systems with continuous phase transitions, a gradual change of such an external field, at fixed temperature, may induce jumps between metastable states, similar to those in systems with first-order phase transitions (see, e.g. the dashed arrows in figure 4.2), with nonvanishing decreases of the appropriate free energy.

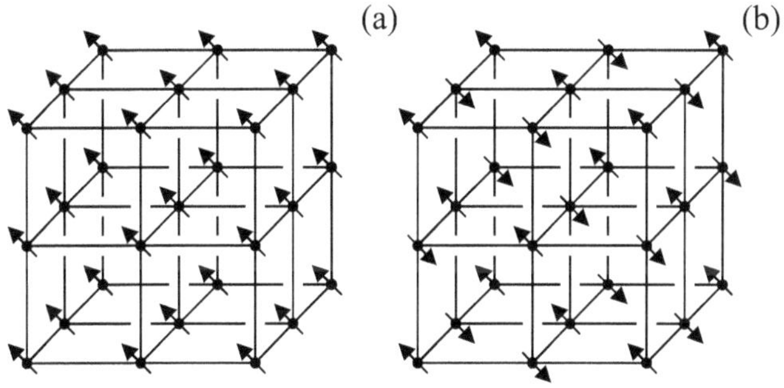

Figure 4.5. Classical images of fully ordered phases: (a) a ferromagnet, and (b) an antiferromagnet.

[15] For ferromagnets, this point is usually referred to at the *Curie temperature*, and for antiferromagnets, as the *Néel temperature*.

Besides these standard examples, some other threshold phenomena, such as formation of a coherent optical field in a laser, and even the self-excitation of oscillators with negative damping (see, e.g. *Part CM* section 5.4), may be treated, at certain conditions, as continuous phase transitions[16].

The general feature of all these transitions is the gradual formation, at $T < T_c$, of certain *ordering*, which may be characterized by some *order parameter* $\eta \neq 0$. The simplest example of such order parameter is the magnetization at the ferromagnetic transitions, and this is why the continuous phase transitions are usually discussed on certain models of ferromagnetism. (I will follow this tradition, while mentioning in passing other important cases that require a substantial modification of theory.) Most of such models are defined on an infinite 3D cubic lattice (see, e.g. figure 4.5), with evident generalizations to lower dimensions. For example, the *Heisenberg model* of a ferromagnet is defined by the following Hamiltonian:

$$\hat{H} = -J \sum_{\{k,k'\}} \hat{\boldsymbol{\sigma}}_k \cdot \hat{\boldsymbol{\sigma}}_{k'} - \sum_k \mathbf{h} \cdot \hat{\boldsymbol{\sigma}}_k, \tag{4.21}$$

where $\hat{\boldsymbol{\sigma}}_k$ is the Pauli vector operator[17] acting on the kth spin, and $\mathbf{h}$ is the normalized external magnetic field $\mathscr{H}$:

$$\mathbf{h} \equiv \mathscr{m}_0 \mu_0 \mathscr{H}. \tag{4.22}$$

(Here $\mathscr{m}_0$ is the magnitude of the spin's magnetic moment; for the Heisenberg model to be realistic, is should be of the order of the *Bohr magneton* $\mu_B \equiv e\hbar/2m_e \approx 0.927 \times 10^{-23}$ J T^{-1}.) The figure brackets $\{j, j'\}$ in Eq. (4.21) denote the summation over the pairs of adjacent lattice sites, so that the magnitude of the constant J may be interpreted as the maximum coupling energy per 'bond' between two adjacent particles. At $J > 0$, the coupling tries to keep spins aligned, i.e. to install the ferromagnetic ordering[18]. The second term in Eq. (4.21) describes the effect of the external magnetic field, which tries to turn all spins, with their magnetic moments, along its direction[19].

However, even the Heisenberg model, while being rather approximate (in particular because its standard form (4.21) is only valid for spins-½), is still rather complex for analysis. This is why most theoretical results have been obtained for its classical twin, the *Ising model*[20]:

$$E_m = -J \sum_{\{k,k'\}} s_k s_{k'} - h \sum_k s_k. \tag{4.23}$$

[16] Unfortunately, I will have no time/space for these interesting (and practically important) generalizations, and have to refer the interested reader to the famous monograph by R Stratonovich [2, 3] and/or the influential review by H Haken [4].

[17] See, e.g. *Part QM* section 4.4. In the standard z-basis, this operator is represented by the set of three Pauli matrices.

[18] At $J < 0$, the first term of Eq. (4.21) gives a reasonable model of an antiferromagnet, but in this case the external magnetic field effects are more subtle, so I will not have time to discuss it.

[19] See, e.g. *Part QM* Eq. (4.163).

[20] Named after E Ising who explored the 1D version of the model in detail in 1925, though a similar model was discussed earlier (in 1920) by W Lenz.

Here E_m are the values of system's energy, the constant h models an external magnetic field's magnitude, and s_k are classical scalar variables that may take only two values, $s_k = \pm 1$. (Despite its classical character, the variable s_k modeling the real spin of an electron, is usually called 'spin' for brevity, and I will follow this tradition.) The index m in the notation E_m numbers all possible combinations of the binary variables s_k; there are evidently 2^N of them in a system of N Ising 'spins'. Somewhat shockingly, even for this toy model, no exact analytical 3D solution that would be valid at arbitrary temperature and $N \to \infty$ has been found yet, and the solution of its 2D version by L Onsager in 1944 (see section 4.5 below) is still considered one of the top intellectual achievements of statistical physics. Still, Eq. (4.23) is very useful for the introduction of basic notions of continuous phase transitions, and methods of their analysis, so that for my brief discussion I will mostly use this model[21].

Evidently, if $T = 0$ and $h = 0$, the lowest possible energy,

$$E_{\min} = -JNd, \tag{4.24}$$

where d is the lattice dimensionality, is achieved in the 'ferromagnetic' phase in which all spins s_k are equal to either $+1$ or -1. On the other hand, at $J = 0$ and $h = 0$, the spins are independent, and in the absence of external field their signs are completely random, with the 50% probability to have either of values ± 1, so that $\langle s_k \rangle = 0$. Hence in the case of arbitrary parameters we may use the average

$$\eta \equiv \langle s_k \rangle \tag{4.25}$$

as a good measure of spin ordering, i.e. as the order parameter. Since in a real ferromagnet, each spin carries a magnetic moment, the order parameter η is proportional to the Cartesian component of the system's magnetization, in the direction of the applied magnetic field.

Now that the Ising model gave us a very clear illustration of the order parameter, let me discuss the general characterization of continuous phase transitions, using this notion. Due to the difficulty of theoretical analyses of most models of such transitions at arbitrary temperatures, their theoretical discussions are focused mostly on a close vicinity of the critical point T_c. Both experiment and theory show that (in the absence of external field), for continuous phase transitions in most systems, the function $\eta(T)$ is close to a certain power,

$$\eta \propto \tau^{\beta}, \quad \text{for } \tau > 0, \quad \text{i.e. } T < T_c, \tag{4.26}$$

of the small deviation from the critical temperature—which is conveniently normalized as

[21] For more detailed discussions of phase transition theories (including other popular models of the ferromagnetic phase transition, e.g. the *Potts model*), see, e.g. either [5], or [6], or [7]. For a very concise text, I can recommend [8].

$$\tau \equiv \frac{T_c - T}{T_c}. \qquad (4.27)$$

Remarkably, most other key variables follow a similar temperature behavior, with a *critical exponent* being the same for both signs of τ. In particular, the heat capacity at fixed magnetic field behaves as[22]

$$c_h \propto |\tau|^{-\alpha}. \qquad (4.28)$$

Similarly, the (normalized) low-field *susceptibility*[23]

$$\chi \equiv \frac{\partial \eta}{\partial h}\Big|_{h=0} \propto |\tau|^{-\gamma}. \qquad (4.29)$$

Two more important critical exponents, ζ and ν, describe the temperature behavior of the *correlation function* $\langle s_k s_{k'} \rangle$ whose dependence on distance $r_{kk'}$ between two spins may be well fitted by the following law,

$$\langle s_k s_{k'} \rangle \propto \frac{1}{r_{kk'}{}^{d-2+\zeta}} \exp\left\{-\frac{r_{kk'}}{r_c}\right\}, \qquad (4.30)$$

with the *correlation radius*

$$r_c \propto |\tau|^{-\nu}. \qquad (4.31)$$

Finally, three more critical exponents, usually denoted ε, δ, and μ, describe the external field dependences of, respectively, c, η and r_c at $\tau = 0$. For example, δ is defined as

$$\eta \propto h^{1/\delta}. \qquad (4.32)$$

(Other field exponents are used less frequently, and for their discussion I have to refer the interested reader to the special literature that was cited above.)

The leftmost column of table 4.1 shows the ranges of experimental values of the critical exponents for various 3D physical systems featuring continuous phase transitions. One can see that their values vary from system to system, leaving no hope for a universal theory that would describe them all exactly. However, certain combinations of the exponents are much more reproducible—see the four bottom lines of the table.

Historically the first (and perhaps the most fundamental) of these *universal relations* was derived in 1963 by J Essam and M Fisher:

$$\alpha + 2\beta + \gamma = 2. \qquad (4.33)$$

It may be proved, for example, by finding the temperature dependence of the magnetic field value, h_τ, that changes the order parameter by the amount similar to

[22] The forms of this and following temperature functions is selected to make all critical exponents non-negative.

[23] In the models of ferromagnetic phase transitions, this variable is proportional to the genuine low-field magnetic susceptibility χ_m of the material—see, e.g. *Part EM* Eq. (5.111).

Table 4.1. Major critical exponents of continuous phase transitions.

Exponents and combinations	Experimental range (3D)[(a)]	Landau's theory	2D Ising model	3D Ising model	3D Heisenberg model[(d)]
α	0 – 0.14	0[(b)]	[(c)]	0.12	−0.14
β	0.32 – 0.39	1/2	1/8	0.31	0.3
γ	1.3 – 1.4	1	7/4	1.25	1.4
δ	4 – 5	3	15	5	?
ν	0.6 – 0.7	1/2	1	0.64	0.7
ζ	0.05	0	1/4	0.05	0.04
$(\alpha + 2\beta + \gamma)/2$	1.00 ± 0.005	1	1	1	1
$\delta - \gamma/\beta$	0.93 ± 0.08	1	1	1	?
$(2 - \zeta)\nu/\gamma$	1.02 ± 0.05	1	1	1	1
$(2 - \alpha)/\nu d$	?	$4/d$	1	1	1

[a] Experimental data are from the monograph by A Patashinskii and V Pokrovskii, cited above.
[b] Discontinuity at $\tau = 0$—see below.
[c] Instead of following Eq. (4.28), in this case c_h diverges as $\ln|\tau|$.
[d] With the order parameter η defined as $\langle \sigma_j \cdot \mathscr{B} \rangle / \mathscr{B}$.

that already existing at $h = 0$ due to a finite temperature deviation $\tau > 0$. First, we may compare Eqs. (4.26) and (4.29), to get

$$h_\tau \propto \tau^{\beta+\gamma}. \tag{4.34}$$

By the physical sense of h_τ we may expect that such a field has to affect a system's free energy[24] F by the amount comparable to the effect of a bare temperature change τ. Ensemble-averaging the last term of Eq. (4.23) and using the definition (4.25) of the order parameter η, we see that the change of F (per particle) due to the field equals $-h_\tau \eta$ and, according to Eq. (4.26), scales as $h_\tau \tau^{\beta} \propto \tau^{(2\beta+\gamma)}$.

In order to estimate the thermal effect on F, let us first elaborate a bit more on the useful thermodynamic formulas already mentioned in section 1.3:

$$C_X = T\left(\frac{\partial S}{\partial T}\right)_X, \tag{4.35}$$

where X means the variable(s) maintained constant at the temperature variation. In the standard 'P–V' thermodynamics, we may use Eqs. (1.35) for $X = V$, and Eqs. (1.39) for $X = P$, to write

$$C_V = T\left(\frac{\partial S}{\partial T}\right)_{V,N} = -T\left(\frac{\partial^2 F}{\partial T^2}\right)_{V,N}, \quad C_P = T\left(\frac{\partial S}{\partial T}\right)_{P,N} = -T\left(\frac{\partial^2 G}{\partial T^2}\right)_{P,N}. \tag{4.36}$$

[24] As was already discussed in sections 1.4 and 2.4, there is some dichotomy of terminology (and notation) in literature on this topic. In the models (4.21) and (4.23), the magnetic field effects are accounted at the microscopic level, by the inclusion of the corresponding term into each particular value of energy E_m. From this point of view, the list of variables in these systems does not include pressure and volume, and we may take $G \equiv F + PV = F + \text{const}$, so that their equilibrium (at fixed h, T and N) corresponds to the minimum of the *Helmholtz* free energy F.

As was just discussed, in the ferromagnetic models of the type (4.21) or (4.23), at a constant field h, the role of G is played by F, so that Eq. (4.35) yields

$$C_h = T\left(\frac{\partial S}{\partial T}\right)_{h,N} = -T\left(\frac{\partial^2 F}{\partial T^2}\right)_{h,N}. \tag{4.37}$$

The last form of this relation means that F may be found by double integration of $(-C_h/T)$ over temperature. With Eq. (4.28) for $c_h \propto C_h$, this means that near T_c, the free energy scales as the double integral of $c_h \propto \tau^{-\alpha}$ over τ. In the limit $\tau \ll 1$, the factor T may be treated as a constant; as a result, the change of F due to $\tau > 0$ alone scales as $\tau^{(2-\alpha)}$. Requiring this change to be proportional to the same power of τ as the field-induced part of energy, we finally get the Essam–Fisher relation (4.33).

Using similar reasoning, it is straightforward to derive a few other universal relations of critical exponents, including the *Widom relation*,

$$\delta - \frac{\gamma}{\beta} = 1, \tag{4.38}$$

very similar relations for other high-field exponents ε and μ (which I do not have time to discuss), and the *Fisher relation*

$$\nu(2 - \zeta) = \gamma. \tag{4.39}$$

A slightly more complex reasoning, involving the so-called *scaling hypothesis*, yields the dimensionality-dependent *Josephson relation*

$$\nu d = 2 - \alpha. \tag{4.40}$$

The second column of table 4.1 shows that at least three of these relations are in a very reasonable agreement with experiment, so that we may use their set as a testbed for various theoretical approaches to continuous phase transitions.

4.3 Landau's mean-field theory

The highest-level approach to continuous phase transitions, formally not based on any particular model (though in fact implying either the Ising model (4.23) or one of it siblings), is the *mean-field theory* developed in 1937 by L Landau, on the basis of prior ideas by P Weiss—to be discussed in the next section. The main approximation of this phenomenological approach is to represent the free energy's change ΔF at the phase transition as an explicit function of the order parameter η (4.25). Since at $T \to T_c$, η has to tend to zero, this change,

$$\Delta F \equiv F(T) - F(T_c), \tag{4.41}$$

may be expanded into the Taylor series in η, and only a few, most important first terms of that expansion retained. In order to keep the symmetry between two possible signs of the order parameter (i.e. between two possible spin directions in the Ising model) in the absence of external field, at $h = 0$ this expansion should include only even powers of η:

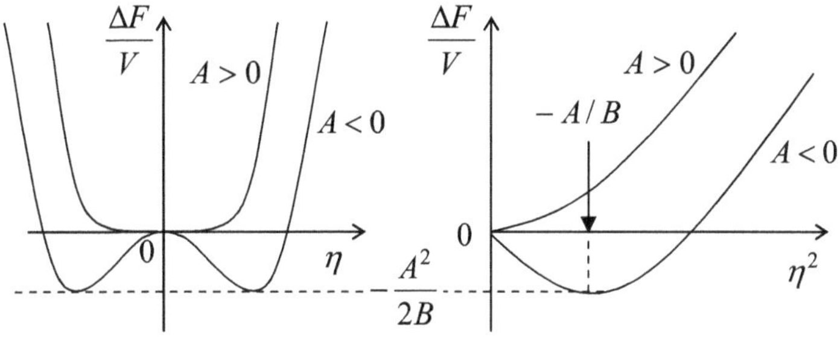

Figure 4.6. The Landau free energy (4.42) as a function of (a) η and (b) η^2, for two signs of the coefficient $A(T)$, both for $B(T) > 0$.

$$\Delta f\,|_{h=0} \equiv \frac{\Delta F}{V}\,|_{h=0} = A(T)\eta^2 + \frac{1}{2}B(T)\eta^4 + \ldots, \quad \text{at } T \approx T_c. \tag{4.42}$$

As figure 4.6 shows, at $A(T) < 0$, and $B(T) > 0$, these two terms are sufficient to describe the minimum of the free energy at $\eta^2 > 0$, i.e. to calculate stationary values of the order parameter; this is why Landau's theory ignores higher terms of the Taylor expansion—which are much smaller at $\eta \to 0$.

Now let us discuss temperature dependences of the coefficients A and B. As Eq. (4.42) shows, first of all, the coefficient $B(T)$ has to be positive for any sign of $\tau \propto (T_c - T)$, to ensure the equilibrium at a finite value of η^2. Thus, it is reasonable to ignore the temperature dependence of B near the critical temperature altogether and use the approximation

$$B(T) = b > 0. \tag{4.43}$$

On the other hand, as figure 4.6 shows, the coefficient $A(T)$ has to change sign at $T = T_c$, to be positive at $T > T_c$ and negative at $T < T_c$, to ensure the transition from $\eta = 0$ at $T > T_c$ to a certain nonvanishing value at $T < T_c$. Since A should be a smooth function of temperature, we may approximate it by the leading term of its Taylor expansion in τ:

$$A(T) = -a\tau, \quad \text{with } a > 0, \tag{4.44}$$

so that Eq. (4.42) becomes

$$\Delta f\,|_{h=0} = -a\tau\eta^2 + \frac{1}{2}b\eta^4. \tag{4.45}$$

In this rudimentary form, the Landau theory may look almost trivial, and its main strength is the possibility of its straightforward extension to the effects of the external field and of spatial variations of the order parameter. First, as the averaging of the field term in Eqs. (4.21) or (4.23) shows, the applied field gives such systems an energy addition of $-h\eta$ per particle, i.e.—$nh\eta$ per unit volume, where n is the particle density. Second, since according to Eq. (4.31) (with $\nu > 0$, see table 4.1) the correlation radius diverges at $\tau \to 0$, spatial variations of the order parameter should be slow, $|\nabla\eta| \to 0$. Hence, the effects of the gradient on ΔF may be approximated by the first nonvanishing term of its expansion into the Taylor series in $(\nabla\eta)^2$.[25] As a result, Eq. (4.45) may be generalized as

$$\Delta F = \int \Delta f d^3r, \quad \text{with} \quad \Delta f = -a\tau\eta^2 + \frac{1}{2}b\eta^4 - nh\eta + c(\nabla\eta)^2, \tag{4.46}$$

where c is a factor independent of η. In order to avoid the unphysical effect of spontaneous formation of spatial variations of the order parameter, that factor has to be positive at all temperatures, and hence may be taken for constant in a small vicinity of T_c—the only region where Eq. (4.46) may be expected to provide quantitatively correct results.

Let us find out what critical exponents are predicted by this phenomenological approach. First of all, we may find the equilibrium values of the order parameter from the condition of F having a minimum, $\partial F/\partial\eta = 0$. At $h = 0$, it is easier to use the equivalent equation $\partial F/\partial(\eta^2) = 0$, where F is given by Eq. (4.45)—see figure 4.6b. This immediately yields

$$|\eta| = \begin{cases} (a\tau/b)^{1/2}, & \text{for } \tau > 0, \\ 0, & \text{for } \tau < 0. \end{cases} \tag{4.47}$$

Comparing this result with Eq. (4.26), we see that in the Landau theory, $\beta = ½$. Next, plugging the result (4.47) back into Eq. (4.45), for the equilibrium (minimal) value of the free energy, we get

$$\Delta f = \begin{cases} -a^2\tau^2/2b, & \text{for } \tau > 0, \\ 0, & \text{for } \tau < 0. \end{cases} \tag{4.48}$$

From here and Eq. (4.37), the specific heat,

$$\frac{C_h}{V} = \begin{cases} a^2/bT_c, & \text{for } \tau > 0, \\ 0, & \text{for } \tau < 0, \end{cases} \tag{4.49}$$

has, at the critical point, a discontinuity rather than a singularity, so that we need to prescribe the zero value to the critical exponent α.

In the presence of a uniform field, the equilibrium order parameter should be found from the condition $\partial f/\partial\eta = 0$ applied to Eq. (4.46) with $\nabla\eta = 0$, giving

$$\frac{\partial f}{\partial\eta} \equiv -2a\tau\eta + 2b\eta^3 - nh = 0. \tag{4.50}$$

In the limit of small order parameter, $\eta \to 0$, the term with η^3 is negligible, and Eq. (4.50) gives

$$\eta = -\frac{nh}{2a\tau}, \tag{4.51}$$

[25] Historically, the last term belongs to the later (1950) extension of the theory by V Ginzburg and L Landau—see below.

so that according to Eq. (4.29), $\gamma = 1$. On the other hand, at $\tau = 0$ (or at relatively high fields at other temperatures), the cubic term in Eq. (4.50) is much larger than the linear one, and this equation yields

$$\eta = \left(\frac{nh}{2b}\right)^{1/3}, \tag{4.52}$$

so that comparison with Eq. (4.32) yields $\delta = 3$.

Finally, according to Eq. (4.30), the last term in Eq. (4.46) scales as $c\eta^2/r_c^2$. (If $r_c \neq \infty$, the effects of the pre-exponential factor in Eq. (4.30) are negligible.) As a result, the gradient term's contribution is comparable[26] with the two leading terms in Δf (which, according to Eq. (4.47), are of the same order), if

$$r_c \approx \left(\frac{c}{a\,|\tau|}\right)^{1/2}, \tag{4.53}$$

so that according to the definition (4.31) of the critical exponent ν, in the Landau theory it is equal to ½.

The third column in table 4.1 summarizes the critical exponents and their combinations in Landau's theory. It shows that these values are somewhat out of the expcrimcntal ranges, and while some of their universal relations are correct, some are not; for example, the Josephson relation would be only correct at $d = 4$ (not the most realistic spatial dimensionality :-) The main reason for this disappointing result is that describing the spin interaction with the field, the Landau mean-field theory neglects spin randomness, i.e. fluctuations. Though a quantitative theory of fluctuations will be discussed only in the next chapter, we can readily perform their crude estimate. Looking at Eq. (4.46), we see that its first term is a quadratic function of the effective 'half-degree of freedom', η. Hence in accordance with the equipartition theorem (2.28) we may expect that the average square of its thermal fluctuations, within a d-dimensional volume with a linear size of the order of r_c, should be of the order of $T/2$ (close to the critical temperature, $T_c/2$ is a good enough approximation):

$$a\,|\tau|\,\langle\tilde{\eta}^2\rangle r_c^d \sim \frac{T_c}{2}. \tag{4.54}$$

In order to be negligible, the variance has to be negligible in comparison with the average $\eta^2 \sim a\tau/b$—see Eq. (4.47). Plugging in the τ-dependences of the operands of this relation, and values of the critical exponents in the Landau theory, for $\tau > 0$ we get the so-called *Levanyuk–Ginzburg criterion* of its validity:

$$\frac{T_c}{2a\tau}\left(\frac{a\tau}{c}\right)^{\frac{d}{2}} \ll \frac{a\tau}{b}. \tag{4.55}$$

[26] According to Eq. (4.30), the correlation radius may be interpreted as the length distance at which the order parameter η relaxes to its equilibrium value, if it is deflected from it at some point. Since the law of such spatial change may be obtained by a variational differentiation of F, for the actual relaxation law, all major terms of (4.46) have to be comparable.

We see that for any realistic dimensionality, $d < 4$, at $\tau \to 0$ the order parameter's fluctuations grow faster than the its average value, and hence the theory becomes invalid.

Thus the Landau mean-field theory is not a perfect approach to finding critical indices at continuous phase transitions in Ising-type systems with their next-neighbor interactions between the particles. Despite that fact, this theory is very much valued for the following reason. Any long-range interactions between particles increase the correlation radius r_c, and hence suppress the order parameter fluctuations. For an example, at laser self-excitation, the emerging coherent optical field couples essentially *all* photon-emitting particles in the electromagnetic cavity (resonator). As another example, in superconductors the role of the correlation radius is played by the Cooper-pair size ξ_0, which is typically of the order of 10^{-6} m, i.e. much larger than the average distance between the pairs ($\sim 10^{-8}$ m). As a result, the mean-field theory remains valid at all temperatures besides an extremely small temperature interval near T_c—for bulk superconductors, of the order of 10^{-6} K.

Another strength of Landau's *classical* mean-field theory (4.46) is that it may be readily generalized for description of Bose–Einstein condensates, i.e. *quantum* fluids. Of those generalizations, the most famous is the *Ginzburg–Landau theory* of superconductivity developed in 1950, i.e. even before the 'microscopic' explanation of this phenomenon by Bardeen, Cooper and Schrieffer in 1956–57. In this theory, the real order parameter η is replaced with the modulus of a complex function ψ, physically the wavefunction of the coherent Bose–Einstein condensate of Cooper pairs. Since each pair carries the electric charge $q = -2e$, and has zero spin s, it interacts with the magnetic field in a way different from that described by the Heisenberg or Ising models. Namely, as was already discussed in section 3.4, the del operator ∇ in Eq. (4.46) has to be complemented with the term $-i(q/\hbar)\mathbf{A}$, where $\mathbf{A}$ is the vector-potential of the total magnetic field $\mathscr{B} = \nabla \times \mathbf{A}$, including not only the external magnetic field $\mathscr{H}$, but also the field induced by the supercurrent itself. With the account for the well-known formula for the magnetic field energy, Eq. (4.46) is now replaced with

$$\Delta f = -a\tau \,|\psi|^2 + \frac{1}{2}b\,|\psi|^4 - \frac{\hbar^2}{2m}\left|\left(\nabla - i\frac{q}{\hbar}\mathbf{A}\right)\psi\right|^2 + \frac{\mathscr{B}^2}{2\mu_0}, \tag{4.56}$$

where m is a phenomenological coefficient rather than the actual particle's mass.

The variational minimization of the resulting Gibbs energy density $\Delta g = \Delta f - \mu_0 \mathscr{H} \cdot \mathscr{M} \equiv \Delta f - \mathscr{H} \cdot \mathscr{B} + \text{const}$[27] over the variables ψ and $\mathscr{B}$ (which is suggested for the reader's exercise) yields two differential equations:

[27] As an immediate elementary sanity check of this relation, resulting from the analogy of Eqs. (1.1) and (1.3), the minimization of Δg in the absence of superconductivity ($\psi = 0$) gives the correct result $\mathscr{B} = \mu_0 \mathscr{H}$. Note that this account of this difference between Δf and Δg is necessary because (unlike Eqs. (4.21) and (4.23)), the Ginzburg–Landau free energy (4.56) does not take into account the effect of the field on each particle directly.

$$\frac{\nabla \times \mathscr{B}}{\mu_0} = q\frac{i\hbar}{2m}\left[\psi\left(\nabla - i\frac{q}{\hbar}\mathbf{A}\right)\psi^* - \text{c.c.}\right], \tag{4.57a}$$

$$a\tau\psi = b\,|\psi|^2\,\psi - \frac{\hbar^2}{2m}\left(\nabla - i\frac{q}{\hbar}\mathbf{A}\right)^2\psi. \tag{4.57b}$$

The first of these *Ginzburg–Landau equations* should be no big surprise for the reader, because according to the Maxwell equations, in magnetostatics the left-hand side of Eq. (4.57*a*) has to be equal to the electric current density, while its right-hand side is the usual quantum-mechanical probability current density multiplied by q, i.e. the electric current density $\mathbf{j}$ of the Cooper pair condensate. (Indeed, after plugging $\psi = n^{1/2}\exp\{i\varphi\}$ into that expression, we come back to Eq. (3.84) which, as we already know, explains such macroscopic quantum phenomena as the magnetic flux quantization and the Meissner–Ochsenfeld effect.)

However, Eq. (4.57*b*) is new—at least for this course[28]. Since the last term on its right-hand side is the standard wave-mechanical expression for the kinetic energy of a particle in the presence of magnetic field[29], if this term dominates that side of the equation, Eq. (4.57*b*) is reduced to the stationary Schrödinger equation, $E\psi = \hat{H}\psi$, for the ground state of free Cooper pairs, with the total energy $E = a\tau$. However, in contrast to the usual (single-particle) Schrödinger equation, in which $|\psi|$ is determined by the normalization condition, the Cooper pair condensate density $n = |\psi|^2$ is determined by the thermodynamic balance of the condensate with the ensemble of 'normal' (unpaired) electrons, which plays the role of the uncondensed part of the particles in the usual Bose–Einstein condensate—see section 3.4. In Eq. (4.57*b*), such balance is enforced by the first term $b|\psi|^2\psi$ on the right-hand side. As we have already seen, in the absence of magnetic field and spatial gradients, such a term yields $|\psi| \propto \tau^{1/2} \propto (T_c - T)^{1/2}$—see Eq. (4.47).

As a parenthetic remark, from the mathematics standpoint, the term $b|\psi|^2\psi$, nonlinear in ψ, makes Eq. (4.57*b*) a member of the family of the so-called *nonlinear Schrödinger equations*. Another member of this family, important for physics, is the *Gross–Pitaevskii equation*,

$$a\tau\psi = b\,|\psi|^2\,\psi - \frac{\hbar^2}{2m}\nabla^2\psi + U(\mathbf{r})\psi, \tag{4.58}$$

which gives a very reasonable (albeit approximate) description of gradient and field effects on Bose–Einstein condensates of electrically neutral atoms at $T \approx T_c$. The differences between Eqs. (4.58) and (4.57) reflect, first, the zero electric charge q of the atoms (so that Eq. (4.57*a*) becomes trivial) and, second, the fact that the atoms forming the condensates may be readily placed in external potentials $U(\mathbf{r}) \neq \text{const}$ (including the time-averaged potentials of optical traps—see *Part EM* chapter 7), while in superconductors such potential profiles are much harder to create due to the

[28] It was discussed in *Part EM* section 6.5.

[29] See, e.g. *Part QM* section 3.1.

screening of external electric and optical fields by good conductors—see, e.g. *Part EM* section 2.1.

Returning to the discussion of Eq. (4.57*b*), it is easy to see that its last term increases as either the external magnetic field or the density of current passed through a superconductor are increased, increasing the vector-potential. In the Ginzburg–Landau equation, this *increase* is matched by a corresponding *decrease* of $|\psi|^2$, i.e. of the condensate density n, until it is completely suppressed. This balance describes the well documented effect of superconductivity suppression by an external magnetic field and/or supercurrent passed through the sample. Moreover, together with Eq. (4.57*a*), describing the flux quantization (see section 3.4), Eq. (4.57*b*) explains the existence of the so-called *Abrikosov vortices*—thin tubes of magnetic field, each carrying one quantum Φ_0 of magnetic flux—see Eq. (3.86). At the core part of the vortex, $|\psi|^2$ is suppressed (down to zero at its central line) by the persistent, dissipation-free current of the superconducting condensate, which circulates around the core and screens the rest of superconductor from the magnetic field carried by the vortex[30]. The penetration of such vortices into the so-called type-II superconductors enables them to sustain zero dc resistance up to very high magnetic fields of the order of 20 T, and as a result, to be used in very compact magnets—including those used for beam bending in particle accelerators.

Moreover, generalizing Eqs. (4.57) to the time-dependent case, just as is done with the usual Schrödinger equation, one can describe other fascinating quantum macroscopic phenomena such as the *Josephson effects*, including the generation of oscillations with frequency $\omega_J = (q/\hbar)\mathscr{V}$ by weak links between two superconductors, biased by dc voltage $\mathscr{V}$. Unfortunately, time/space restrictions do not allow me to discuss these effects in any detail in this course, and I have to refer the reader to special literature[31]. Let me only note that at $T \approx T_c$, and not extremely pure superconductors (in which the so-called non-local transport phenomena may be important), the Ginzburg–Landau equations are exact, and may be derived (and their parameters T_c, a, b, q, and m determined) from the 'microscopic' theory of superconductivity based on the initial work by Bardeen, Cooper and Schrieffer[32]. Most importantly, such derivation proves that $q = -2e$—the electric charge of a single Cooper pair.

4.4 Ising model: the Weiss' molecular-field theory

The Landau mean-field theory is phenomenological in the sense that even within the range of its validity, it tells us nothing about the value of the critical temperature T_c and other parameters (in Eq. (4.46), the coefficients a, b, and c), so that they have to be found from a particular 'microscopic' model of the system under analysis. In this course, we would have time to discuss only the Ising model (4.23) for various dimensionalities d.

The most simplistic way to map this model on a mean-field theory is to assume that all spins are exactly equal, $s_k = \eta$, with an additional condition $\eta^2 \leqslant 1$, forgetting

[30] See, e.g. *Part EM* section 6.5.

[31] See, e.g. [9]. A short discussion of the Josephson effects and Abrikosov vortices may be found in *Part QM* section 1.6 and *Part EM* section 6.5.

[32] See, e.g. section 4.45 in [10].

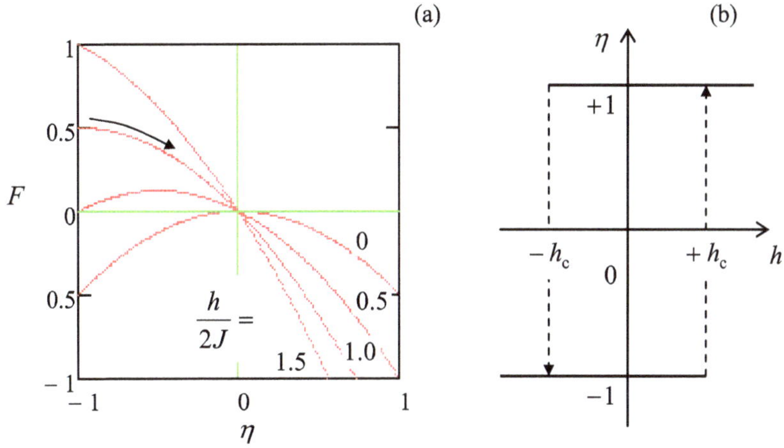

Figure 4.7. Field dependences of (a) the free energy profile and (b) the order parameter (i.e. magnetization) in the crudest mean-field approach to the Ising model.

for a minute that in the genuine Ising model, s_k may equal only +1 or −1. Plugging this relation into Eq. (4.23), we get[33]

$$F = -(NJd)\eta^2 - Nh\eta. \tag{4.59}$$

This energy is plotted in figure 4.7a as a function of η, for several values of h. The plots show that at $h = 0$, the system may be in either of two stable states, with $\eta = \pm 1$, corresponding to two different directions of spins (i.e. two different directions of magnetization), with equal energy[34]. (Formally, the state with $\eta = 0$ is also stationary, because at this point $\partial F/\partial\eta = 0$, but it is unstable, because for the ferromagnetic interaction, $J > 0$, the second derivative $\partial^2 F/\partial\eta^2$ is always negative.) As the external field is increased, it tilts the potential profile, and finally at a critical field,

$$h = h_c \equiv 2Jd, \tag{4.60}$$

the state with $\eta = -1$ becomes unstable, leading to the system's jump into the only remaining state with opposite magnetization, $\eta = +1$—see the arrow in figure 4.7a. Application of the similar external field of the opposite polarity leads to the similar switching, at the field $h = -h_c$, back to $\eta = -1$, so that the full field dependence of η follows the hysteretic pattern shown in figure 4.7b.[35] Such a pattern is the most visible experimental feature of actual ferromagnetic materials, with the *coercive magnetic field* $\mathscr{H}_c$ of the order of 10^3 A m^{-1}, and the *saturated* (or 'remnant') *magnetization* corresponding to fields $\mathscr{B}$ of the order of a few tesla. The most important property of

[33] Since in this naïve approach we neglect the fluctuations of spin, i.e. their disorder, this assumption of full ordering implies $S = 0$, so that $F \equiv E - TS = E$, and we may use either notation for the system's energy.
[34] The fact that the stable states always correspond to $\eta = \pm 1$, partly justifies the treatment, in this crude approximation, of the order parameter η as a continuous variable.
[35] Since these magnetization jumps are accompanied by (negative) jumps of the free energy F, they are sometimes called the first-order phase transitions. Note, however, that in this simple theory, these transitions are between two *fully-ordered* phases.

these materials, also called *permanent magnets*, is their stability, i.e. the ability to retain the history-determined direction of magnetization in the absence of external field, for a very long time. In particular, this property is the basis of all magnetic systems for data recording, including the ubiquitous hard disk drives with their incredible information density—currently approaching 1 Terabit per square inch[36].

So, this simplest mean-field theory (4.59) gives a crude description of the ferromagnetic ordering. However, this theory grossly overestimates the stability of these states with respect to thermal fluctuations. Indeed, in this theory, there is no thermally-induced randomness at all, until T becomes comparable with the height of the energy barrier separating two stable states,

$$\Delta F \equiv F(\eta = 0) - F(\eta = \pm 1) = NJd, \tag{4.61}$$

which is proportional to the number of particles. At $N \to \infty$, this value diverges, and in this sense the critical temperature is infinite, while numerical experiments and more refined theories of the Ising model show that actually the ferromagnetic phase is suppressed at $T_c \sim Jd$—see below.

The accuracy of such a theory may be dramatically improved by even an approximate account for thermally-induced randomness. In this approach (suggested in 1907 by P Weiss), called the *molecular-field theory*[37], random deviations of individual spin values from the lattice average,

$$\tilde{s}_k \equiv s_k - \eta, \quad \text{with } \eta \equiv \langle s_k \rangle, \tag{4.62}$$

are allowed, but considered small, $|\tilde{s}_k| \ll \eta$. This assumption allows us, after plugging the resulting expression $s_k = \eta + \tilde{s}_k$ into the first term on the right-hand side of Eq. (4.23),

$$\begin{aligned} E_m &= - J \sum_{\{k,k'\}} (\eta + \tilde{s}_k)(\eta + \tilde{s}_{k'}) - h \sum_k s_k \\ &\equiv - J \sum_{\{k,k'\}} [\eta^2 + \eta\, (\tilde{s}_k + \tilde{s}_{k'}) + \tilde{s}_k \tilde{s}_{k'}] - h \sum_k s_k, \end{aligned} \tag{4.63}$$

ignore the last term in the square brackets. Making the replacement (4.62) in the terms proportional to $\tilde{s}_k$, we may rewrite the result as

$$E_m \to E_m' \equiv (NJd)\eta^2 - h_{\text{ef}} \sum_k s_k, \tag{4.64}$$

[36] For me, it was always shocking how little my students knew about this fascinating (and very important) field of modern engineering, which involves so much interesting physics and a fantastic electromechanical technology. For getting acquainted with it, I may recommend, for example, the monograph [11].

[37] In some texts, this approximation is called the mean-field theory. This terminology may lead to confusion, because the molecular-field theory belongs to a different, deeper level of the theoretical hierarchy than, say, the (more phenomenological) Landau-style mean-field theories. For example, for a given microscopic model, the molecular-field approach may be used for the calculation of the parameters a, b, and T_c participating in Eq. (4.46)—the starting point of Landau's theory.

where h_{ef} is defined as the sum

$$h_{\text{ef}} \equiv h + (2Jd)\,\eta. \tag{4.65}$$

The physical interpretation of h_{ef} is the *effective* external field, which takes into account (besides the real external field h) the effect that *would be* exerted on spin s_k by its $2d$ next neighbors if they all had non-fluctuating (but possibly continuous) spin values $s_{k'} = \eta$. Such an addition to the external field,

$$h_{\text{mol}} \equiv h_{\text{ef}} - h = (2Jd)\,\eta, \tag{4.66}$$

is called the *molecular field*—giving its name to the Weiss theory.

From the point of view of statistical physics, at fixed parameters of the system (including the order parameter η), the first term on the right-hand side of Eq. (4.64) is merely a constant energy offset, and h_{ef} is just another constant, so that

$$E_m' = \text{const} + \sum_k \varepsilon_k, \quad \text{with} \quad \varepsilon_k = -h_{\text{ef}} s_k \equiv \begin{cases} -h_{\text{ef}}, & \text{for } s_k = +1, \\ +h_{\text{ef}}, & \text{for } s_k = -1. \end{cases} \tag{4.67}$$

Such separability of the energy means that in the molecular-field approximation the fluctuations of different spins are independent of each other, and their statistics may be examined individually, using the energy spectrum ε_k. But this is exactly the two-level system which was the subject of problems 2.2–2.4. Actually, its statistics is so simple that it is easier to redo this fundamental problem starting from scratch, rather than to use the results of those exercises (which would require changing notation). Indeed, according to the Gibbs distribution (2.58)–(2.59), the equilibrium probabilities of the states $s_k = \pm 1$ may be found as

$$W_\pm = \frac{1}{Z} e^{\pm h_{\text{ef}}/T}, \quad \text{with} \quad Z = \exp\left\{+\frac{h_{\text{ef}}}{T}\right\} + \exp\left\{-\frac{h_{\text{ef}}}{T}\right\} \equiv 2\cosh\frac{h_{\text{ef}}}{T}. \tag{4.68}$$

From here, we may readily calculate $F = -T\ln Z$ and all other thermodynamic variables, but let us immediately use Eq. (4.68) to calculate the statistical average of s_j, i.e. the order parameter:

$$\eta \equiv \langle s_j \rangle = (+1)W_+ + (-1)W_- = \frac{e^{+h_{\text{ef}}/T} - e^{-h_{\text{ef}}/T}}{2\cosh(h_{\text{ef}}/T)} \equiv \tanh\frac{h_{\text{ef}}}{T}. \tag{4.69}$$

Now comes the punch line of the Weiss' approach: plugging this result back into Eq. (4.65), we may write the condition of *self-consistency* of the molecular-field theory:

$$h_{\text{ef}} - h = 2Jd\tanh\frac{h_{\text{ef}}}{T}. \tag{4.70}$$

This is a transcendental equation, which evades an explicit analytical solution, but whose properties may be readily analyzed by plotting both its sides as functions of their argument, so that the stationary state(s) of the system corresponds to the intersection point(s) of these plots.

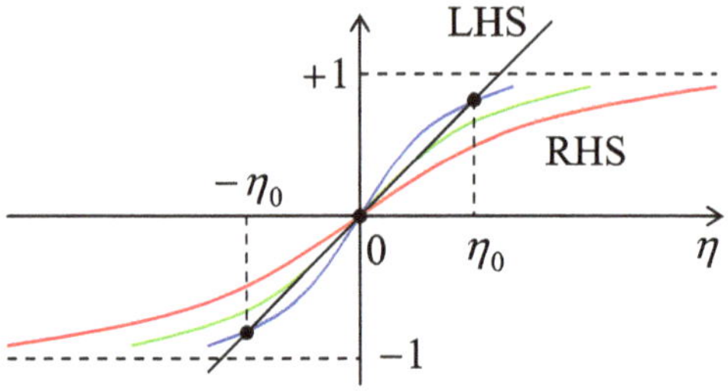

Figure 4.8. The ferromagnetic phase transition in Weiss' molecular-field theory: two sides of Eq. (4.71) plotted as functions of η for 3 temperatures: above T_c (red), below T_c (blue), and equal to T_c (green).

First of all, let us explore the field-free case ($h = 0$), when $h_{ef} = h_{mol} \equiv 2dJ\eta$, so that Eq. (4.70) is reduced to

$$\eta = \tanh\left(\frac{2Jd}{T}\eta\right), \tag{4.71}$$

giving one of the patterns sketched in figure 4.8, depending on the dimensionless parameter $2Jd/T$. If this parameter is small, the right-hand side of Eq. (4.71) grows slowly with η (see the red line in figure 4.8), and there is only one intersection point with the left-hand side plot, at $\eta = 0$. This means that the spin system has no spontaneous magnetization; this is the so-called *paramagnetic phase*. However, if the parameter $2Jd/T$ exceeds 1, i.e. T is decreased below the following critical value,

$$T_c = 2Jd, \tag{4.72}$$

the right-hand side of Eq. (4.71) grows, at small η, faster than its left-hand side, so that their plots intersect it at three points: $\eta = 0$ and $\eta = \pm\eta_0$—see the blue line in figure 4.8. It is almost evident that the former stationary point is unstable, while the two latter points are stable. (This fact may be readily verified by using Eq. (4.68) to calculate F. Now the condition $\partial F/\partial\eta|_{h=0} = 0$ returns us to Eq. (4.71), while calculating the second derivative, for $T < T_c$ we get $\partial^2 F/\partial\eta^2 > 0$ at $\eta = \pm\eta_0$, and $\partial^2 F/\partial\eta^2 < 0$ at $\eta = 0$.) Thus, below T_c the system is in the ferromagnetic phase, with one of two possible directions of the average spontaneous magnetization, so that the critical (*Curie*) temperature, given by Eq. (4.72), marks the transition between the paramagnetic and ferromagnetic phases. (Since the stable minimum value of the free energy F is a continuous function of temperature at $T = T_c$, this phase transition is continuous.)

Now let us repeat this graphics analysis to examine how each of these phases responds to an external magnetic field $h \neq 0$. According to Eq. (4.70), the effect of h is just a horizontal shift of the straight-line plot of its left-hand side—see figure 4.9. (Note a different, here more convenient, normalization of both axes.) In the paramagnetic case (figure 4.9a) the resulting dependence $h_{ef}(h)$ is evidently continuous, but the coupling effect ($J > 0$) makes it more steep than it would be without spin interaction. This effect may be characterized by the low-field susceptibility defined by

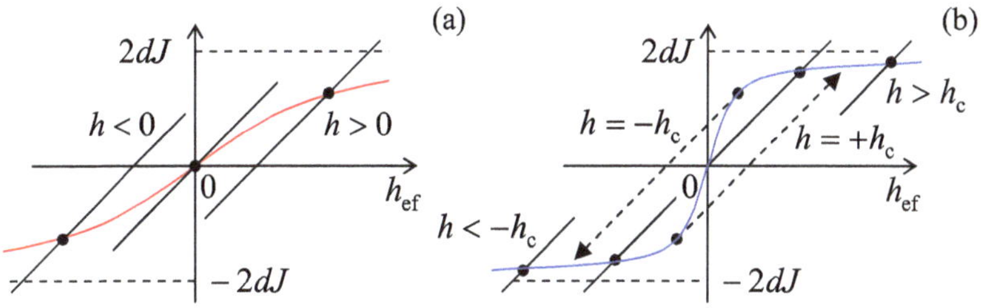

Figure 4.9. External field effects on: (a) a paramagnet ($T > T_c$), and (b) a ferromagnet ($T < T_c$).

Eq. (4.29). To calculate it, let us notice that for small h, and hence h_{ef}, the function tanh in Eq. (4.70) is approximately equal to its argument, so that Eq. (4.70) is reduced to

$$h_{ef} - h = \frac{2Jd}{T}h_{ef}, \quad \text{for} \quad \left|\frac{2Jd}{T}h_{ef}\right| \ll 1. \tag{4.73}$$

Solving this equation for h_{ef}, and then using Eq. (4.72), we get

$$h_{ef} = \frac{h}{1 - 2Jd/T} \equiv \frac{h}{1 - T_c/T}. \tag{4.74}$$

Recalling Eq. (4.66), we can rewrite this result for the order parameter:

$$\eta = \frac{h_{ef} - h}{T_c} = \frac{h}{T - T_c}, \tag{4.75}$$

so that the low-field susceptibility

$$\chi \equiv \frac{\partial \eta}{\partial h}\Big|_{h=0} = \frac{1}{T - T_c}, \quad \text{for } T > T_c. \tag{4.76}$$

This is the famous *Curie–Weiss law*, which shows that the susceptibility diverges at the approach to the Curie temperature T_c.

In the ferromagnetic case, the graphic solution (figure 4.9b) of Eq. (4.70) gives a qualitatively different result. A field increase leads, depending on the spontaneous magnetization, either to the further saturation of h_{mol} (with the order parameter η gradually approaching 1), or, if the initial η was negative, to a jump to positive η at some critical (coercive) field h_c. In contrast with the crude approximation (4.59), at $T > 0$ the coercive field is smaller than that given by Eq. (4.60), and the magnetization saturation is gradual, in good (semi-qualitative) accordance with experiment.

To summarize, the Weiss' molecular-field theory gives an approximate, but realistic description of the ferromagnetic and paramagnetic phases in the Ising model, and a very simple prediction (4.72) of the temperature of the phase transition between them, for an arbitrary dimensionality d of the cubic lattice. It also allows finding all other parameters of the Landau's mean-field theory for this model—an easy exercise, left for the reader.

4.5 Ising model: exact and numerical results

In order to evaluate the main prediction (4.72) of the Weiss theory, let us now discuss the exact (analytical) and quasi-exact (numerical) results obtained for the Ising model, going from the lowest value of dimensionality, $d = 0$, to its higher values.

Zero dimensionality means that a spin has no nearest neighbors at all, so that the first term of Eq. (4.23) vanishes. Hence Eq. (4.64) is exact, with $h_{\text{ef}} = h$, and so is its solution (4.69). Now we can repeat the calculations that have led us to Eq. (4.76), with $J = 0$, i.e. $T_c = 0$, and reduce this result to the so-called *Curie law*:

$$\chi = \frac{1}{T}. \tag{4.77}$$

It shows that the system is paramagnetic at any temperature. One may say that for this case the Weiss molecular-field theory is exact—or even trivial. (However, in some sense it is more general than the Ising model, because as we know from chapter 2, it gives the exact result for a fully quantum-mechanical treatment of any two-level system, including spin-½, at its negligible interaction with its environment.) Experimentally, the Curie law is approximately valid for many so-called *paramagnetic materials*, i.e. 3D systems with sufficiently weak interaction between particle spins.

The case $d = 1$ is more complex, but has an exact analytical solution. Probably the simplest way to obtain it is to use the so-called *transfer matrix approach*[38]. For this, first of all, we may argue that most properties of a 1D system of $N \gg 1$ spins (say, put at equal distances on a straight line) should not change noticeably if we bend that line gently into a closed ring (figure 4.10), assuming that the spins s_1 and s_N interact exactly as all other next-neighbor pairs. Then the energy (4.23) is

$$E_m = -(Js_1s_2 + Js_2s_3 + \ldots + Js_Ns_1) - (hs_1 + hs_2 + \ldots + hs_N). \tag{4.78}$$

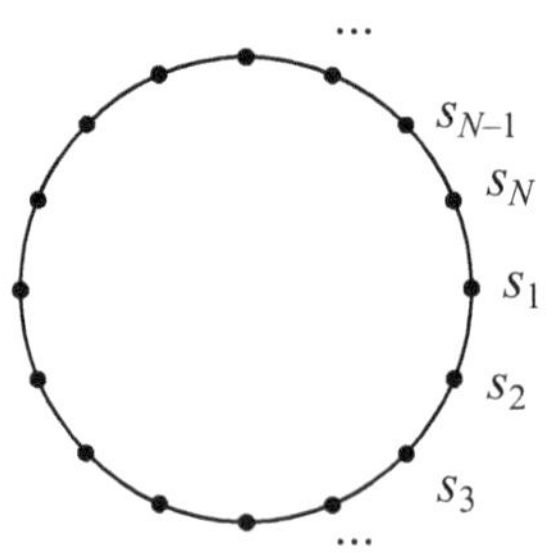

Figure 4.10. The closed-ring 1D Ising system.

[38] It was developed in 1941 by H Kramers and G Wannier. Note that the approach is very close to the one used in quantum mechanics—see, e.g. *Part QM* section 2.5.

Let us regroup terms of this sum in the following way:

$$E_m = -\left[\left(\frac{h}{2}s_1 + Js_1s_2 + \frac{h}{2}s_2\right) + \left(\frac{h}{2}s_2 + Js_2s_3 + \frac{h}{2}s_3\right) + \dots \right. \\ \left. + \left(\frac{h}{2}s_N + Js_Ns_1 + \frac{h}{2}s_1\right)\right], \tag{4.79}$$

so that the group in each parentheses depends only on the state of two adjacent spins. The corresponding statistical sum,

$$Z = \sum_{\substack{s_k=\pm1,\text{ for}\\ k=1,2,\dots N}} \exp\left\{h\frac{s_1}{2T} + J\frac{s_1s_2}{T} + h\frac{s_2}{2T}\right\}\exp\left\{h\frac{s_2}{2T} + J\frac{s_2s_3}{T} + h\frac{s_3}{2T}\right\}\dots \\ \dots \exp\left\{h\frac{s_N}{2T} + J\frac{s_Ns_1}{T} + h\frac{s_1}{2T}\right\}, \tag{4.80}$$

has 2^N terms, each corresponding to a certain combination of signs of N spins. Each operand of the product under the sum may take four values for four different combinations of its two arguments:

$$\exp\left\{h\frac{s_k}{2T} + J\frac{s_ks_{k+1}}{T} + h\frac{s_{k+1}}{2T}\right\} = \begin{cases}\exp\{(J+h)/T\}, & \text{for } s_k = s_{k+1} = +1,\\ \exp\{(J-h)/T\}, & \text{for } s_k = s_{k+1} = -1,\\ \exp\{-J/T\}, & \text{for } s_k = -s_{k+1}.\end{cases} \tag{4.81}$$

These values do not depend on the index k,[39] and may be represented as elements $M_{j,j'}$ (with j, j' = 1, 2) of the so-called *transfer matrix*

$$\mathrm{M} \equiv \begin{pmatrix}\exp\{(J+h)/T\} & \exp\{-J/T\}\\ \exp\{-J/T\} & \exp\{(J-h)/T\}\end{pmatrix}, \tag{4.82}$$

so that the whole statistical sum (4.80) may be recast as a product:

$$Z = \sum_{j_k=1,2} M_{j_1j_2}M_{j_2j_3}\dots M_{j_{N-1}j_N}M_{j_Nj_1}. \tag{4.83}$$

According to the basic rule of the matrix multiplication, this sum is just

$$Z = \mathrm{Tr}(\mathrm{M}^N). \tag{4.84}$$

Linear algebra tells us that this trace may be represented just as

$$Z = \lambda_+^N + \lambda_-^N \tag{4.85}$$

[39] This is of course a result of the 'translational' (or rather rotational) symmetry of the system, i.e. its invariance to the index replacement $k \to k + 1$ in all terms of the energy (4.78).

where $\lambda_\pm$ are the eigenvalues of the transfer matrix M, i.e. the roots of its characteristic equation,

$$\begin{vmatrix} \exp\{(J+h)/T\} - \lambda & \exp\{-J/T\} \\ \exp\{-J/T\} & \exp\{(J-h)/T\} - \lambda \end{vmatrix} = 0. \tag{4.86}$$

A straightforward calculation yields

$$\lambda_\pm = \exp\left\{\frac{J}{T}\right\}\left[\cosh\frac{h}{T} \pm \left(\sinh^2\frac{h}{T} + \exp\left\{-\frac{4J}{T}\right\}\right)^{1/2}\right]. \tag{4.87}$$

The last simplification comes from the condition $N \gg 1$—which we need anyway, to make the ring model sufficiently close to an infinite 1D system. In this limit, even a small difference of the exponents, $\lambda_+ > \lambda_-$, makes the second term in Eq. (4.85) negligible, so that we finally get

$$Z = \lambda_+^N = \exp\left\{\frac{NJ}{T}\right\}\left[\cosh\frac{h}{T} + \left(\sinh^2\frac{h}{T} + \exp\left\{-\frac{4J}{T}\right\}\right)^{1/2}\right]^N. \tag{4.88}$$

From here, we can find the free energy per particle

$$\frac{F}{N} = \frac{T}{N}\ln\frac{1}{Z} = -J - T\ln\left[\cosh\frac{h}{T} + \left(\sinh^2\frac{h}{T} + \exp\left\{-\frac{4J}{T}\right\}\right)^{1/2}\right], \tag{4.89}$$

and then use thermodynamics to calculate such variables as entropy—see the first of Eqs. (1.35).

However, we are mostly interested in the order parameter defined by Eq. (4.25): $\eta \equiv \langle s_j \rangle$. The conceptually simplest approach to the calculation of this statistical average would be to use the sum (2.7), with the Gibbs probabilities $W_m = Z^{-1}\exp\{-E_m/T\}$. However, the number of terms in this sum is 2^N, so that for $N \gg 1$ this approach is completely impracticable. Here the analogy between the canonical pair $\{-P, V\}$ and other generalized force-coordinate pairs $\{\mathscr{F}, q\}$, in particular $\{\mu_0\mathscr{H}(\mathbf{r}_k), \mathscr{m}_k\}$ for the magnetic field, discussed in sections 1.1 and 1.4, becomes invaluable—see in particular Eq. (1.3*b*). (In our normalization (4.22), and for a uniform field, the pair $\{\mu_0\mathscr{H}(\mathbf{r}_k), \mathscr{m}_k\}$ becomes $\{h, s_k\}$.) Indeed, in this analogy the last term of Eq. (4.23), i.e. the sum of N products $(-hs_k)$ for all spins, with the statistical average $(-Nh\eta)$, is similar to the product PV, i.e. the difference between the thermodynamic potentials F and $G \equiv F + PV$ in the usual 'P–V thermodynamics'. Hence, the free energy F, given by Eq. (4.89), may be understood as the Gibbs energy of the Ising system in the external field, and the equilibrium value of the order parameter may be found from the last of Eqs. (1.39), with the replacements $-P \to h$, $V \to N\eta$:

$$N\eta = -\left(\frac{\partial F}{\partial h}\right)_T. \tag{4.90}$$

Note that this formula is valid for any model of ferromagnetism, of any dimensionality, if it has the same form of the interaction with the external field as the Ising model.

For the 1D Ising ring with $N \gg 1$, Eqs. (4.89) and (4.90) yield

$$\eta = \sinh \frac{h}{T} \Big/ \left(\sinh^2 \frac{h}{T} + \exp\left\{ -\frac{4J}{T} \right\} \right)^{1/2}, \qquad \text{giving} \quad \chi \equiv \frac{\partial \eta}{\partial h}\Big|_{h=0} = \frac{1}{T} \exp\left\{ \frac{2J}{T} \right\}. \tag{4.91}$$

This result means that the 1D Ising model does *not* exhibit a phase transition, i.e. in this model $T_c = 0$. However, its susceptibility grows, at $T \to 0$, much faster than the Curie law (4.77). This gives us a hint that at low temperatures the system is 'virtually ferromagnetic', i.e. has the ferromagnetic order with some rare violations. (In physics, such violations are called low-temperature *excitations*.) This perception may be confirmed by the following approximate calculation.

It is almost evident that the lowest-energy excitation of the ferromagnetic state of an open-end 1D Ising chain at $h = 0$ is the reversal of signs of all spins in one of its parts—see figure 4.11. Indeed, such an excitation (called the *Bloch wall*) involves the change of sign of just one product $s_k s_{k'}$, so that according to Eq. (4.23), its energy E_W (defined as the difference between the values of E_m with and without the excitation) equals $2J$, regardless of the wall position[40]. Since in a ferromagnetic Ising model, the parameter J is positive, $E_W > 0$. If the system tried to minimize its internal energy, having any wall in the system would be energy-disadvantageous. However, thermodynamics tells us that at $T \neq 0$, the system's thermal equilibrium corresponds to the minimum of the free energy $F \equiv E - TS$, rather than just energy E.[41] Hence, we have to calculate Bloch wall's contribution F_W to the free energy. Since in an open-end linear chain of $N \gg 1$ spins, the wall can take $(N - 1) \approx N$ positions with the same energy E_W, we may claim that the entropy S_W associated with an excitation of this type is $\ln N$, so that

$$F_W \equiv E_W - TS_W \approx 2J - T \ln N. \tag{4.92}$$

This result tells us that in the limit $N \to \infty$, and at $T \neq 0$, walls are always free-energy-beneficial, thus explaining the absence of the perfect ferromagnetic order in the 1D Ising system. Note, however, that since the logarithmic function changes

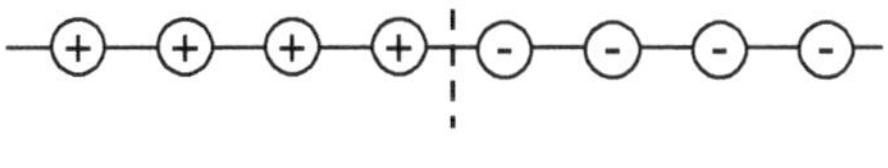

Figure 4.11. A Bloch wall in an open-end 1D Ising system.

[40] On this issue the closed-ring model (figure 4.10) gives a slightly different prediction than the open-end model (figure 4.11), because in the former system, the Bloch walls may appear only in pairs, each with the energy $2E_W = 4J$.

[41] This is a very vivid application of one of the core results of thermodynamics. If the reader is still uncomfortable with it, he or she is strongly encouraged to revisit Eq. (1.42) and its discussion.

extremely slowly at large values of its argument, one may argue that a large but finite 1D system should still feature a quasi-critical temperature

$$\text{“}T_c\text{”} = \frac{2J}{\ln N}, \tag{4.93}$$

below which it would be in a virtually complete ferromagnetic order. (The exponentially large susceptibility (4.91) is a manifestation of this fact.)

Now let us apply a similar approach to estimate T_c of a 2D Ising model, with open borders. Here the Bloch wall is a line of certain length L—see figure 4.12. (For the example presented in that figure, counting from the left to the right, $L = 2 + 1 + 4 + 2 + 3 = 12$ lattice periods.) Evidently, the additional energy associated with such a wall is $E_W = 2JL$, while the wall's entropy S_W may be estimated using the following reasoning. Let the wall be formed by the path of a 'Manhattan pedestrian' traveling between its nodes. (The dashed line in figure 4.12 is an example of such a path.) At each junction, the pedestrian may select three choices of four possible directions (except the one that leads backward), so that there are approximately $3^{(L-1)} \approx 3^L$ options for a walk starting from a certain point. Now taking into account that the open borders of a square-shaped lattice with N spins, have the length of the order of $N^{1/2}$, and the Bloch wall may start from any of them, there are approximately $M \sim N^{1/2}3^L$ different walks between two borders (of the linear size $N^{1/2}$). Again calculating S_W as $\ln M$, we get

$$\begin{aligned} F_W = E_W - TS_W &\approx 2JL - T\ln(N^{1/2}3^L) \\ &\equiv L(2J - T\ln 3) - (T/2)\ln N. \end{aligned} \tag{4.94}$$

(Actually, since L scales as $N^{1/2}$ or higher, at $N \to \infty$ the last term in Eq. (4.94) is negligible.) We see that the sign of $\partial F_W/\partial L$ depends on whether the temperature is higher or lower than the following critical value

$$T_c = \frac{2}{\ln 3}J \approx 1.82\ J. \tag{4.95}$$

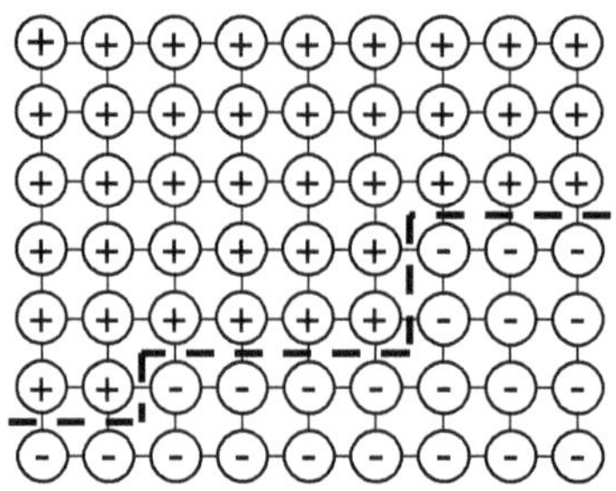

Figure 4.12. A Bloch wall in a 2D Ising system.

At $T < T_c$, the free energy minimum corresponds to $L \to 0$, i.e. Bloch walls are free-energy-detrimental, and the system is in the ferromagnetic phase.

So, for $d = 2$ the estimates predict a non-zero critical temperature of the same order as the Weiss' theory (according to Eq. (4.72), $T_c = 4J$). The major approximation in the calculation leading to Eq. (4.95) is disregarding possible self-crossings of the 'Manhattan walk'. An accurate counting of such self-crossings is rather difficult. It had been carried out in 1944 by L Onsager; since then his calculations have been redone in several easier ways, but even they are rather cumbersome, and I will not have time to discuss them[42]. The final result, however, is surprisingly simple:

$$\tanh\frac{J}{T_c} = \sqrt{2} - 1, \quad \text{i.e.}\ T_c \approx 2.269\,J, \tag{4.96}$$

i.e. showing that the simple estimate (4.95) is off the mark by only ~20%.

The Onsager solution, as well as all alternative solutions of the problem that were found later, are so 'artificial' (2D-specific) that they do not give a clear clue to their generalization to other (higher) dimensions. As a result, the 3D Ising problem is still unsolved analytically. Nevertheless, we do know T_c for it with an extremely high precision—at least to the 6th decimal place. This has been achieved by numerical methods; they deserve a thorough discussion, because they are important for the solution of other similar problems as well. Conceptually, this task is rather simple: just compute, to the desired precision, the statistical sum of the system (4.23):

$$Z = \sum_{\substack{s_k = \pm 1,\ \text{for} \\ k = 1, 2, \ldots, N}} \exp\left\{\frac{J}{T}\sum_{\{k,k'\}} s_k s_{k'} + \frac{h}{T}\sum_k s_k\right\}. \tag{4.97}$$

As soon as this has been done for a sufficient number of values of the dimensionless parameters J/T and h/T, everything becomes easy; in particular, we can compute the dimensionless function

$$F/T = -\ln Z, \tag{4.98}$$

and then find the ratio J/T_c as the smallest value of the parameter J/T at which the ratio F/T (as a function of h/T) has a minimum at zero field. However, for any system of a reasonable size N, the 'exact' computation of the statistical sum (4.97) is impossible, because it contains too many terms for any supercomputer to handle. For example, let us take a relatively small 3D lattice with $N = 10 \times 10 \times 10 = 10^3$ spins, which still feature substantial boundary artifacts even using the periodic boundary conditions, so that its phase transition is smeared about T_c by ~ 3%. Still, even for such a crude model, Z would include $2^{1000} \equiv (2^{10})^{100} \approx (10^3)^{100} \equiv 10^{300}$ terms. Let us suppose we are using a modern exaflops-scale supercomputer performing 10^{18} floating-point operations per second, i.e. ~10^{26} such operations per year.

[42] For that, the interested reader may be referred to either section 4.151 in the textbook by Landau and Lifshitz, or chapter 15 in the text by Huang, both cited above.

With those resources, the computation of just one statistical sum would require $\sim 10^{(300-26)} = 10^{274}$ years. To call such a number 'astronomic' would be a strong understatement. (As a reminder, the age of our Universe is estimated to be close to 1.3×10^{10} years—a very humble number in comparison.)

This situation may be improved dramatically by noticing that any statistical sum,

$$Z = \sum_m \exp\left\{-\frac{E_m}{T}\right\}, \tag{4.99}$$

is dominated by terms with lower values of E_m. In order to find those lowest-energy states, we may use the following powerful approach (belonging to a broad class of numerical *Monte-Carlo techniques*), which essentially mimics one (randomly selected) path of a system's evolution in time. One could argue that for that we would need to know the exact laws of evolution of statistical systems[43], which may differ from one system to another, even if their energy spectra E_m are the same. This is true, but since the genuine value of Z should be independent of these details, it may be evaluated using *any* reasonable kinetic model that satisfies certain general rules. In order to reveal these rules, let us start from a system with just two states, E_m and $E_{m'} = E_m + \Delta$—see figure 4.13.

In the absence of quantum coherence between the states (see section 2.1), the equations for time evolution of the corresponding probabilities W_m and $W_{m'}$ should depend only on the probabilities (plus certain constant coefficients). Moreover, since equations of quantum mechanics are linear, the equations of probability evolution should be also linear. Hence, it is natural to expect them to have the following form,

$$\frac{dW_m}{dt} = W_{m'}\Gamma_\downarrow - W_m\Gamma_\uparrow, \qquad \frac{dW_{m'}}{dt} = W_m\Gamma_\uparrow - W_{m'}\Gamma_\downarrow, \tag{4.100}$$

where the constant coefficients $\Gamma_\uparrow$ and $\Gamma_\downarrow$ have the physical sense of the *rates* of the corresponding transitions (see figure 4.13); for example, $\Gamma_\uparrow dt$ is the probability of the system's transition into the state m' during an infinitesimal time interval dt, provided that in the beginning of that interval it was in the state m with full certainty: $W_m = 1$, $W_{m'} = 0$.[44] Since for the system with just two energy levels, the time derivatives of the probabilities have to be equal and opposite, Eqs. (4.100) describe an (irreversible) redistribution of the probabilities while keeping their sum $W = W_m + W_{m'}$ constant.

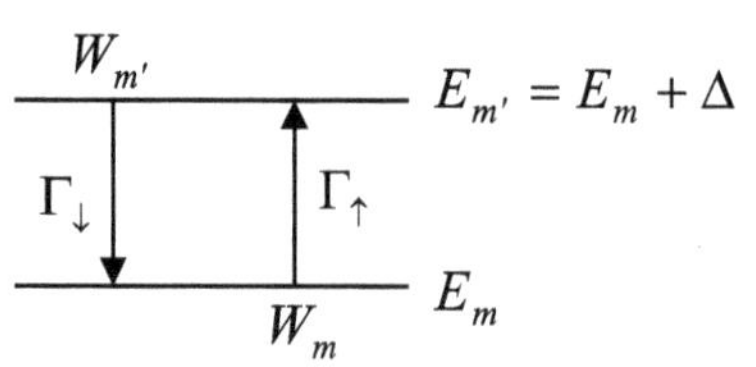

Figure 4.13. Deriving the detailed balance equation.

[43] Discussion of such laws in the task of *physical kinetics*, which will be briefly reviewed in chapter 6.

[44] The calculation of these rates for several particular cases is described in *Part QM* sections 6.6, 6.7, and 7.6—see, e.g. Eq. (7.196) for a very general model of a quantum system with a factorable coupling to its environment.

According to Eqs. (4.100), at $t \to \infty$, the probabilities settle to their stationary values related as

$$\frac{W_{m'}}{W_m} = \frac{\Gamma_\uparrow}{\Gamma_\downarrow}. \qquad (4.101)$$

Now let us require these stationary values to obey the Gibbs distribution (2.58); from it

$$\frac{W_{m'}}{W_m} = \exp\left\{\frac{E_m - E_{m'}}{T}\right\} = \exp\left\{-\frac{\Delta}{T}\right\} < 1. \qquad (4.102)$$

Comparing these two expressions, we see that the rates have to satisfy the following *detailed balance relation*:

$$\frac{\Gamma_\uparrow}{\Gamma_\downarrow} = \exp\left\{-\frac{\Delta}{T}\right\}. \qquad (4.103)$$

Now comes the final step: since the rates of transition between two particular states should not depend on other states and their occupation, Eq. (4.103) has to be valid for *each* pair of states of any multi-state system. (By the way, this relation may serve as an important sanity check: the rates calculated using any reasonable model of a quantum system have to satisfy it.) The detailed balance yields only one equation for two rates $\Gamma_\uparrow$ and $\Gamma_\downarrow$; if our only goal is the calculation of Z, the choice of the other equation is not too important. Perhaps the simplest choice is

$$\Gamma(\Delta) \propto \gamma(\Delta) \equiv \begin{cases} 1, & \text{if } \Delta < 0, \\ \exp\{-\Delta/T\}, & \text{otherwise,} \end{cases} \qquad (4.104)$$

where Δ is the energy change resulting from the transition. This model, which evidently satisfies the detailed balance relation (4.103), is very popular for its simplicity (despite the uphysical cusp this function has at $\Delta = 0$), and enables the following *Metropolis algorithm* (figure 4.14).

The calculation starts from setting a certain initial state of the system. At relatively high temperatures, the state may be generated randomly; for example, in the Ising system the initial state of each spin s_k may be selected independently, with the 50% probability. At low temperatures, starting the calculations from the lowest-energy state (in particular, for the Ising model, from the ferromagnetic state $s_k = \text{sgn}(h) = \text{const}$) may give the fastest convergence of the sum (4.97).

Now one spin is flipped at random, and the corresponding change Δ of energy is calculated[45], and plugged into Eq. (4.104) to calculate $\gamma(\Delta)$. Next, a pseudo-random number generator is used to generate a random number ξ, with the probability density uniformly distributed on the segment [0, 1]. (Such functions, typically called RND, are available in virtually any numerical library.) If the resulting ξ is less than

---

[45] Note that the flip changes signs of only $(2d + 1)$ terms in the sum (4.23), i.e. does not require re-calculation of all $(2d + 1)N$ terms of the sum, so that the computation of Δ takes just a few add-multiply operations even at $N \gg 1$.

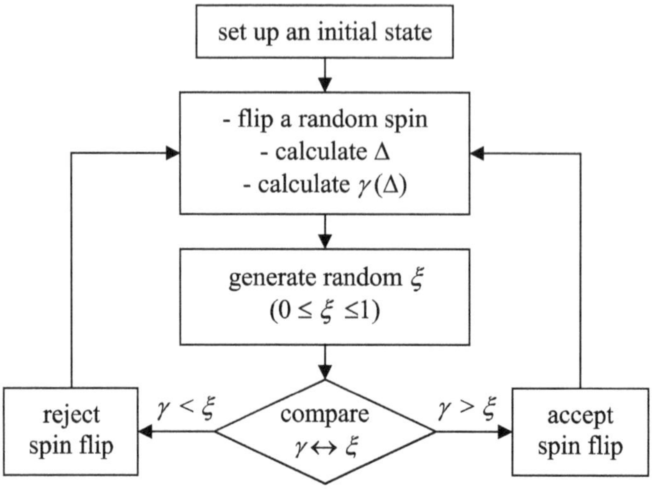

Figure 4.14. A crude scheme of the Metropolis algorithm for the Ising model simulation.

$\gamma(\Delta)$, the transition is accepted, while if $\xi > \gamma(\Delta)$, it is rejected. In the view of Eq. (4.104), this means that any transition down the energy spectrum ($\Delta < 0$) is always accepted, while those up the energy profile ($\Delta > 0$) are accepted with the probability proportional to $\exp\{-\Delta/T\}$. The latter feature is necessary to avoid the system's trapping in local minima of its multidimensional energy profile $E_m(s_1,s_2,\ldots,s_N)$. Now the statistical sum may be calculated approximately as a partial sum over the states already passed by the system. (It is better to discard the contributions from a few first steps to avoid effects of the initial state choice.)

This algorithm is extremely efficient. Even with modest computers available in the 1980s, it has allowed one to simulate a 3D Ising system of $(128)^3$ spins to get the following result: $J/T_c \approx 0.221\,650 \pm 0.000\,005$. For all practical purposes, this result is exact (so that perhaps the largest benefit of the possible analytical solution for the infinite 3D Ising system would be a virtually certain Nobel Prize for the author :-). Table 4.2 summarizes values of T_c for the Ising model. Very visible is the fast improvement of the prediction accuracy of the molecular-field theory—which is asymptotically correct at $d \to \infty$.

Finally, I need to mention the *renormalization-group* ('RG') approach[46], despite its low efficiency for the Ising-type problems. The basic idea of this approach stems from the scaling law (4.30)–(4.31): at $T = T_c$ the correlation radius r_c diverges. Hence, the critical temperature may be found from the requirement for the system to be *spatially self-similar*. Namely, let us form larger and larger groups ('blocks') of adjacent spins, and require that all properties of the resulting system of the blocks approach those of the initial system, as T approaches T_c.

Let us see how does this idea work for the simplest nontrivial (1D) case, described by the statistical sum (4.80). Assuming N to be even (which does not matter at

[46] Initially developed in the quantum field theory in the 1950s, it was adapted to statistics by L Kadanoff in 1966, with a spectacular solution of the so-called *Kubo problem* by K Wilson in 1972, later rewarded with a Nobel Prize.

Table 4.2. The critical temperature T_c (in the units of J) of the Ising model of a ferromagnet ($J > 0$), for several values of dimensionality d.

d	Molecular-field theory—Eq. (4.72)	Exact value	Exact value's source
0	0	0	Gibbs distribution
1	2	0	Transfer matrix theory
2	4	2.269...	Onsager's solution
3	6	4.513...	Numerical simulation

$N \to \infty$), and adding an inconsequential constant C to each exponent (for the purpose that will be clear soon), we may rewrite this expression as

$$Z = \sum_{s_k=\pm 1} \prod_{k=1,2,...N} \exp\left\{\frac{h}{2T}s_k + \frac{J}{T}s_k s_{k+1} + \frac{h}{2T}s_{k+1} + C\right\}. \tag{4.105}$$

Let us group each pair of adjacent exponents to recast this expression as a product over only *even* numbers k,

$$Z = \sum_{s_k=\pm 1} \prod_{k=2,4,...N} \exp\left\{\frac{h}{2T}s_{k-1} + s_k\left[\frac{J}{T}(s_{k-1} + s_{k+1}) + \frac{h}{T}\right] + \frac{h}{2T}s_{k+1} + 2C\right\}, \tag{4.106}$$

and carry out the summation over two possible states of the internal spin s_k explicitly:

$$\begin{aligned} Z = \sum_{s_k=\pm 1} \prod_{k=2,4,...N} &\left[\exp\left\{\frac{h}{2T}s_{k-1} + \frac{J}{T}(s_{k-1} + s_{k+1}) + \frac{h}{T} + \frac{h}{2T}s_{k+1} + 2C\right\}\right. \\ &\left. + \exp\left\{\frac{h}{2T}s_{k-1} - \frac{J}{T}(s_{k-1} + s_{k+1}) - \frac{h}{T} + \frac{h}{2T}s_{k+1} + 2C\right\}\right] \\ \equiv \sum_{s_k=\pm 1} \prod_{k=2,4,...N} & 2\cosh\left\{\frac{J}{T}(s_{k-1} + s_{k+1}) + \frac{h}{T}\right\} \\ &\times \exp\left\{\frac{h}{2T}(s_{k-1} + s_{k+1}) + 2C\right\}. \end{aligned} \tag{4.107}$$

Now let us require this statistical sum (and hence all statistical properties of the system of 2-spin blocks) to be identical to that of the Ising system of $N/2$ spins, numbered by odd k:

$$Z' = \sum_{s_k=\pm 1} \prod_{k=2,4,...,N} \exp\left\{\frac{J'}{T}s_{k-1}s_{k+1} + \frac{h'}{T}s_{k+1} + C'\right\}, \tag{4.108}$$

with some different parameters h', J', and C', for all four possible values of $s_{k-1} = \pm 1$ and $s_{k+1} = \pm 1$. Since the right-hand side of Eq. (4.107) depends only on the sum $(s_{k-1} + s_{k+1})$, this requirement yields only three (rather than four) independent equations for finding h', J', and C'. Of them, the equations for h' and J' depend only on h and J (but not on C)[47], and may be represented in an especially simple form,

$$x' = \frac{x(1+y)^2}{(x+y)(1+xy)}, \qquad y' = \frac{y(x+y)}{1+xy}, \tag{4.109}$$

if the following notation is used:

$$x \equiv \exp\left\{-4\frac{J}{T}\right\}, \qquad y \equiv \exp\left\{-2\frac{h}{T}\right\}. \tag{4.110}$$

Now the grouping procedure may be repeated, with the same result (4.109)–(4.110). Hence these equations may be considered as recurrence relations describing repeated doubling of the spin block size. Figure 4.15 shows (schematically) the *trajectories* of this dynamic system on the phase plane $[x, y]$. (Each trajectory is defined by the following property: for each of its points $\{x, y\}$, the point $\{x', y'\}$ defined by the 'mapping' Eq. (4.109) is also on the same trajectory.) For ferromagnetic coupling ($J > 0$) and $h > 0$, we may limit the analysis to the unit square $0 \leqslant x, y \leqslant 1$. If this *flow diagram* had a stable fixed point with $x' = x = x_\infty \neq 0$ (i.e. $T/J < \infty$) and $y' = y = 1$ (i.e. $h = 0$), then the first of Eqs. (4.110) would immediately give us the critical temperature of the phase transition in the field-free system:

$$T_c = \frac{4J}{\ln(1/x_\infty)}. \tag{4.111}$$

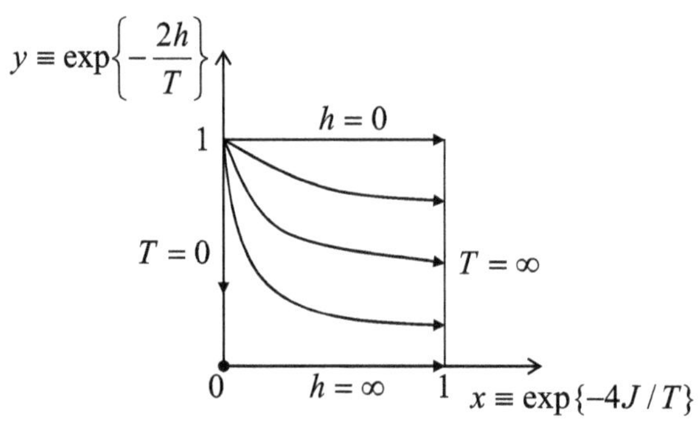

Figure 4.15. The RG flow diagram of the 1D Ising system (schematically).

[47] This might be expected, because physically C is just a certain constant addition to the system's energy. However, the introduction of that constant is mathematically necessary, because Eqs. (4.107) and (4.108) may be reconciled only if $C' \neq C$.

However, figure 4.15 shows that the only fixed point of the 1D system is $x = y = 0$, which (at a finite coupling J) should be interpreted as $T_c = 0$. This is of course in agreement with the exact result of the transfer-matrix analysis, but does not give any additional information.

Unfortunately, for higher dimensionalities the renormalization-group approach rapidly becomes rather cumbersome, and requires certain approximations, whose accuracy cannot be easily controlled. For the 2D Ising system, such approximations lead to the prediction $T_c/J \approx 2.55$, i.e. to a substantial difference from the exact result (4.96).

4.6 Problems

Problem 4.1. Compare the third virial coefficient $C(T)$ that follows from the van der Waals equation, with its value for the hardball model of particle interactions (whose calculation was the subject of problem 3.28), and comment.

Problem 4.2. Calculate the entropy and the internal energy of the van der Waals gas, and discuss the results.

Problem 4.3. Use two different approaches to calculate the so-called *Joule–Thomson coefficient* $(\partial E/\partial V)_T$ for the van der Waals gas, and the change of temperature of such a gas, with a temperature-independent C_V, at its fast expansion.

Problem 4.4. Calculate the difference $C_P - C_V$ for the van der Waals gas, and compare it with that for an ideal classical gas.

Problem 4.5. Calculate the temperature dependence of the phase-equilibrium pressure $P_0(T)$ and the latent heat $\Lambda(T)$, for the van der Waals model, in the low-temperature limit $T \ll T_c$.

Problem 4.6. Perform the same tasks as in the previous problem in the opposite limit—in a close vicinity of the critical point T_c.

Problem 4.7. Calculate the critical values P_c, V_c, and T_c for the so-called *Redlich–Kwong model* of the real gas, with the following equation of state[48]:

$$P + \frac{a}{V(V + Nb)\,T^{1/2}} = \frac{NT}{V - Nb},$$

with constant parameters a and b.

Hint: Be prepared to solve a cubic equation with particular (numerical) coefficients.

[48] This equation of state, suggested in 1948, describes most real gases better than not only the original van der Waals model, but also other two-parameter alternatives, such as the *Berthelot*, *modified-Berthelot*, and *Dieterici* models, though some approximations with more fitting parameters (such as the *Soave–Redlich–Kwong model*) work even better.

Problem 4.8. Calculate the critical values P_c, V_c, and T_c for the phenomenological *Dieterici model*, with the following equation of state[49]:

$$P = \frac{NT}{V - b} \exp\left\{ -\frac{a}{NTV} \right\},$$

with constant parameters a and b. Compare the value of the dimensionless factor P_cV_c/NT_c with those given by the van der Waals and Redlich–Kwong models.

Problem 4.9. In the crude sketch shown in figure 4.3b, the values of derivatives dP/dT of the phase transitions liquid–gas ('vaporization') and solid–gas ('sublimation') at the triple point are different, with

$$\left(\frac{dP_v}{dT}\right)_{T=T_t} < \left(\frac{dP_s}{dT}\right)_{T=T_t}.$$

Is this occasional? What relation between these derivatives can be obtained from thermodynamics?

Problem 4.10. Use the Clapeyron–Clausius formula (4.17) to calculate the latent heat Λ of the Bose–Einstein condensation, and compare the result with that obtained in the solution of problem 3.21.

Problem 4.11.

(i) Write the effective Hamiltonian for which the usual single-particle stationary Schrödinger equation coincides with the Gross–Pitaevskii equation (4.58).
(ii) Use this *Gross–Pitaevskii Hamiltonian*, with the particular trapping potential $U(\mathbf{r}) = m\omega^2 r^2/2$, to calculate the energy E of $N \gg 1$ trapped particles, assuming the approximate solution $\psi \propto \exp\{-r^2/2r_0^2\}$, as a function of the parameter r_0.[50]
(iii) Explore the function $E(r_0)$ for positive and negative values of the constant b, and interpret the results.
(iv) For small $b < 0$, estimate the largest number N of particles that may form a metastable Bose–Einstein condensate.

Problem 4.12. Superconductivity may be suppressed by a sufficiently strong magnetic field. In the simplest case of a bulk, long cylindrical sample of a type-I superconductor, placed into an external magnetic field $\mathscr{H}_{\text{ext}}$ parallel to its surface, this suppression takes a simple form of a simultaneous transition of the whole sample from the superconducting state to the 'normal' (non-superconducting) state at a certain value $\mathscr{H}_c(T)$ of the field's magnitude. This *critical field* gradually decreases with temperature from its maximum value $\mathscr{H}_c(0)$ at $T \to 0$ to zero at the

[49] This model is currently less popular than the Redlich–Kwong model (also with two fitting parameters), whose analysis was the task of the previous problem.
[50] This task is essentially the first step of the variational method of quantum mechanics—see, e.g. *Part QM* section 2.9.

critical temperature T_c. Assuming that the function $\mathscr{H}_c(T)$ is known, calculate the latent heat of this phase transition as a function of temperature, and spell out its values at $T \to 0$ and $T = T_c$.

Hint: In this context, the 'bulk sample' means a sample of size much larger than the intrinsic length scales of the superconductor (such as the London penetration depth δ_L and the coherence length ξ)[51]. For such bulk samples of type-I superconductors, magnetic properties of the superconducting phase may be well described just as the perfect diamagnetism, with the magnetic induction $\mathscr{B} = 0$ inside it.

Problem 4.13. In some textbooks, the discussion of thermodynamics of superconductivity is started with displaying, as self-evident, the following formula:

$$F_n(T) - F_s(T) = \frac{\mu_0 \mathscr{H}_c^2(T)}{2} V,$$

where F_s and F_n are the free energy values in the superconducting and non-superconducting ('normal') phases, and $\mathscr{H}_c(T)$ is the critical value of the magnetic external field. Is this formula correct, and if not, what qualification is necessary to make it valid? Assume that all conditions of the simultaneous field-induced phase transition in the whole sample, spelled out in the previous problem, are satisfied.

Problem 4.14. In section 4.4, we have discussed Weiss' molecular-field approach to the Ising model, in which the average $\langle s_j \rangle$ plays the role of the order parameter η. Use the results of that analysis to calculate the coefficients a and b in the corresponding Landau expansion (4.46) of the free energy. List the critical exponents α and β, defined by Eqs. (4.26) and (4.28), within this approach.

Problem 4.15. Consider a ring of $N = 3$ Ising 'spins' ($s_k = \pm 1$), with similar ferromagnetic coupling J between all sites, in thermal equilibrium.

(i) Calculate the order parameter and the low-field susceptibility χ of the system.
(ii) Use the low-temperature limit of the result for χ to predict the result for a ring with an arbitrary N, and verify it by a direct calculation (in this limit).
(iii) Discuss the relation between the last result, in the limit $N \to \infty$, and Eq. (4.91).

Problem 4.16. Calculate the average energy, entropy, and heat capacity of a three-site ring of Ising-type 'spins' ($s_k = \pm 1$), with *anti*-ferromagnetic coupling (of magnitude J) between the sites, in thermal equilibrium at temperature T, with no external magnetic field. Find the asymptotic behavior of its heat capacity for low and high temperatures, and give an interpretation of the results.

Problem 4.17. Using the results discussed in section 4.5, calculate the average energy, free energy, entropy, and heat capacity (all per spin) as functions of

[51] A discussion of these parameters, as well as of the difference between the type-I and type-II superconductivity, may be found in *Part EM* sections 6.4–6.5. However, those details are not needed for the solution of this problem.

temperature T and external field h, for the infinite 1D Ising model. Sketch the temperature dependence of the heat capacity for various values of ratio h/J, and give a physical interpretation of the result.

Problem 4.18. Use the molecular-field theory to calculate the critical temperature and the low-field susceptibility of a d-dimensional cubic lattice of spins, described by the so-called *classical Heisenberg model*[52]:

$$E_m = -J \sum_{\{k,k'\}} \mathbf{s}_k \cdot \mathbf{s}_{k'} - \sum_k \mathbf{h} \cdot \mathbf{s}_k.$$

Here, in contrast to the (otherwise, very similar) Ising model (4.23), the spin of each site is modeled by a classical 3D vector $\mathbf{s}_k = \{s_{xk}, s_{yk}, s_{zk}\}$ of the unit length: $s_k^2 = 1$.

References

[1] Schwabl F 2000 *Statistical Mechanics* (Springer)
[2] Stratonovich R 1963 *Topics in the Theory of Random Noise* vol 1 (Gordon and Breach)
[3] Stratonovich R 1967 *Topics in the Theory of Random Noise* vol 2 (Gordon and Breach)
[4] Haken H 1970 *Ferstkörperprobleme* **10** 351
[5] Stanley H 1971 *Introduction to Phase Transitions and Critical Phenomena* (Oxford University Press)
[6] Patashinskii A and Pokrovskii V 1979 *Fluctuation Theory of Phase Transitions* (Pergamon)
[7] McCoy B 2010 *Advanced Statistical Mechanics* (Oxford University Press)
[8] Yeomans J 1992 *Statistical Mechanics of Phase Transitions* (Clarendon)
[9] Tinkham M 1996 *Introduction to Superconductivity* 2nd edn (McGraw-Hill)
[10] Lifshitz E and Pitaevskii L 1980 *Statistical Physics* Part 2 (Pergamon)
[11] Mee C and Daniel E 1996 *Magnetic Recording Technology* 2nd edn (McGraw-Hill)

[52] This classical model is formally similar to the generalization of the genuine (quantum) Heisenberg model (4.21) to arbitrary spin s, and serves as its infinite-spin limit.

IOP Publishing

Statistical Mechanics
Lecture notes
Konstantin K Likharev

Chapter 5

Fluctuations

This chapter discusses fluctuations of macroscopic variables, mostly at thermodynamic equilibrium. In particular, it describes the intimate connection between fluctuations and dissipation (damping) in dynamic systems weakly coupled to multi-particle environments, which culminates in the Einstein relation between the diffusion coefficient and mobility, the Nyquist formula, and their quantum-mechanical generalization—the fluctuation–dissipation theorem. An alternative approach to the same problem, based on the Smoluchowski and Fokker–Planck equations, is also discussed in brief.

5.1 Characterization of fluctuations

In the beginning of chapter 2, we have discussed the notion of averaging, $\langle f \rangle$, of a variable f over a statistical ensemble—see Eqs. (2.7) and (2.10). Now, the variable's *fluctuation* may be defined simply as its deviation from such an average:

$$\tilde{f} \equiv f - \langle f \rangle; \tag{5.1}$$

this deviation is, evidently, also a random variable. The most important property of any fluctuation is that its average (over the same statistical ensemble) equals zero:

$$\langle \tilde{f} \rangle = \langle f - \langle f \rangle \rangle = \langle f \rangle - \langle \langle f \rangle \rangle = \langle f \rangle - \langle f \rangle = 0. \tag{5.2}$$

As a result, such an average cannot characterize fluctuations' *intensity*, and the simplest characteristic of the intensity is the *variance* (also called 'dispersion'):

$$\langle \tilde{f}^2 \rangle = \langle (f - \langle f \rangle)^2 \rangle. \tag{5.3}$$

The following simple property of the variance is frequently convenient for its calculation:

$$\langle \tilde{f}^2 \rangle = \langle (f - \langle f \rangle)^2 \rangle = \langle f^2 - 2f\langle f \rangle + \langle f \rangle^2 \rangle = \langle f^2 \rangle - 2\langle f \rangle^2 + \langle f \rangle^2, \tag{5.4a}$$

so that, finally:

$$\langle \tilde{f}^2 \rangle = \langle f^2 \rangle - \langle f \rangle^2. \tag{5.4b}$$

As the simplest example, consider a variable that takes only two values, ±1, with equal probabilities $W_j = ½$. For such a variable,

$$\begin{aligned} \langle f \rangle &= \sum_j W_j f_j = \frac{1}{2}(+1) + \frac{1}{2}(-1) = 0, \quad \text{but} \\ \langle f^2 \rangle &= \sum_j W_j f_j^2 = \frac{1}{2}(+1)^2 + \frac{1}{2}(-1)^2 = 1, \\ \text{so that} \quad &\langle \tilde{f}^2 \rangle = \langle f^2 \rangle - \langle f \rangle^2 = 1. \end{aligned} \tag{5.5}$$

The square root of the variance,

$$\delta f \equiv \langle \tilde{f}^2 \rangle^{1/2} \tag{5.6}$$

is called the *root-mean-square (rms) fluctuation*. An advantage of this measure is that it has the same dimensionality as the variable itself, so that the ratio $\delta f/\langle f \rangle$ is dimensionless, and may be used to characterize the *relative intensity* of fluctuations. As has been mentioned in chapter 1, all results of thermodynamics are valid only if the fluctuations of thermodynamic variables (internal energy E, entropy S, etc) are relatively small[1], so let us make a simple estimate of the relative intensity of fluctuations by considering a system of N independent, similar particles, and an extensive variable

$$\mathscr{F} \equiv \sum_{k=1}^{N} f_k. \tag{5.7}$$

where f_k depends on the state of just one (kth) particle.

The statistical average of such $\mathscr{F}$ is evidently

$$\langle \mathscr{F} \rangle = \sum_{k=1}^{N} \langle f \rangle = N\langle f \rangle, \tag{5.8}$$

while the variance of its fluctuations is

$$\langle \tilde{\mathscr{F}}^2 \rangle \equiv \langle \tilde{\mathscr{F}} \tilde{\mathscr{F}} \rangle \equiv \left\langle \sum_{k=1}^{N} \tilde{f}_k \sum_{k'=1}^{N} \tilde{f}_{k'} \right\rangle = \left\langle \sum_{k,k'=1}^{N} \tilde{f}_k \tilde{f}_{k'} \right\rangle = \sum_{k,k'=1}^{N} \langle \tilde{f}_k \tilde{f}_{k'} \rangle. \tag{5.9}$$

---

[1] Let me remind the reader that up to this point, the averaging signs $\langle \ldots \rangle$ were dropped in most formulas, for the sake of notation simplicity. In this chapter I have to restore these signs to avoid confusion. The only exception will be temperature—whose average, following (bad :-) tradition, will be still call just T everywhere, besides the last part of section 5.3, where temperature fluctuations are discussed explicitly.

Now we may use the fact that for two independent variables

$$\left\langle \tilde{f}_k \tilde{f}_{k'} \right\rangle = 0, \quad \text{for } k' \neq k; \tag{5.10}$$

indeed, this relation may be considered as the mathematical definition of the independence. Hence, in the sum (5.9), only the terms with $k' = k$ survive, and

$$\langle \tilde{\mathscr{F}}^2 \rangle = \sum_{k,k'=1}^{N} \left\langle \tilde{f}_k^2 \right\rangle \delta_{k,k'} = N \langle \tilde{f}^2 \rangle. \tag{5.11}$$

Comparing Eqs. (5.8) and (5.11), we see that the relative intensity of fluctuations of the variable $\mathscr{F}$,

$$\frac{\delta \mathscr{F}}{\langle \mathscr{F} \rangle} = \frac{1}{N^{1/2}} \frac{\delta f}{\langle f \rangle}, \tag{5.12}$$

tends to zero as the system size grows ($N \to \infty$). It is this fact that justifies the thermodynamic approach to typical physical systems, with the number N of particles of the order of the Avogadro number $N_A \sim 10^{24}$. Nevertheless, in many situations even small fluctuations of variables are important, and in this chapter we will calculate their basic properties, starting from the variance.

It should be comforting for the reader to notice that for some simple (but important) cases, such a calculation has already been done in our course. For example, for any generalized coordinate q and generalized momentum p that give quadratic contributions of the type (2.46) to system's Hamiltonian, we have derived the equipartition theorem (2.48), valid in the classical limit. Since the average values of these variables, in the thermodynamic equilibrium, equal zero, Eq. (5.6) immediately yields their rms fluctuations:

$$\delta p = (mT)^{1/2}, \quad \delta q = \left(\frac{T}{\kappa}\right)^{1/2} \equiv \left(\frac{T}{m\omega^2}\right)^{1/2}, \quad \text{where} \quad \omega = \left(\frac{\kappa}{m}\right)^{1/2}. \tag{5.13}$$

The generalization of these classical relations to the quantum-mechanical case ($T \sim \hbar\omega$) is provided by Eqs. (2.78) and (2.81):

$$\delta p = \left[\frac{\hbar m\omega}{2} \coth \frac{\hbar\omega}{2T}\right]^{1/2}, \quad \delta q = \left[\frac{\hbar}{2m\omega} \coth \frac{\hbar\omega}{2T}\right]^{1/2,}. \tag{5.14}$$

However, the intensity of fluctuations in other systems requires special calculations. Moreover, only a few cases allow for general, model-independent results. Let us review some of them.

5.2 Energy and the number of particles

First of all, note that fluctuations of macroscopic variables depend on particular conditions[2]. For example, in a mechanically- and thermally-insulated system, i.e. a

[2] Unfortunately, even in some popular textbooks, some formulas pertaining to fluctuations are either incorrect, or given without specifying the conditions of their applicability, so the reader's caution is advised.

member of a microcanonical ensemble, there are no fluctuations of internal energy: $\delta E = 0$.

However, if a system is in thermal contact with the environment, for example it is a member of a canonical ensemble (figure 2.6), the situation is different. Indeed, for such a system we may apply the general Eq. (2.7), with W_m given by the Gibbs distribution (2.58)–(2.59), not only to E, but also to E^2. As we already know from section 2.4, the first average,

$$\langle E \rangle = \sum_m W_m E_m, \quad W_m = \frac{1}{Z}\exp\left\{-\frac{E_m}{T}\right\}, \quad Z = \sum_m \exp\left\{-\frac{E_m}{T}\right\}, \tag{5.15}$$

yields Eq. (2.61*b*), which may be rewritten in the form

$$\langle E \rangle = \frac{1}{Z}\frac{\partial Z}{\partial(-\beta)}, \quad \text{with} \quad \beta \equiv \frac{1}{T}, \tag{5.16}$$

which is more convenient for our current purposes. Let us carry out a similar calculation for E^2:

$$\langle E^2 \rangle = \sum_m W_m E_m^2 = \frac{1}{Z}\sum_m E_m^2 \exp\{-\beta E_m\}. \tag{5.17}$$

It is straightforward to verify, by double differentiation, that the last expression may be rewritten in a form similar to Eq. (5.16):

$$\langle E^2 \rangle = \frac{1}{Z}\frac{\partial^2}{\partial(-\beta)^2}\sum_m \exp\{-\beta E_m\} \equiv \frac{1}{Z}\frac{\partial^2 Z}{\partial(-\beta)^2}. \tag{5.18}$$

Now it is easy to use Eq. (5.4) to calculate the variance of energy fluctuations:

$$\begin{aligned}\langle \tilde{E}^2 \rangle = \langle E^2 \rangle - \langle E \rangle^2 &= \frac{1}{Z}\frac{\partial^2 Z}{\partial(-\beta)^2} - \left(\frac{1}{Z}\frac{\partial Z}{\partial(-\beta)}\right)^2 \\ &\equiv \frac{\partial}{\partial(-\beta)}\left(\frac{1}{Z}\frac{\partial Z}{\partial(-\beta)}\right) = \frac{\partial\langle E \rangle}{\partial(-\beta)}.\end{aligned} \tag{5.19}$$

Since Eqs. (5.15)–(5.19) are valid only if system's volume V is fixed (because its change may affect the energy spectrum E_m), it is customary to rewrite this important result as follows:

$$\langle \tilde{E}^2 \rangle = \frac{\partial\langle E \rangle}{\partial(-1/T)} = T^2\left(\frac{\partial\langle E \rangle}{\partial T}\right)_V \equiv C_V T^2. \tag{5.20}$$

This is a remarkably simple, fundamental result. As a sanity check, for a system of N similar, independent particles, $\langle E \rangle$ and hence C_V and are proportional to N, so that $\delta E \propto N^{1/2}$ and $\delta E/\langle E \rangle \propto N^{-1/2}$, in agreement with Eq. (5.12). Let me emphasize that the classically-looking Eq. (5.20) is based on the general Gibbs distribution, and hence is valid for any system (either classical or quantum) in thermal equilibrium.

The corollaries of this result will be discussed in the next section, and now let us carry out a very similar calculation for a system whose number N of particles in a system is not fixed, because they may go to, and come from its environment at will. If the chemical potential μ of the environment and its temperature T are fixed, i.e. we are dealing with the grand canonical ensemble (figure 2.13), we may use the grand canonical distribution (2.106)–(2.107):

$$W_{m,N} = \frac{1}{Z_G}\exp\left\{\frac{\mu N - E_{m,N}}{T}\right\}, \quad Z_G = \sum_{N,m}\exp\left\{\frac{\mu N - E_{m,N}}{T}\right\}. \tag{5.21}$$

Acting exactly as we did above for the internal energy, we get

$$\langle N\rangle = \frac{1}{Z_G}\sum_{m,N} N\exp\left\{\frac{\mu N - E_{m,N}}{T}\right\} = \frac{T}{Z_G}\frac{\partial Z_G}{\partial \mu}, \tag{5.22}$$

$$\langle N^2\rangle = \frac{1}{Z_G}\sum_{m,N} N^2\exp\left\{\frac{\mu N - E_{m,N}}{T}\right\} = \frac{T^2}{Z_G}\frac{\partial^2 Z_G}{\partial \mu^2}, \tag{5.23}$$

so that the particle number variance is

$$\begin{aligned}\langle \tilde{N}^2\rangle = \langle N^2\rangle - \langle N\rangle^2 &= \frac{T^2}{Z_G}\frac{\partial Z_G}{\partial \mu} - \frac{T^2}{Z_G^2}\left(\frac{\partial Z_G}{\partial \mu}\right)^2 \\ &= T\frac{\partial}{\partial \mu}\left(\frac{T}{Z_G}\frac{\partial Z_G}{\partial \mu}\right) = T\frac{\partial \langle N\rangle}{\partial \mu},\end{aligned} \tag{5.24}$$

in the full analogy with Eq. (5.19).

In particular, for the ideal classical gas we may combine the last result with Eq. (3.32). As was already emphasized in section 3.2, though that result has been obtained for the canonical ensemble, in which the number of particles N is fixed, at $N \gg 1$ the fluctuations of N in the grand canonical ensemble should be relatively small, so that the same relation should be valid for the *average* $\langle N\rangle$ in that ensemble. Easily solving Eq. (3.32*b*) for $\langle N\rangle$, we get

$$\langle N\rangle = \text{const} \times \exp\left\{\frac{\mu}{T}\right\}, \tag{5.25}$$

where 'const' means a factor constant at the differentiation of $\langle N\rangle$ over μ, required by Eq. (5.24). Performing the differentiation and then using Eq. (5.25) again,

$$\frac{\partial \langle N\rangle}{\partial \mu} = \text{const} \times \frac{1}{T}\exp\left\{\frac{\mu}{T}\right\} = \frac{\langle N\rangle}{T}, \tag{5.26}$$

we get from Eq. (5.24) a very simple result:

$$\langle \tilde{N}^2\rangle = \langle N\rangle, \quad \text{i.e.} \quad \delta N = \langle N\rangle^{1/2}. \tag{5.27}$$

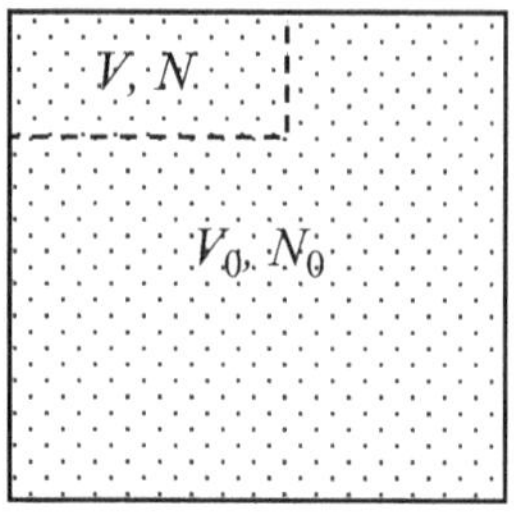

Figure 5.1. Deriving the binomial and Poisson distributions.

This relation is so important that I will now show how it may be derived in a different way. As a by-product, we will prove that this result is valid for systems with an arbitrary (say, small) N, and also get more detailed information about the statistics of fluctuations of that number. Let us consider an ideal classical gas of N_0 particles in its volume V_0, and calculate the probability W_N to have exactly $N \leqslant N_0$ of these particles in its part of volume $V \leqslant V_0$—see figure 5.1. For one particle such probability is of course $W = V/V_0 \leqslant 1$, while the probability of one particle being in the remaining part of the volume is $W' = 1 - W = 1 - V/V_0$. If all particles were distinguishable, the probability of having $N \leqslant N_0$ *specific* particles in volume V, and $(N - N_0)$ *specific* particles in volume $(V - V_0)$, would be $W^N W'^{(N_0 - N)}$. However, if we do not want to distinguish the particles, we should multiply this probability by the number of possible particle combinations keeping the numbers N and N_0 constant, i.e. by the binomial coefficient $N_0!/N!(N_0 - N)!$.[3] As the result, the required probability is

$$W_N = W^N W'^{(N_0-N)} \frac{N_0!}{N!(N_0 - N)!} = \left(\frac{\langle N \rangle}{N_0}\right)^N \left(1 - \frac{\langle N \rangle}{N_0}\right)^{N_0-N} \frac{N_0!}{N!(N_0 - N)!}, \quad (5.28)$$

where at the last step I have used the expression $\langle N \rangle = WN_0 = (V/V_0)N_0$ for the average number of particles in volume V. This is the so-called *binomial probability distribution*, valid for *any* $\langle N \rangle$ and N_0.

Still keeping $\langle N \rangle$ arbitrary, we can simplify the binomial distribution by assuming that the whole volume V_0, and hence N_0, are very large:

$$N_0 \gg N, \quad (5.29)$$

where N means all values of interest, including $\langle N \rangle$. Indeed, in this limit we can neglect N in comparison with N_0 in the second exponent of Eq. (5.28), and also approximate the fraction $N_0!/(N_0 - N)!$, i.e. the product of N terms, $(N_0 - N + 1)\,(N_0 - N + 2)\ldots(N_0 - 1)N_0$, as just N_0^N. As a result, we get

$$\begin{aligned} W_N &\approx \left(\frac{\langle N \rangle}{N_0}\right)^N \left(1 - \frac{\langle N \rangle}{N_0}\right)^{N_0} \frac{N_0^N}{N!} \equiv \frac{\langle N \rangle^N}{N!}\left(1 - \frac{\langle N \rangle}{N_0}\right)^{N_0} \\ &= \frac{\langle N \rangle^N}{N!}\left[(1 - W)^{\frac{1}{W}}\right]^{\langle N \rangle}, \end{aligned} \quad (5.30)$$

[3] See, e.g. Eq. (A.5).

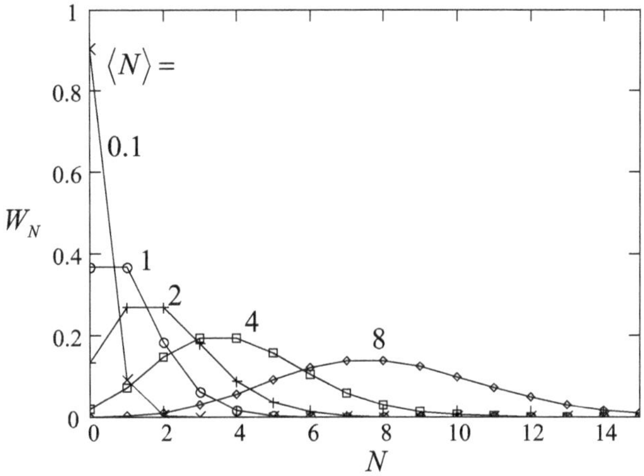

Figure 5.2. The Poisson distribution for several values of $\langle N \rangle$. In contrast to that average, the argument N may take only integer values, so that lines in these plots are only guides for the eye.

where, as before, $W = \langle N \rangle / N_0$. In the limit (5.29), $W \to 0$, and the factor inside the square brackets tends to $1/e$, the reciprocal of the natural logarithm base[4]. Thus, we get an expression independent of N_0:

$$W_N = \frac{\langle N \rangle^N}{N!} e^{-\langle N \rangle}. \tag{5.31}$$

This is the much celebrated *Poisson distribution*, which describes a very broad family of random phenomena. Figure 5.2 shows this distribution for several values of $\langle N \rangle$—which, in contrast to N, are not necessarily integer. In the limit of very small $\langle N \rangle$, the function $W_N(N)$ is close to an exponential one, $W_N \approx W^N \propto \langle N \rangle^N$, while in the opposite limit, $\langle N \rangle \gg 1$, it rapidly approaches the *Gaussian* (or 'normal') *distribution*

$$W_N = \frac{1}{(2\pi)^{1/2}\delta N} \exp\left\{-\frac{(N - \langle N \rangle)^2}{2(\delta N)^2}\right\}. \tag{5.32}$$

(Note that the Gaussian distribution is also valid if both N and N_0 are large, regardless of the relation (5.29) between them—see figure 5.3.)

A major property of the Poisson (and hence of the Gaussian) distribution is that it has the same variance as given by Eq. (5.27):

$$\langle \tilde{N}^2 \rangle \equiv \langle (N - \langle N \rangle)^2 \rangle = \langle N \rangle. \tag{5.33}$$

(This is not true for the general binomial distribution.) For our current purposes, this means that for the ideal classical gas, Eq. (5.27) is valid for *any* number of particles.

[4] Indeed, this is just the most popular definition of this major mathematical constant—see, e.g. Eq. (A.2*a*) with $n = -1/W$.

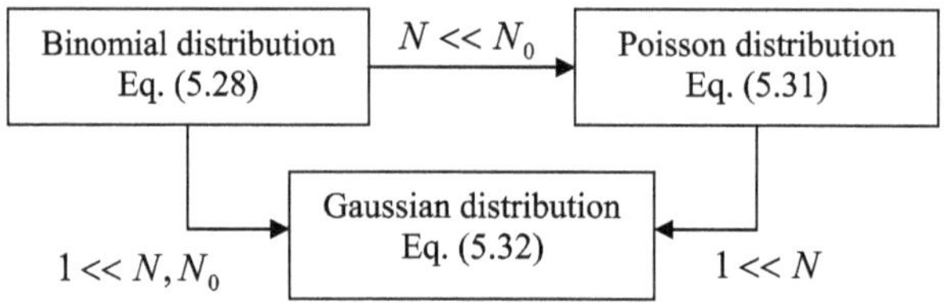

Figure 5.3. The hierarchy of three major probability distributions.

5.3 Volume and temperature

What are the rms fluctuations of other thermodynamic variables—like V, T, etc? Again, the answer depends on specific conditions. For example, if the volume V occupied by a gas is externally fixed (say, by rigid walls), it evidently does not fluctuate at all: $\delta V = 0$. On the other hand, the volume may fluctuate in the situation when the average pressure is fixed—see, e.g. figure 1.5. A formal calculation of these fluctuations, using the approach applied in the last section, is complicated by the fact that it is physically impracticable to fix its conjugate variable, P, i.e. suppress its fluctuations. For example, the force $\mathscr{F}(t)$ exerted by an ideal classical gas on a container's wall (whose measure the pressure is) is the result of individual, independent hits of the wall by particles (figure 5.4), with the time scale $\tau_c \sim r_B/\langle v^2\rangle^{1/2} \sim r_B/(T/m)^{1/2} \sim 10^{-16}$ s, so that its frequency spectrum extends to very high frequencies, virtually impossible to control.

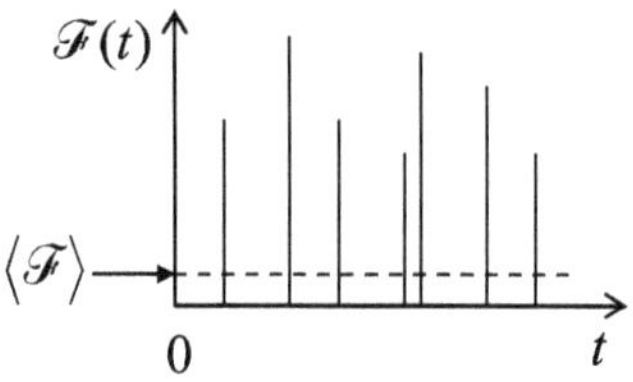

Figure 5.4. The force exerted by gas particles on container's wall, as a function of time (schematically).

However, we can use the following trick, very typical for the theory of fluctuations. It is almost evident that the rms fluctuations of volume are independent of the shape of the container. Let us consider the particular situation similar to that shown in figure 1.5, with the container of a cylindrical shape, with the base area A.[5] Then the coordinate of the piston is just $q = V/A$, while the average force exerted by the gas on the cylinder is $\mathscr{F} = PA$—see figure 5.5. Now if the piston is sufficiently massive, its free oscillation frequency ω near the equilibrium position is small enough to satisfy the following three conditions.

First, besides balancing the average force $\langle \mathscr{F} \rangle$, and thus sustaining the average pressure $\langle P \rangle = \langle \mathscr{F} \rangle / A$ of the gas, the interaction between the heavy piston and light particles of the gas is weak because of a relatively short duration of the particle hits

[5] As a reminder, the term 'cylinder' does not necessarily means the 'circular cylinder'; the shape of the base A may be arbitrary; it just should not change with height.

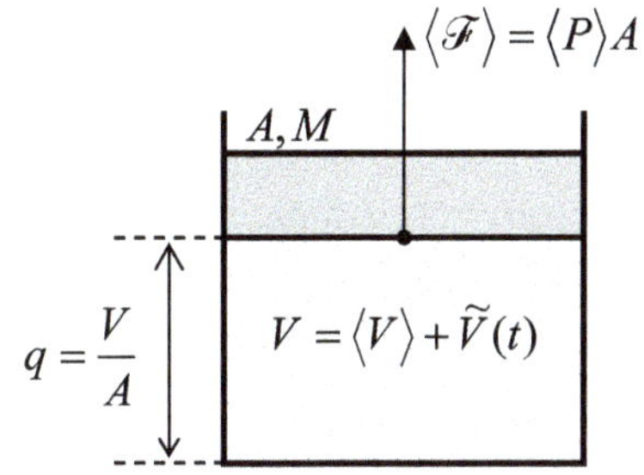

Figure 5.5. Deriving Eq. (5.37).

(figure 5.4). Because of that, the full energy of the system may be represented as a sum of those of the particles and the piston, with a quadratic contribution to the piston's potential energy by small deviations from the equilibrium:

$$U_p = \frac{\kappa}{2}\tilde{q}^2, \quad \text{where} \quad \tilde{q} \equiv q - \langle q \rangle = \frac{\tilde{V}}{A}, \tag{5.34}$$

and κ is the effective spring constant arising from the finite compressibility of the gas.

Second, at $\omega \to 0$, that spring constant may be calculated just as for constant variations of the volume, with the gas remaining in quasi-equilibrium at all times:

$$\kappa = -\frac{\partial \langle \mathscr{F} \rangle}{\partial q} = A^2\left(-\frac{\partial \langle P \rangle}{\partial \langle V \rangle}\right). \tag{5.35}$$

This partial derivative[6] should be calculated at whatever the given thermal conditions are, e.g. with S = const for adiabatic conditions (i.e. a thermally insulated gas), or with T = const for isothermal conditions (including a good thermal contact between the gas and a heat bath), etc. With that constant denoted as X, Eqs. (5.34) and (5.35) give

$$U_p = \frac{1}{2}\left(-A^2\frac{\partial \langle P \rangle}{\partial \langle V \rangle}\right)_X \left(\frac{\tilde{V}}{A}\right)^2 = \frac{1}{2}\left(-\frac{\partial \langle P \rangle}{\partial \langle V \rangle}\right)_X \tilde{V}^2. \tag{5.36}$$

Finally, making $\omega = (\kappa/M)^{1/2}$ sufficiently small (namely, $\hbar\omega \ll T$) by a sufficiently large piston mass M, we can apply, to the piston's fluctuations, the classical equipartition theorem: $\langle U_p \rangle = T/2$, giving[7]

---

[6] As was already discussed in section 4.1 in the context of the van der Waals equation, for the mechanical stability of a gas (or liquid), the derivative $\partial P/\partial V$ has to be negative, so that κ is positive.

[7] One may meet statements that a similar formula,

$$\left\langle \tilde{P}^2 \right\rangle = T\left(-\frac{\partial \langle P \rangle}{\partial \langle V \rangle}\right)_X, \quad \text{(WRONG!)}$$

is valid for pressure fluctuations. However, such a statement does not take into account a different physical nature of pressure (figure 5.4), with its very broad frequency spectrum. This issue will be discussed later in this chapter.

$$\langle \tilde{V}^2 \rangle_X = T\left(-\frac{\partial \langle V \rangle}{\partial \langle P \rangle} \right)_X. \tag{5.37a}$$

Since this result is valid for any A and ω, it should not depend on the system's geometry and the piston's mass, provided that it is large in comparison with the effective mass of a single system component (say, a gas molecule)—the condition that is naturally fulfilled in most experiments. For the particular case of fluctuations at constant temperature ($X = T$),[8] we may use the definition (3.58) of the isothermal bulk compressibility K_T of the gas to rewrite Eq. (5.37*a*) as

$$\langle \tilde{V}^2 \rangle_T = \frac{TV}{K_T}. \tag{5.37b}$$

For an ideal classical gas of N particles, with the equation of state $\langle V \rangle = NT/\langle P \rangle$, it is easier to use directly Eq. (5.37*a*), again with $X = T$, to get

$$\langle \tilde{V}^2 \rangle_T = -T\left(-\frac{NT}{\langle P \rangle^2} \right) = \frac{\langle V \rangle^2}{N}, \quad \text{i.e.} \quad \frac{\delta V_T}{\langle V \rangle} = \frac{1}{N^{1/2}}, \tag{5.38}$$

in full agreement with the general trend given by Eq. (5.12).

Now let us proceed to fluctuations of temperature, for simplicity focusing on the case $V = \text{const}$. Let us again assume that the system we are considering is weakly coupled to a heat bath of temperature T_0, in the sense that the time τ of temperature equilibration between the two is much larger than the internal temperature relaxation (*thermalization*) time. Then we may assume that, on the former time scale, T changes virtually simultaneously in the whole system, and consider it a function of time alone:

$$T = \langle T \rangle + \tilde{T}(t). \tag{5.39}$$

Moreover, due to the (relatively) large τ, we may use the stationary relation between small fluctuations of temperature and the internal energy of the system:

$$\tilde{T}(t) = \frac{\tilde{E}(t)}{C_V}, \quad \text{so that} \quad \delta T = \frac{\delta E}{C_V}. \tag{5.40}$$

With those assumptions, Eq. (5.20) immediately yields the famous expression for the so-called *thermodynamic fluctuations of temperature*:

$$\delta T = \frac{\delta E}{C_V} = \frac{\langle T \rangle}{C_V^{1/2}}. \tag{5.41}$$

The most straightforward application of this result is to analyse so-called *bolometers*—broadband detectors of electromagnetic radiation in microwave and infrared frequency bands, in particular the CMB radiation (discussed in section 2.6).

[8] In this case we may also use the second of Eqs. (1.39) to rewrite Eq. (5.37) via the second derivative $(\partial^2 G/\partial P^2)_T$.

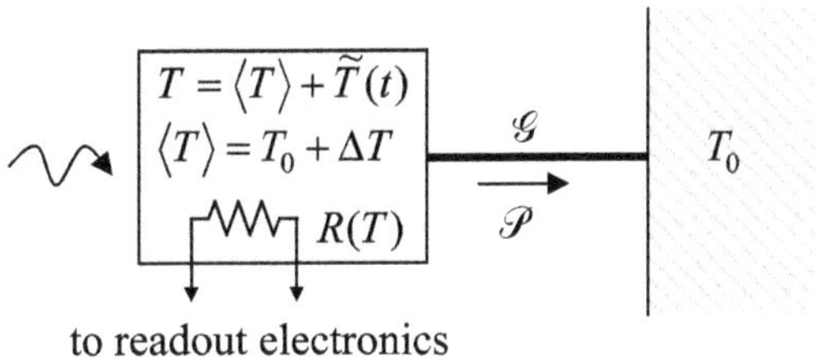

Figure 5.6. The conceptual scheme of a bolometer.

In such a detector (figure 5.6), the incoming radiation it focused on a small sensor (e.g. either a small piece of a germanium crystal, or a superconductor thin film at temperature $T \approx T_c$, etc), which is well isolated thermally from the environment. As a result, the absorption of even small radiation power $\mathscr{P}$ leads to a noticeable change ΔT of the sensor's average temperature $\langle T \rangle$ and hence of its electric resistance R, which is probed up by low-noise external electronics[9]. If the power does not change in time too fast, ΔT is a certain function of $\mathscr{P}$, turning to 0 at $\mathscr{P} = 0$. Hence, if ΔT is much lower than the environment temperature T_0, we may keep only the main, linear term in its Taylor expansion in $\mathscr{P}$:

$$\Delta T \equiv \langle T \rangle - T_0 = \frac{\mathscr{P}}{\mathscr{G}}, \tag{5.42}$$

where the coefficient $\mathscr{G} \equiv \partial \mathscr{P} / \partial T$ is called the *thermal conductance* of the (unavoidable) thermal coupling between the sensor and the heat bath—see figure 5.6.

The power may be detected if the electric signal from the sensor, which results from the change ΔT, is not drowned in spontaneous fluctuations. In practical systems, these fluctuations are contributed by several sources including electronic amplifiers, sensor, etc. However, in modern systems these 'technical' contributions to noise are successfully suppressed, and the dominating noise source is the fundamental sensor temperature fluctuations, described by Eq. (5.41). In this case the so-called *noise-equivalent power* ('NEP'), defined as the level of $\mathscr{P}$ that produces the signal equal to the rms value of noise, may be calculated by equating the expressions (5.41) (with $\langle T \rangle = T_0$) and (5.42):

$$\text{NEP} \equiv \mathscr{P}|_{\Delta T = \delta T} = \frac{T_0 \mathscr{G}}{C_V^{1/2}}. \tag{5.43}$$

This expression shows that in order to decrease the NEP, i.e. improve the detector's sensitivity, both the environment temperature T_0 and the thermal conductance $\mathscr{G}$ should be reduced. In modern receivers of radiation, their typical values are of the order of 0.1 K and 10^{-10} W K^{-1}, respectively.

On the other hand, Eq. (5.43) implies that in order to increase the bolometer's sensitivity, i.e. reduce NEP, the C_V of the sensor, and hence its mass, should be *increased*. This conclusion is valid only to a certain extent, because due to technical

[9] Besides low internal electric noise, the sensor should have a sufficiently large *temperature responsivity* dR/dT, making the noise contribution by the readout electronics insignificant—see below.

reasons (parameter drifts and the so-called $1/f$ *noise* of the sensor and external electronics), the incoming power has to be modulated with as high frequency ω as technically possible (in most practical cases, the cyclic frequency $\nu = \omega/2\pi$ of the modulation is between 10 and 1000 Hz), so that the electrical signal might be picked up from the sensor at that frequency. As a result, the C_V may be increased only until the thermal constant of the sensor,

$$\tau = \frac{C_V}{\mathscr{G}}, \tag{5.44}$$

becomes close to $1/\omega$, because at $\omega\tau \gg 1$ the useful signal drops faster than noise. As a result, the lowest (i.e. the best) value of the NEP,

$$\frac{(\text{NEP})_{\min}}{\nu^{1/2}} = \alpha T_0 \mathscr{G}^{1/2}, \quad \text{with} \quad \alpha \sim 1, \tag{5.45}$$

is reached at $\nu\tau \approx 1$. (The exact values of the optimal product $\omega\tau$, and the numerical constant $\alpha \sim 1$ in Eq. (5.45) depend on the exact law of the power modulation in time, and the readout signal processing procedure.) With the parameters cited above, this estimate yields $(\text{NEP})_{\min}/\nu^{1/2} \sim 3 \times 10^{-17}$ W Hz$^{-1/2}$—a very low power indeed.

However, surprisingly enough, the power modulation allows the bolometric (and other broadband) receivers to register radiation with a power much lower than this NEP! Indeed, picking up the sensor signal at the modulation frequency ω, we can use the following electronics stages to filter out all the noise besides its components within a very narrow band, of width $\Delta\nu \ll \nu$, around the modulation frequency (figure 5.7). This is the idea of a *microwave radiometer*[10], currently used in all sensitive broadband receivers of radiation. In order to analyze this opportunity, we need to develop theoretical tools for a quantitative description of the spectral distribution of fluctuations. Another motivation for that description is the need in

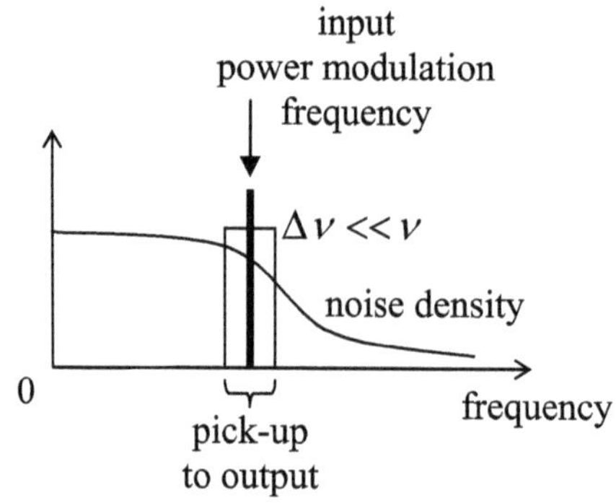

Figure 5.7. The basic idea of the Dicke radiometer.

[10] It was pioneered in the 1950s by R Dicke, so that the device is frequently called the *Dicke radiometer*. Note that the optimal strategy of using essentially the same devices for time- and energy-resolved detection of single high-energy photons is different—though even it is essentially based on Eq. (5.41). For a recent brief review of such detectors see, e.g. [1], and references therein.

analysis of variables dominated by fast (high-frequency) components, such as pressure—please have one more look at figure 5.4. Finally, during such analysis, we will run into the fundamental relation between fluctuations and dissipation, which is one of the main results of statistical physics as a whole.

5.4 Fluctuations as functions of time

In the previous sections the averaging $\langle\ldots\rangle$ of any function was assumed to be over an appropriate statistical ensemble of many similar systems. However, as was discussed in section 2.1, most physical systems of interest are ergodic. If such a system is also *stationary*, i.e. statistical averages of its variables do not change with time, the averaging may be also understood as that over a sufficiently long time interval. In this case, we may think about fluctuations of any variable f as of a random process taking place in just one system, but developing in time: $\tilde{f} = \tilde{f}(t)$.

There are two mathematically-equivalent approaches to the description of such random functions of time, called the *time-domain* picture and the *frequency-domain* picture, their relative convenience depending on the particular problem to be solved. In the time domain, we need to characterize a *random* fluctuation $\tilde{f}(t)$ by some *deterministic* function of time. Evidently, the average $\langle\tilde{f}(t)\rangle$ cannot be used for this purpose, because it equals zero—see Eq. (5.2). Of course, the variance (5.3) does not equal zero, but if the system is stationary, this average cannot depend on time either. Because of that, let us consider the following average:

$$\langle\tilde{f}(t)\tilde{f}(t')\rangle. \tag{5.46}$$

Generally, this is a function of two arguments. Moreover, in a stationary system, the average like (5.46) may depend only on the difference,

$$\tau \equiv t' - t, \tag{5.47}$$

between the two observation times. In this case, the average (5.46) is called the *correlation function* of the variable f:

$$K_f(\tau) \equiv \langle\tilde{f}(t)\tilde{f}(t+\tau)\rangle. \tag{5.48}$$

Again, here the averaging may be understood as that either over a statistical ensemble of macroscopically similar systems, or over a sufficiently long interval of the time argument t, with the argument τ kept constant. The correlation function's name[11] catches the idea of this notion very well: $K_f(\tau)$ characterizes the mutual relation between the fluctuation of the variable f at two times separated by the given interval τ. Let us list the basic properties of this function[12].

First of all, $K_f(\tau)$ has to be an even function of the time delay τ. Indeed, we may write

[11] Another term, the *autocorrelation function*, is sometimes used for the average (5.48) to distinguish it from the *mutual correlation function*, $\langle f_1(t)f_2(t+\tau)\rangle$, of two different stationary processes.

[12] Please notice that this correlation function is the direct temporal analog of the spatial correlation function briefly discussed in section 4.2—see Eq. (4.30).

$$K_f(-\tau) = \langle \tilde{f}(t)\tilde{f}(t-\tau)\rangle \equiv \langle \tilde{f}(t-\tau)\tilde{f}(t)\rangle = \langle \tilde{f}(t')\tilde{f}(t'+\tau)\rangle, \tag{5.49}$$

with $t' \equiv t - \tau$. For stationary processes, this average cannot depend on the common shift of two observation times, so that the averages (5.48) and (5.49) have to be equal:

$$K_f(-\tau) = K_f(\tau). \tag{5.50}$$

Second, at $\tau \to 0$ the correlation function tends to the variance:

$$K_f(0) = \langle \tilde{f}(t)\tilde{f}(t)\rangle = \langle \tilde{f}^2\rangle. \tag{5.51}$$

In the opposite limit, when τ is much larger than certain characteristic *correlation time* τ_c of the system[13], the correlation function has to tend to zero, because the fluctuations separated by such a time interval are virtually independent (*uncorrelated*). As a result, the correlation function typically looks like one of the plots sketched in figure 5.8. Note that on a time scale much longer than τ_c, any physically-realistic correlation function may be well approximated with a delta-function of τ. (For example, for a process which is a sum of independent very short pulses, e.g. the gas pressure force exerted on the container wall (figure 5.4), such an approximation is legitimate on time scales much longer than the single pulse duration, e.g. the time of particle's interaction with on the wall at the impact.)

In the reciprocal, frequency domain, the same process $\tilde{f}(t)$ is represented as a Fourier integral,

$$\tilde{f}(t) = \int_{-\infty}^{+\infty} f_\omega e^{-i\omega t} d\omega, \tag{5.52}$$

with the reciprocal transform being[14]

$$f_\omega = \frac{1}{2\pi}\int_{-\infty}^{+\infty} \tilde{f}(t) e^{i\omega t} dt. \tag{5.53}$$

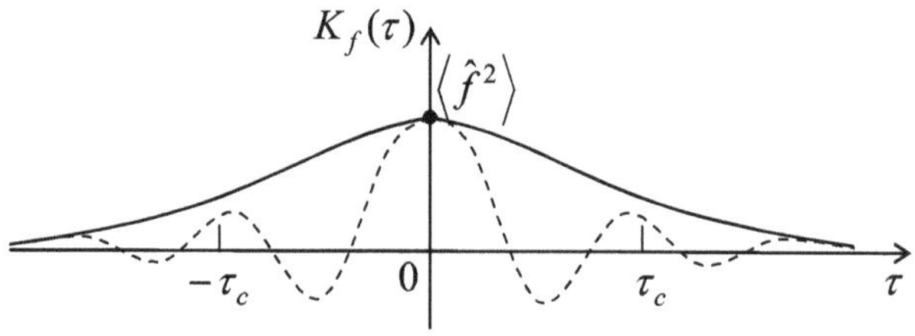

Figure 5.8. The correlation function of fluctuations: two typical examples.

[13] Note that the correlation time τ_c is the direct temporal analog of the correlation radius r_c which was discussed in section 4.2—see the same Eq. (4.30).

[14] The argument of the function f_ω is represented as its index with a purpose to emphasize that this function is different from $\tilde{f}(t)$, while (very conveniently) still using the same letter for the same variable.

If the initial function $\tilde{f}(t)$ is random (as it is in the case of fluctuations), with zero average, its Fourier transform f_ω is also a random function (now of frequency), and also with a vanishing statistical average. Indeed, now thinking of the operation $\langle \ldots \rangle$ as an ensemble averaging, we have

$$\langle f_\omega \rangle = \left\langle \frac{1}{2\pi} \int_{-\infty}^{+\infty} \tilde{f}(t) e^{i\omega t} dt \right\rangle = \frac{1}{2\pi} \int_{-\infty}^{+\infty} \langle \tilde{f}(t) \rangle e^{i\omega t} dt = 0. \tag{5.54}$$

The simplest nonvanishing average may be formed similarly to Eq. (5.46), but with a due respect to the complex-variable character of the Fourier images:

$$\left\langle f_\omega f_{\omega'}^* \right\rangle = \frac{1}{(2\pi)^2} \int_{-\infty}^{+\infty} dt' \int_{-\infty}^{+\infty} dt \, \langle \tilde{f}(t) \tilde{f}(t') \rangle \, e^{i(\omega' t' - \omega t)}. \tag{5.55}$$

It turns out that for a stationary process, the averages (5.46) and (5.55) are directly related. Indeed, since the integration over t' in Eq. (5.55) is in infinite limits, we may replace it with the integration over $\tau \equiv t' - t$ (at fixed t), also in infinite limits. Replacing t' with $t + \tau$ in the expressions under the integral, we see that the average is just the correlation function $K_f(\tau)$, while the time exponent is equal to $\exp\{i(\omega' - \omega)t\}$ $\exp\{i\omega'\tau\}$. As a result, changing the order of integration, we get

$$\begin{aligned} \left\langle f_\omega f_{\omega'}^* \right\rangle &= \frac{1}{(2\pi)^2} \int_{-\infty}^{+\infty} dt \int_{-\infty}^{+\infty} d\tau K_f(\tau) e^{i(\omega - \omega')t} e^{i\omega'\tau} \\ &\equiv \frac{1}{(2\pi)^2} \int_{-\infty}^{+\infty} K_f(\tau) e^{i\omega'\tau} d\tau \int_{-\infty}^{+\infty} e^{i(\omega - \omega')t} dt. \end{aligned} \tag{5.56}$$

But the last integral is just $2\pi\delta(\omega - \omega')$,[15] so that we finally get

$$\left\langle f_\omega f_{\omega'}^* \right\rangle = S_f(\omega) \delta(\omega - \omega'), \tag{5.57}$$

where the real function of frequency,

$$S_f(\omega) \equiv \frac{1}{2\pi} \int_{-\infty}^{+\infty} K_f(\tau) e^{i\omega\tau} d\tau = \frac{1}{\pi} \int_0^{\infty} K_f(\tau) \cos \omega\tau \, d\tau, \tag{5.58}$$

is called the *spectral density of fluctuations at frequency* ω. According to Eq. (5.58), the spectral density is just the Fourier image of the correlation function, and hence the reciprocal Fourier transform is[16,17]:

---

[15] See, e.g. Eq. (A.88).

[16] The second form of Eq. (5.59) uses the fact that, according to Eq. (5.58), $S_f(\omega)$ is an even function of frequency—just as $K_f(\tau)$ is an even function of time.

[17] Although Eqs. (5.58) and (5.59) look not much more than straightforward corollaries of the Fourier transform, they bear a special name of the *Wiener–Khinchin theorem*—after the mathematicians N Wiener and A Khinchin who have proved that these relations are valid even for the functions $f(t)$ that are not square-integrable, so that from the point of view of rigorous mathematics, their Fourier transforms are not well defined.

$$K_f(\tau) = \int_{-\infty}^{+\infty} S_f(\omega) e^{-i\omega\tau} d\omega = 2\int_0^{\infty} S_f(\omega)\cos\omega\tau \; d\omega. \tag{5.59}$$

In particular, for the fluctuation variance, Eq. (5.59) yields

$$\langle \tilde{f}^2 \rangle \equiv K_f(0) = \int_{-\infty}^{+\infty} S_f(\omega) d\omega \equiv 2\int_0^{\infty} S_f(\omega) d\omega. \tag{5.60}$$

The last relation shows that the term 'spectral density' describes the physical sense of the function $S_f(\omega)$ very well. Indeed, if a random signal $f(t)$ had been passed through a frequency filter with a small bandwidth $\Delta\nu \ll \nu$ of positive cyclic frequencies, the integral in Eq. (5.60) could be limited to the interval $\Delta\omega = 2\pi\Delta\nu$, i.e. the variance of the filtered signal would become

$$\langle \tilde{f}^2 \rangle_{\Delta\nu} = 2S_f(\omega)\Delta\omega \equiv 4\pi S_f(\omega)\Delta\nu. \tag{5.61}$$

(A popular alternative definition of the spectral density is $\mathscr{S}_f(\nu) \equiv 4\pi S_f(\omega)$, making the average (5.61) equal to just $\mathscr{S}_f(\nu)\Delta\nu$.)

To conclude this introductory (mostly mathematical) section, let me note an important particular case. If the spectral density of some process is nearly constant within all the frequency range of interest, $S_f(\omega) = \text{const} = S_f(0)$,[18] Eq. (5.59) shows that its correlation function may be well approximated with a delta-function:

$$K_f(\tau) = S_f(0)\int_{-\infty}^{+\infty} e^{-i\omega\tau} d\omega = 2\pi S_f(0)\delta(\tau). \tag{5.62}$$

From this relation stems another popular name of the white noise, the *delta-correlated process*. We have already seen that this is a very reasonable approximation, for example, for the gas pressure force fluctuations (figure 5.4). Of course, for the spectral density of a realistic, limited physical variable the approximation of constant spectral density cannot be true for *all* frequencies (otherwise, for example, the integral (5.60) would diverge, giving an unphysical, infinite value of the variance), and may be valid only at frequencies much lower than $1/\tau_c$.

5.5 Fluctuations and dissipation

Now we are well equipped mathematically to address one of the most important issues of statistical physics, the relation between fluctuations and dissipation. This relation is especially simple for the following hierarchical situation: a relatively 'heavy', slowly moving system, weakly interacting with an environment consisting of rapidly moving, 'light' components. A popular theoretical term for such a system is the *Brownian particle*, named after botanist R Brown who first noticed in 1827 the random motion of pollen grains, caused by their random hits by fluid's molecules, under a microscope. However, the family of such systems is much broader than that

[18] Such process is frequently called the *white noise*, because it consists of all frequency components with equal amplitudes, reminding the white light, which consists of many monochromatic components with close amplitudes.

of mechanical particles. Just for a few examples, such description is valid for an atom interacting with electromagnetic field modes of the surrounding space, a clock pendulum interacting with molecules of the air around it, the probe masses of the now-famous LIGO detectors, weakly interacting with the acoustic modes of the Earth via the suspension system, etc[19].

One more important assumption of this theory is that the system's motion does not violate the thermal equilibrium of the environment—well fulfilled in many cases. (Think, for example, about a mechanical pendulum whose motion does not overheat the air around it to any noticeable extent.) In this case, the averaging over a statistical ensemble of similar environments, at a fixed motion of the system of interest, may be performed assuming them to be in thermal equilibrium[20]. I will denote such a 'primary' averaging by the usual angle brackets $\langle\ldots\rangle$. At a later stage we may carry out another, 'secondary' averaging, over an ensemble of many similar systems of interest, coupled to similar environments. When we go after such additional averaging, it will be denoted by double angle brackets $\langle\langle\ldots\rangle\rangle$.

Let me start from a simple classical system, a 1D harmonic oscillator whose equation of evolution may be represented as

$$m\ddot{q} + \kappa q = \mathscr{F}_{\text{det}}(t) + \mathscr{F}_{\text{env}}(t) \equiv \mathscr{F}_{\text{det}}(t) + \langle\mathscr{F}\rangle + \tilde{\mathscr{F}}(t), \quad \text{with } \langle\tilde{\mathscr{F}}(t)\rangle = 0, \quad (5.63)$$

where q is the (generalized) coordinate of the oscillator, $\mathscr{F}_{\text{det}}(t)$ is the deterministic external force, while both components of the force $\mathscr{F}_{\text{env}}(t)$ represent the impact of the environment on the oscillator's motion. Again, from the point of view of the fast-moving environmental components, the oscillator's motion is slow. The average component $\langle\mathscr{F}\rangle$ of the force exerted by environment on such a slowly moving object may have a part depending on not only q, but on the velocity $\dot{q}$ as well. For most systems, the Taylor expansion of the force in small velocity would have a non-vanishing leading, linear term:

$$\langle\mathscr{F}\rangle = -\eta\,\dot{q}, \quad (5.64)$$

where the constant η is usually called the *drag* (or 'kinematic friction', or 'damping') *coefficient*, so that Eq. (5.63) may be rewritten as

$$m\ddot{q} + \eta\,\dot{q} + \kappa q = \mathscr{F}_{\text{det}}(t) + \tilde{\mathscr{F}}(t). \quad (5.65)$$

This way of describing the effects of environment on an otherwise Hamiltonian system is called the *Langevin equation*[21]. Due to the linearity of the differential equation (5.65), its general solution may be represented as a sum of two parts: the

[19] To emphasize this generality, in the forthcoming discussion of the 1D case, I will use letter q rather than x for the system's displacement.

[20] For a usual (ergodic) environment, the primary averaging may be interpreted as that over relatively short time intervals, $\tau_c \ll \Delta t \ll \tau$, where τ_c is the correlation time of the environment, while τ is the characteristic time scale of motion of our 'heavy' system of interest.

[21] After P Langevin, whose 1908 work was the first systematic development of A Einstein's ideas on Brownian motion (see below) using this formalism. A detailed discussion of this approach, with numerical examples of its application, may be found, e.g. in the monograph [2].

deterministic motion of the damped linear oscillator due to the external force $\mathscr{F}_{\text{det}}(t)$, and random fluctuations due to the random force $\tilde{\mathscr{F}}(t)$ exerted by the environment. The former effects are well known from classical dynamics[22], so let us focus on the latter part by taking $\mathscr{F}_{\text{det}}(t) = 0$. The remaining term in the right-hand part of Eq. (5.65) describes the fluctuating part of the environmental force; in contrast to the average component (5.64), its intensity (read: its spectral density at relevant frequencies $\omega \sim \omega_0 \equiv (\kappa/m)^{1/2}$) does not vanish at $q(t) = 0$, and hence may be evaluated ignoring the system's motion[23].

Plugging into Eq. (5.65) the representation of both variables in the form similar to Eq. (5.52), and requiring the coefficients before the same $\exp\{-i\omega t\}$ to be equal on both sides of the equation, for their Fourier images we get the following relation:

$$-m\omega^2 q_\omega - i\omega\eta\, q_\omega + \kappa q_\omega = \mathscr{F}_\omega. \tag{5.66}$$

which immediately gives us q_ω, i.e. the (random) complex amplitude of oscillations:

$$q_\omega = \frac{\mathscr{F}_\omega}{(\kappa - m\omega^2) - i\eta\omega} \equiv \frac{\mathscr{F}_\omega}{m\left(\omega_0^2 - \omega^2\right) - i\eta\omega}. \tag{5.67}$$

Now multiplying Eq. (5.67) by its complex conjugate for a different frequency, averaging both parts of the resulting equation, and using the formulas similar to Eq. (5.57) for each of them[24], we get the following relation between spectral densities of the oscillations and the random force:

$$S_q(\omega) = \frac{1}{m^2\left(\omega_0^2 - \omega^2\right)^2 + \eta^2\omega^2} S_{\mathscr{F}}(\omega). \tag{5.68}$$

In the so-called low-damping limit ($\eta \ll m\omega_0$), the fraction on the right-hand side of Eq. (5.68) has a sharp peak near the oscillator's own frequency ω_0 (describing the well-known effect of high-Q *resonance*)[25], and may be approximated in that vicinity as

$$\frac{1}{m^2\left(\omega_0^2 - \omega^2\right)^2 + (\eta\omega)^2} \approx \frac{1}{\eta^2\omega_0^2(\xi^2 + 1)}, \quad \text{with} \quad \xi \equiv \frac{2m(\omega - \omega_0)}{\eta}. \tag{5.69}$$

In contrast, the spectral density $S_{\mathscr{F}}(\omega)$ of fluctuations of a typical environment is changing relatively slowly near that frequency, so that for the purpose of integration over frequencies near ω_0 we may replace $S_{\mathscr{F}}(\omega)$ with $S_{\mathscr{F}}(\omega_0)$. As a result, the variance

[22] See, e.g. *Part CM* section 5.1. Here I assume that the variable $f(t)$ is classical, with the discussion of the quantum case postponed until the end of the section.

[23] Note that the direct secondary statistical averaging of Eq. (5.65) with $\mathscr{F}_{\text{det}} = 0$ yields $\langle\langle q \rangle\rangle = 0$! This, perhaps a bit counter-intuitive result becomes less puzzling if we recognize that this is the averaging over a large statistical ensemble of random sinusoidal oscillations with all values of their phase, and that the (equally probable) oscillations with opposite phases give mutually canceling contributions to the sum in Eq. (2.6).

[24] At this stage we restrict our analysis to random, stationary processes $q(t)$, so that Eq. (5.57) is valid for this variable as well, if the averaging in it is understood in the $\langle\langle \ldots \rangle\rangle$ sense.

[25] Regardless of the physical sense of such a function of ω, and of whether its maximum is situated at a finite frequency ω_0 as in Eq. (5.68) or at $\omega = 0$, it is often referred to as the *Lorentzian* (or 'Breit–Wigner') *line*.

of the environment-imposed random oscillations may be calculated, using Eq. (5.60), as

$$\begin{aligned}\langle\langle\tilde{q}^2\rangle\rangle &= 2\int_0^{\infty} S_q(\omega)d\omega \approx 2\int_{\omega\approx\omega_0} S_q(\omega)d\omega \\ &= 2S_{\mathscr{F}}(\omega_0)\frac{1}{\eta^2\omega_0^2}\frac{\eta}{2m}\int_{-\infty}^{+\infty}\frac{d\xi}{\xi^2+1}.\end{aligned} \quad (5.70)$$

This is a well-known table integral[26], equal to π, so that, finally:

$$\langle\langle\tilde{q}^2\rangle\rangle = 2S_{\mathscr{F}}(\omega_0)\frac{1}{\eta^2\omega_0^2}\frac{\eta}{2m}\pi \equiv \frac{\pi}{m\omega_0^2\eta}S_{\mathscr{F}}(\omega_0) \equiv \frac{\pi}{\kappa\eta}S_{\mathscr{F}}(\omega_0). \quad (5.71)$$

But on the other hand, the weak interaction with the environment should keep the oscillator in thermodynamic equilibrium at the same temperature T. Since our analysis has been based on the classical Langevin equation (5.65), we may only use it in the classical limit $\hbar\omega_0 \ll T$, in which we may use the equipartition theorem (2.48). In our current notation, it yields

$$\frac{\kappa}{2}\langle\langle\tilde{q}^2\rangle\rangle = \frac{T}{2}. \quad (5.72)$$

Comparing Eqs. (5.71) and (5.72), we see that the spectral density of the random force exerted by environment has to be fundamentally related to the damping it provides:

$$S_{\mathscr{F}}(\omega_0) = \frac{\eta}{\pi}T. \quad (5.73a)$$

Now we may argue (rather convincingly :-) that since this relation does not depend on the oscillator's parameters m and κ, and hence its eigenfrequency $\omega_0 = (\kappa/m)^{1/2}$, it should be valid at any (but sufficiently low, $\omega\tau_c \ll 1$) frequency. Using Eq. (5.58) with $\omega\rightarrow 0$, it may be rewritten as a formula for the effective low-frequency drag coefficient:

$$\eta = \frac{1}{T}\int_0^{\infty} K_{\mathscr{F}}(\tau)d\tau \equiv \frac{1}{T}\int_0^{\infty}\langle\tilde{\mathscr{F}}(0)\tilde{\mathscr{F}}(\tau)\rangle d\tau. \quad (5.73b)$$

Relations (5.73) reveal an intimate, fundamental connection between fluctuations and dissipation provided by a thermally-equilibrium environment. Verbally, 'there is no dissipation without fluctuations'—and vice versa[27]. To the best of my knowledge, this fact was first recognized in 1905 by A Einstein[28], for the following particular

---

[26] See, e.g. Eq. (A.32*a*).

[27] This means in particular that the phenomenological description of dissipation barely by the drag force in classical mechanics (see, e.g. *Part CM* section 5.1) is only valid approximately, when the energy scale of the process is much larger than T.

[28] It was published in one of the three papers of Einstein's celebrated 1905 'triad'. As a reminder, another paper started the (special) relativity, and one more was the quantum description of the photoelectric effect, essentially starting the quantum mechanics. Not too bad for one year, one person!

case. Let us apply our result (5.73) to a free 1D Brownian particle, by taking $\kappa = 0$. In this case both relations (5.71) and (5.72) give infinities. In order to understand the reason for that divergence, let us go back to the Langevin equation (5.65) with not only $\kappa = 0$, but (just for the sake of simplicity) $m \to 0$ as well. (The latter approximation, frequently called the *overdamping limit*, is quite appropriate for the motion of a small particle in a very viscous fluid.) In this approximation, Eq. (5.65) is reduced to a simple equation,

$$\eta \dot{q} = \mathscr{F}_{\text{det}}(t) + \tilde{\mathscr{F}}(t), \quad \text{with} \quad \langle \tilde{\mathscr{F}}(t) \rangle = 0, \tag{5.74}$$

which may be readily integrated to give the particle displacement during a finite time interval t. In the absence of the deterministic force,

$$\Delta q(t) \equiv q(t) - q(0) = \frac{1}{\eta} \int_0^t \tilde{\mathscr{F}}(t')dt'. \tag{5.75}$$

Evidently, at the full statistical averaging of the displacement, the fluctuation effects vanish, but this does not mean that the particle does not move—just that it has equal probabilities to be shifted in either of two possible directions. To see that, let us calculate the variance of the displacement:

$$\langle\langle \Delta \tilde{q}^2(t) \rangle\rangle = \frac{1}{\eta^2} \int_0^t dt' \int_0^t dt'' \langle \tilde{\mathscr{F}}(t') \tilde{\mathscr{F}}(t'') \rangle \equiv \frac{1}{\eta^2} \int_0^t dt' \int_0^t dt'' K_{\mathscr{F}}(t' - t''). \tag{5.76}$$

As we already know, at times $\tau \gg \tau_c$, the correlation function may be well approximated by the delta-function—see Eq. (5.62). In this approximation, with $S_{\mathscr{F}}(0)$ expressed by Eq. (5.73*a*), we get

$$\begin{aligned} \langle\langle \Delta \tilde{q}^2(t) \rangle\rangle &= \frac{2\pi}{\eta^2} S_{\mathscr{F}}(0) \int_0^t dt' \int_0^t dt'' \delta(t'' - t') \\ &= \frac{2\pi}{\eta^2} \frac{\eta T}{\pi} \int_0^t dt' = \frac{2T}{\eta} t \equiv 2Dt, \end{aligned} \tag{5.77}$$

with

$$D = \frac{T}{\eta}. \tag{5.78}$$

The final form of Eq. (5.77) describes the well-known law of *diffusion* ('random walk') of a 1D system, with the rms deviation from the point of origin growing as $(2Dt)^{1/2}$. The coefficient D in this relation is called the *coefficient of diffusion*, and Eq. (5.78) describes the extremely simple and important[29] *Einstein's relation* between

[29] In particular, in 1908, i.e. very soon after Einstein's publication, it was used by J Perrin for an accurate determination of the Avogadro number N_A. (It was actually Perrin who graciously suggested to call this constant after A Avogadro, honoring his pioneering studies of gases in the 1810s.)

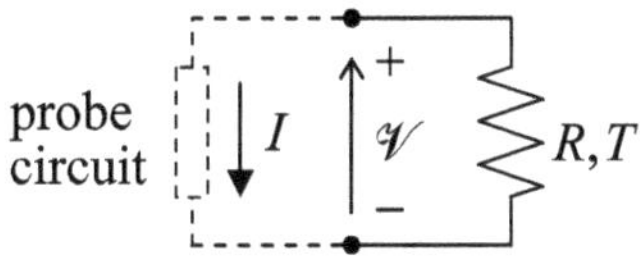

Figure 5.9. A resistive device as a dissipative environment of a two-terminal probe circuit.

that coefficient and the drag coefficient. Often this relation is rewritten, in the SI units of temperature, as $D = \mu\, k_B T_K$, where $\mu \equiv 1/\eta$ is the *mobility* of the particle. The physical sense of μ becomes clear from the expression for the deterministic velocity (particle's 'drift'), which follows from the averaging of both sides of Eq. (5.74):

$$v_{\text{drift}} \equiv \langle\langle \dot{q}(t) \rangle\rangle = \frac{1}{\eta}\mathscr{F}_{\text{det}}(t) \equiv \mu \mathscr{F}_{\text{det}}(t), \tag{5.79}$$

so that the mobility is just the drift velocity given to the particle by a unit force[30].

Another famous manifestation of the general Eq. (5.73) is the *thermal* (or 'Johnson', or 'Johnson–Nyquist', or just 'Nyquist') *noise* in resistive electron devices. Let us consider a two-terminal, dissipation-free 'probe' circuit, playing the role of the harmonic oscillator in our analysis carried out above, connected to a resistive device (figure 5.9), playing the role of the probe circuit's environment. (The noise is generated by the thermal motion of numerous electrons, randomly moving inside the resistive device.) For this system, one convenient choice of the conjugate variables (the generalized coordinate and generalized force) is, respectively, the electric charge $Q \equiv \int I(t)dt$ that has passed through the 'probe' circuit by time t, and the voltage $\mathscr{V}$ across its terminals, with the polarity shown in figure 5.9. (Indeed, the product $\mathscr{V}dQ$ is the elementary work $d\mathscr{W}$ done by the environment on the probe circuit.)

Making the corresponding replacements, $q \to Q$ and $\mathscr{F} \to \mathscr{V}$ in Eq. (5.64), we see that it becomes

$$\langle \mathscr{V} \rangle = -\,\eta\dot{Q} \equiv -\eta I. \tag{5.80}$$

Comparing this relation with Ohm's law, $\mathscr{V} = R(-I)$,[31] we see that in this case, the coefficient η has the physical sense of the usual Ohmic resistance R of our dissipative device[32], so that Eq. (5.73*a*) becomes

$$S_{\mathscr{V}}(\omega) = \frac{R}{\pi}T. \tag{5.81a}$$

---

[30] Note that in solid-state physics and electronics, the charge carrier mobility is usually defined as $|\mathbf{v}_{\text{drift}}/\mathscr{E}| = e\mathbf{v}_{\text{drift}}/|\mathscr{F}_{\text{det}}| \equiv e|\mu|$ (where $\mathscr{E}$ is the applied electric field), and is traditionally measured in cm^2 V^{-1}s^{-1}.

[31] The minus sign is due to the fact that in our notation, the current flowing in the resistor, from the positive terminal to the negative one, is $(-I)$—see figure 5.9.

[32] Due to this fact, Eq. (5.64) is often called the *Ohmic model* of the environment response, even if the physical nature of the variables q and $\mathscr{F}$ is completely different from the electric charge and voltage.

Using the last equality in Eq. (5.61), and transferring to the SI units of temperature ($T = k_B T_K$), we may bring this famous *Nyquist formula*[33] to its most popular form:

$$\langle \tilde{\mathscr{V}}^2 \rangle_{\Delta\nu} = 4 k_B T_K R \Delta\nu. \tag{5.81b}$$

Note that according to Eq. (5.65), this result is only valid at a negligible speed of change of the generalized coordinate q (in this case, negligible current I), i.e. Eq. (5.81) expresses the voltage fluctuations as would be measured by a nearly-ideal *voltmeter*, with its input resistance much higher that R.

On the other hand, using a different choice of generalized coordinate and force, $q \to \Phi$, $\mathscr{F} \to I$ (where $\Phi \equiv \int \mathscr{V}(t)dt$ is the generalized magnetic flux, so that $d\mathscr{W} = Id\Phi$), we get $\eta \to 1/R$, and Eq. (5.73) yields the thermal fluctuations of the current through the resistive device, as measured by a nearly-ideal *ammeter*, i.e. at $\mathscr{V} \to 0$:

$$S_I(\omega) = \frac{1}{\pi R} T, \quad \text{i.e.} \quad \langle \tilde{I}^2 \rangle_{\Delta\nu} = \frac{4 k_B T_K}{R} \Delta\nu. \tag{5.81c}$$

The nature of Eqs. (5.81) is so fundamental that they may be used, in particular, for the so-called *Johnson noise thermometry*[34]. Note, however, that these relations are valid for noise in thermal equilibrium only. In electric circuits that may be readily driven out of equilibrium by an applied voltage $\langle \mathscr{V} \rangle$, other types of noise are frequently important, notably the *shot noise*, which arises in short conductors, e.g. tunnel junctions, at applied voltages $\langle \mathscr{V} \rangle \gg T/q$, due to the discreteness of charge carriers[35]. A straightforward analysis (left for the reader's exercise) shows that this noise may be characterized by current fluctuations with the following low-frequency spectral density:

$$S_I(\omega) = \frac{|q\bar{I}|}{2\pi}, \quad \text{i.e.} \quad \langle \tilde{I}^2 \rangle_{\Delta\nu} = 2|q\bar{I}|\Delta\nu, \tag{5.82}$$

where q is the electric charge of a single current carrier. This is the *Schottky formula*[36], valid for any relation between the average I and $\mathscr{V}$. The comparison of Eqs. (5.81*c*) and (5.82) for a device that obeys the Ohm law shows that the shot noise has the same intensity as the thermal noise with the effective temperature

$$T_{\text{ef}} = \frac{|q\bar{\mathscr{V}}|}{2} \gg T. \tag{5.83}$$

[33] It is named after H Nyquist who derived this formula in 1928 (independently of the prior work by A Einstein, M Smoluchowski, and P Langevin) to describe the noise that had been just discovered experimentally by his Bell Labs colleague J Johnson. The derivation of Eq. (5.73) and hence Eq. (5.81) in these notes is essentially a twist of the derivation used by Nyquist.

[34] See, e.g. [3] and references therein.

[35] Another practically important type of fluctuations in electronic devices is the low-frequency *1/f noise* that was already mentioned in section 5.3 above. I will briefly discuss it in section 5.8.

[36] It was derived by W Schottky as early as in 1918, i.e. before the Nyquist' s work.

This relation may be interpreted as a result of charge carrier overheating by the applied electric field, and explains why the Schottky formula (5.82) is only valid in conductors much shorter than the energy relaxation length l_e of the charge carriers[37]. (Another mechanism of shot noise suppression, which may become noticeable in highly conductive nanoscale devices, is the Fermi–Dirac statistics of electrons[38].)

Now let us return for a minute to the bolometric Dicke radiometer (see figures 5.6–5.7 and their discussion), and use the Langevin formalism to finalize its analysis. For this system, the Langevin equation is the extension of the usual equation of heat balance:

$$C_V \frac{dT}{dt} + \mathscr{G}(T - T_0) = \mathscr{P}_{\text{det}}(t) + \tilde{\mathscr{P}}(t), \tag{5.84}$$

where $\mathscr{P}_{\text{det}} \equiv \langle \mathscr{P} \rangle$ describes the (deterministic) power of the absorbed radiation, and $\tilde{\mathscr{P}}$ represents the effective source of temperature fluctuations. Now we can use Eq. (5.84) to carry out a calculation of the spectral density $S_T(\omega)$ of temperature fluctuations absolutely similarly to how this was done with Eq. (5.65), assuming that the frequency spectrum of the fluctuation source is much broader than the intrinsic bandwidth $1/\tau = \mathscr{G}/C_V$ of the bolometer, so that its spectral density at frequencies $\omega\tau \sim 1$ may be well approximated by its low-frequency value $S_{\mathscr{P}}(0)$:

$$S_T(\omega) = \left| \frac{1}{-i\omega C_V + \mathscr{G}} \right|^2 S_{\mathscr{P}}(0). \tag{5.85}$$

Then, requiring the variance of temperature fluctuations, calculated from this formula and Eq. (5.60),

$$\begin{aligned} (\delta T)^2 \equiv \langle \tilde{T}^2 \rangle &= 2\int_0^\infty S_T(\omega)d\omega = 2S_{\mathscr{P}}(0)\int_0^\infty \left| \frac{1}{-i\omega C_V + \mathscr{G}} \right|^2 d\omega \\ &= 2S_{\mathscr{P}}(0)\frac{1}{C_V^2}\int_0^\infty \frac{d\omega}{\omega^2 + (\mathscr{G}/C_V)^2} = \frac{\pi S_{\mathscr{P}}(0)}{\mathscr{G}C_V}, \end{aligned} \tag{5.86}$$

to coincide with our earlier 'thermodynamic fluctuation' result (5.41), we get

$$S_{\mathscr{P}}(0) = \frac{\mathscr{G}}{\pi} T_0^2. \tag{5.87}$$

The rms value of the 'power noise' within a bandwidth $\Delta\nu \ll 1/\tau$ (see figure 5.7) becomes equal to the deterministic signal power $\mathscr{P}_{\text{det}}$ (or more exactly, the main harmonic of its modulation law) at

$$\mathscr{P} = \mathscr{P}_{\min} \equiv \left(\langle \tilde{\mathscr{P}}^2 \rangle_{\Delta\nu} \right)^{1/2} = (2S_{\mathscr{P}}(0)\Delta\omega)^{1/2} = 2(\mathscr{G}\Delta\nu)^{1/2} T_0. \tag{5.88}$$

[37] See, e.g. [4]. In practically used metals, l_e is of the order of 30 nm even at liquid helium temperatures (and even shorter at room temperature), so that the usual 'macroscopic' resistors do not exhibit the shot noise.

[38] For a review of this effect see, e.g. [5].

This result shows that our earlier prediction (5.45) may be improved by a substantial factor of the order of $(\Delta\nu/\nu)^{1/2}$, where the reduction of the output bandwidth is limited only by the signal accumulation time $\Delta t \sim 1/\Delta\nu$, while the increase of ν is limited by the speed of (typically, mechanical) devices performing the power modulation. In practical systems this factor may improve the sensitivity by a couple orders of magnitude, enabling observation of extremely weak radiation. Maybe the most spectacular example is the recent measurements of the CMB radiation, which corresponds to blackbody temperature $T_K \approx 2.726$ K, with an accuracy $\delta T_K \sim 10^{-6}$ K, using microwave receivers with the physical temperature of all their components much higher than δT. The observed weak ($\sim 10^{-5}$ K) anisotropy of the CMB radiation is a major experimental basis of all modern cosmology[39].

Returning to the discussion of our main result, Eq. (5.73), first let me note that it may be readily generalized to the case when the environment's response is different from the Ohmic model (5.64). This opportunity is virtually evident from Eq. (5.66): by its derivation, the second term on its left-hand side is just the Fourier component of the average response of the environment to the system's displacement:

$$\langle \mathscr{F}_\omega \rangle = i\omega\eta\, q_\omega. \tag{5.89}$$

Now let the response be still linear, but have an arbitrary frequency dispersion,

$$\langle \mathscr{F}_\omega \rangle = \chi(\omega) q_\omega, \tag{5.90}$$

where the function $\chi(\omega)$, called the *generalized susceptibility* (in our case, of the environment) may be complex, i.e. have both the imaginary and real parts:

$$\chi(\omega) = \chi'(\omega) + i\chi''(\omega). \tag{5.91}$$

Then Eq. (5.73) remains valid[40] with the replacement $\eta \rightarrow \chi''(\omega)/\omega$:

$$S_{\mathscr{F}}(\omega) = \frac{\chi''(\omega)}{\pi\omega} T. \tag{5.92}$$

This fundamental relation[41] may be used not only to calculate the fluctuation intensity from the known generalized responsibility (i.e. the deterministic response of the system to a small perturbation), but also in reverse—to calculate such a linear response from the known fluctuations. The latter use is especially attractive at numerical simulations of complex systems, e.g. those based on molecular-dynamics

---

[39] See, e.g. a concise book [6].

[40] Reviewing the calculations leading to Eq. (5.73), we may see that the possible real part $\chi'(\omega)$ of the susceptibility just adds up to $(k - m\omega^2)$ in the denominator of Eq. (5.67), resulting in a change of the oscillator's frequency ω_0. This renormalization is insignificant if the oscillator-to-environment coupling is weak, i.e. if the susceptibility $\chi(\omega)$ small—as had been assumed at the derivation of Eq. (5.69) and hence Eq. (5.73).

[41] It is sometimes called the Green–Kubo (or just 'Kubo') formula. This is hardly fair, because, as the reader could see, Eq. (5.92) is just an elementary generalization of the Nyquist formula (5.81). Moreover, the corresponding works of M Green and R Kubo were published, respectively, in 1954 and 1957, i.e. after the 1951 paper by H Callen and T Welton, where a more general result (5.98) had been derived. More adequately, the Green/Kubo names are associated with Eq. (5.102) below.

approaches, because it allows one to avoid extracting a weak response to a small perturbation from a noisy background.

Now let us discuss what generalization of Eq. (5.92) is necessary to make that fundamental result suitable for arbitrary temperatures, $T \sim \hbar\omega$. The calculations we had performed were based on the apparently classical equation of motion, Eq. (5.63). However, quantum mechanics shows[42] that a similar equation is valid for the corresponding Heisenberg-picture operators, so that repeating all the arguments leading to the Langevin equation (5.65), we may write its quantum-mechanical version

$$m\ddot{\hat{q}} + \eta\dot{\hat{q}} + \kappa\hat{q} = \hat{\mathscr{F}}_{\text{det}} + \hat{\tilde{\mathscr{F}}}. \tag{5.93}$$

This is the so-called the *Heisenberg–Langevin* (or 'quantum Langevin') *equation*—in this particular case, for a harmonic oscillator.

The further operations, however, require certain caution, because the right-hand side of the equation is now an operator, and has some nontrivial properties. For example, the 'values' of the Heisenberg operator, representing the same variable $f(t)$ at different times, do not necessarily commute:

$$\hat{\tilde{f}}(t)\hat{\tilde{f}}(t') \neq \hat{\tilde{f}}(t')\hat{\tilde{f}}(t), \quad \text{if } t' \neq t. \tag{5.94}$$

As a result, the function defined by Eq. (5.46) may not be a symmetric function of the time delay $\tau \equiv t' - t$ even for a stationary process, making it inadequate for representation of the real correlation function—which has to obey Eq. (5.50). This technical difficulty may be overcome by the introduction of the following *symmetrized correlation function*[43]

$$K_f(\tau) \equiv \frac{1}{2}\left\langle \hat{\tilde{f}}(t)\hat{\tilde{f}}(t+\tau) + \hat{\tilde{f}}(t+\tau)\hat{\tilde{f}}(t) \right\rangle \equiv \frac{1}{2}\left\langle \left\{ \hat{\tilde{f}}(t), \hat{\tilde{f}}(t+\tau) \right\} \right\rangle, \tag{5.95}$$

(where $\{\ldots,\ldots\}$ denotes the anticommutator of the two operators), and, similarly, the symmetrical spectral density $S_f(\omega)$, defined by the following relation:

$$S_f(\omega)\delta(\omega - \omega') \equiv \frac{1}{2}\left\langle \hat{f}_\omega \hat{f}^*_{\omega'} + \hat{f}^*_{\omega'}\hat{f}_\omega \right\rangle \equiv \frac{1}{2}\left\langle \left\{ \hat{f}_\omega, \hat{f}^*_{\omega'} \right\} \right\rangle, \tag{5.96}$$

with $K_f(\tau)$ and $S_f(\omega)$ still related by the Fourier transform (5.59).

Now we may repeat all the analysis that was carried out for the classical case, and get Eq. (5.71) again, but now this expression has to be compared not with the equipartition theorem, but with its quantum-mechanical generalization (5.14), which, in our current notation, reads

[42] See, e.g. *Part QM* section 4.6.

[43] Here (and to the end of this section) the averaging $\langle\ldots\rangle$ should be understood in the general quantum-statistical sense—see Eq. (2.12). As was discussed in section 2.1, for the classical-mixture state of the system, this does not create any difference in either mathematical treatment of the averages or their physical interpretation.

$$\langle\langle \tilde{q}^2 \rangle\rangle = \frac{\hbar\omega_0}{2\kappa} \coth \frac{\hbar\omega_0}{2T}. \tag{5.97}$$

As a result, we get the following quantum-mechanical generalization of Eq. (5.92):

$$S_{\mathscr{F}}(\omega) = \frac{\hbar\chi''(\omega)}{2\pi} \coth \frac{\hbar\omega}{2T}. \tag{5.98}$$

This is the much-celebrated *fluctuation–dissipation theorem*, usually referred to just as the FDT, first derived in 1951 by H Callen and T Welton—in a somewhat different way.

As natural as it seems, this generalization of the relation between fluctuations and dissipation poses a very interesting conceptual dilemma. Let, for the sake of clarity, temperature be relatively low, $T \ll \hbar\omega$; then Eq. (5.98) gives a temperature-independent result

$$S_{\mathscr{F}}(\omega) = \frac{\hbar\chi''(\omega)}{2\pi}, \tag{5.99}$$

which describes what is frequently called the *quantum noise*. According to the quantum Langevin equation (5.93), nothing but the random force exerted by the environment, with the spectral density (5.99) proportional to the imaginary part of susceptibility (i.e. damping), is the source of the ground-state 'fluctuations' of the coordinate and momentum of a quantum harmonic oscillator, with the rms values

$$\delta q \equiv \langle\langle \tilde{q}^2 \rangle\rangle^{1/2} = \left(\frac{\hbar}{2m\omega_0}\right)^{1/2}, \quad \delta p \equiv \langle\langle \tilde{p}^2 \rangle\rangle^{1/2} = \left(\frac{\hbar m\omega_0}{2}\right)^{1/2}, \tag{5.100}$$

and the total energy $\hbar\omega_0/2$. On the other hand, the basic quantum mechanics tells us that exactly these formulas describe the ground state of a *dissipation-free* oscillator, not coupled to any environment, and are a direct corollary of the basic commutation relation

$$[\hat{q}, \hat{p}] = i\hbar. \tag{5.101}$$

So, what is the genuine source of the uncertainty described by Eqs. (5.100)?

The best resolution of this paradox I can offer is that *either* interpretation of Eqs. (5.100) is legitimate, with their relative convenience depending on the particular application. One may say that since the right-hand side of the quantum Langevin equation (5.93) is a quantum-mechanical operator, rather than a classical force, it 'carries the uncertainty relation within itself'. However, this (admittedly, opportunistic) resolution leaves the following question open: is the quantum noise (5.99) of the environment observable directly, without any probe oscillator subjected to it? An experimental resolution of this dilemma is not quite simple, because usual scientific instruments have their own ground-state uncertainty, i.e. their own quantum noise, which may be readily confused with those of the system under study. Fortunately, this difficulty may be overcome, for example, using unique frequency-mixing ('down-conversion') properties of Josephson junctions. Special low-temperature experiments using such down-conversion[44] have confirmed that the noise (5.99) is real and measurable.

Finally, let me mention an alternative derivation[45] of the fluctuation-theorem (5.98) from the general quantum mechanics of open systems. This derivation is substantially longer than that given above, but gives an interesting sub-product, the *Green–Kubo formula*

$$\left\langle \left[\hat{\tilde{\mathscr{F}}}(t), \hat{\tilde{\mathscr{F}}}(t+\tau) \right] \right\rangle = i\hbar \mathscr{G}(\tau), \tag{5.102}$$

where $\mathscr{G}(\tau)$ is the temporal Green's function of the environment, defined by the following relation:

$$\langle \mathscr{F}(t) \rangle = \int_0^\infty \mathscr{G}(\tau) q(t-\tau) d\tau = \int_{-\infty}^t \mathscr{G}(t-t') q(t') dt'. \tag{5.103}$$

Plugging the Fourier transforms of all three functions of time participating in Eq. (5.103) into this relation, it is straightforward to check that the Green's function is just the Fourier image of the complex susceptibility $\chi(\omega)$ defined by Eq. (5.90):

$$\int_0^\infty \mathscr{G}(\tau) e^{i\omega\tau} d\tau = \chi(\omega); \tag{5.104}$$

here 0 is used as a lower limit instead of $(-\infty)$ just to emphasize that due to the causality principle, the Green's function has to be equal zero for $\tau < 0$.[46]

In order to reveal the real beauty of Eq. (5.102), we may use the Wiener–Khinchin theorem (5.59) to rewrite the fluctuation–dissipation theorem (5.98) in a similar time-domain form:

$$\left\langle \left\{ \hat{\tilde{\mathscr{F}}}(t), \hat{\tilde{\mathscr{F}}}(t+\tau) \right\} \right\rangle = 2K_{\mathscr{F}}(\tau), \tag{5.105}$$

where the symmetrized correlation function $K_{\mathscr{F}}(\tau)$ is most simply described by its Fourier transform, which is, according to Eq. (5.58), equal to $\pi S_{\mathscr{F}}(\omega)$, so that using the FDT, we get

$$\int_0^\infty K_{\mathscr{F}}(\tau) \cos \omega\tau \, d\tau = \frac{\hbar \chi''(\omega)}{2} \coth \frac{\hbar\omega}{2T}. \tag{5.106}$$

The comparison of Eqs. (5.102) and (5.104) on one hand, and Eqs. (5.105) and (5.106) on the other hand, shows that both the commutation and anticommutation properties of the Heisenberg–Langevin force operator at different moments of time are determined by the same generalized susceptibility $\chi(\omega)$ of the environment. However, the averaged anticommutator also depends on temperature, while the

---

[44] See [7] and references therein.

[45] See, e.g. *Part QM* section 7.4.

[46] See, e.g. *Part CM* section 5.1.

averaged commutator does not—at least explicitly, because the complex susceptibility of the environment may be temperature-dependent as well.

5.6 The Kramers problem and the Smoluchowski equation

Returning to the classical case, it is evident that Langevin equations of the type (5.65) provide means not only for the analysis of stationary fluctuations, but also for the description of an arbitrary time evolution of (classical) dynamic systems coupled to their environment—which, again, may provide both dissipation and fluctuations. However, this approach to evolution analysis suffers from two major handicaps.

First, the Langevin equation does enable a straightforward calculation of the statistical average of the variable q, and its fluctuation variance (i.e. in the common mathematical terminology, the first and second *moments* of the probability distribution) as functions of time, but not of the distribution $w(q, t)$ as such. Admittedly, this is rarely a big problem, because in most cases the distribution is Gaussian—see, e.g. Eq. (2.77).

The second, more painful, drawback of the Langevin approach is that it is instrumental only for the 'linear' systems—i.e. the systems whose dynamics is described by linear differential equations, such as Eq. (5.65). However, as we know from classical dynamics, many important problems (for example, the Kepler problem of planetary motion[47]) are reduced to motion in substantially anharmonic potentials $U_{ef}(q)$, leading to nonlinear equations of motion. If the energy of interaction between the system and its random environment is factorable—i.e. is a product of variables belonging to these subsystems (as is very frequently the case), we may repeat all arguments of the last section to derive the following generalized version of the Langevin equation:

$$m\ddot{q} + \eta\dot{q} + \frac{\partial U(q, t)}{\partial q} = \tilde{\mathscr{F}}(t), \tag{5.107}$$

valid for an arbitrary, possibly time-dependent potential $U(q, t)$.[48] Unfortunately, the solution of this equation may be very hard. Indeed, the Fourier analysis carried out in the last section was essentially based on the linear superposition principle, which is invalid for nonlinear equations.

If the fluctuation intensity is low, $|\delta q| \ll \langle q \rangle$, where $\langle q \rangle(t)$ is the deterministic solution of Eq. (5.107) in the absence of fluctuations, this equation may be linearized[49] with respect to small fluctuations $\tilde{q} \equiv q - \langle q \rangle$ to get a linear equation,

$$m\ddot{\tilde{q}} + \eta\dot{\tilde{q}} + \kappa(t)\tilde{q} = \tilde{\mathscr{F}}(t), \quad \text{with} \quad \kappa(t) \equiv \frac{\partial^2}{\partial q^2}U(\langle q \rangle(t), t). \tag{5.108}$$

---

[47] See, e.g. *Part CM* sections 3.4–3.6.

[48] The generalization of Eq. (5.107) to higher spatial dimensionality is also straightforward, with the scalar variable q replaced with a multi-dimensional vector $\mathbf{q}$, and the scalar derivative dU/dq replaced with the vector ∇U, where ∇ is the del vector-operator in the $\mathbf{q}$-space.

[49] See, e.g. *Part CM* sections 3.2, 5.2, and beyond.

This equation differs from Eq. (5.65) only by the time dependence of the effective spring constant $\kappa(t)$, and may be solved by the Fourier expansion of both the fluctuations and the function $\kappa(t)$. Such calculations may be more cumbersome than have been performed above, but still be doable (especially if the unperturbed motion $\langle q\rangle(t)$ is periodic), and sometimes give useful analytical results[50].

However, some important problems cannot be solved by the linearization. Perhaps, the most apparent (and practically very important) example is the so-called *Kramers problem*[51] of finding the lifetime of a metastable state of a 1D classical system in a potential well separated from the unlimited motion region with a potential barrier—see figure 5.10.

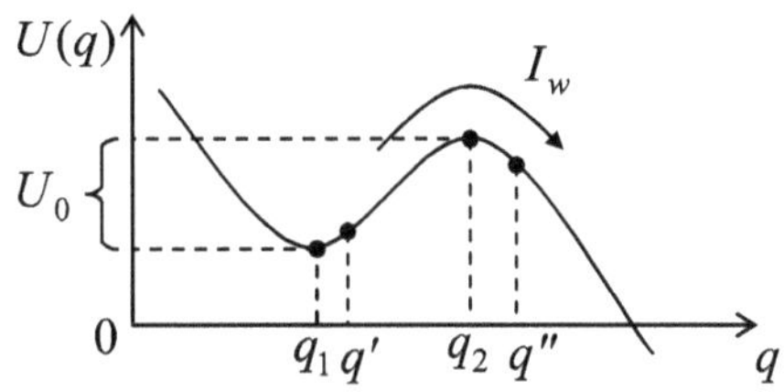

Figure 5.10. The Kramers problem.

In the absence of fluctuations, the system, initially placed close to the well's bottom (in figure 5.10, at $q \approx q_1$), would stay there forever. Fluctuations result not only in a finite spread of the probability density $w(q, t)$ around that point, but also in a gradual decrease of the total probability

$$W(t) = \int_{\substack{\text{well's}\\ \text{bottom}}} w(q, t)dq \tag{5.109}$$

to find the system in the well, because of the growing probability of its escape from it, over the potential barrier, due to thermal activation. What may be immediately expected of the situation is that if the barrier height,

$$U_0 \equiv U(q_2) - U(q_1), \tag{5.110}$$

is much larger than temperature T,[52] the Boltzmann distribution $w \propto \exp\{-U(q)/T\}$ should be still approximately valid in most of the well, so that the probability for the system to overcome the barrier should scale as $\exp\{-U_0/T\}$. From these handwaving arguments, one may reasonably expect that if the probability $W(t)$ that the system is still in the well by time t obeys the usual 'decay law'

$$\dot{W} = -\frac{W}{\tau}, \tag{5.111a}$$

---

[50] See, e.g. *Part QM* problem 7.8, and also chapters 5 and 6 in the monograph by W Coffey *et al*, cited above.

[51] It was named after H Kramers who, besides solving this important problem in 1940, has made several other significant contributions to physics, including the famous Kramers–Kronig dispersion relations (see, e.g. *Part EM* section 7.4) and the WKB approximation in quantum mechanics (see, e.g. *Part QM* section 2.4).

[52] If U_0 is comparable with T, system's behavior also depends substantially on the initial probability distribution, i.e. does not follow the simple law (5.111).

then the lifetime τ has to obey the general Arrhenius law:

$$\tau = \tau_A \exp\left\{\frac{U_0}{T}\right\}. \tag{5.111b}$$

However, these relations need to be proved, and the pre-exponential coefficient τ_A (usually called the *attempt time*) needs to be calculated. This cannot be done by the linearization of Eq. (5.107), because this approximation is equivalent to a quadratic approximation of the potential $U(q)$, which evidently cannot describe the potential well *and* the potential barrier simultaneously—see figure 5.10.

This and other essentially nonlinear problems may be addressed using an alternative approach to fluctuations, dealing directly with the time evolution of the probability density $w(q, t)$. Due to the shortage of time/space, I will review this approach using mostly handwaving arguments, and refer the interested reader to special literature[53] for strict mathematical proofs. Let us start from the diffusion of a free classical 1D particle with inertial effects negligible in comparison with damping. It is described by the Langevin equation (5.74) with $\mathscr{F}_{\text{det}} = 0$. Let us assume that at all times the probability distribution stays Gaussian:

$$w(q, t) = \frac{1}{(2\pi)^{1/2}\delta q(t)} \exp\left\{-\frac{(q - q_0)^2}{2\delta q^2(t)}\right\}, \tag{5.112}$$

where q_0 is the initial position of the particle, and $\delta q(t)$ is the time-dependent distribution width, whose growth in time is described, as we already know, by Eq. (5.77):

$$\delta q(t) = (2Dt)^{1/2}. \tag{5.113}$$

Then it is straightforward to verify, by substitution, that this solution satisfies the following simple partial differential equation[54],

$$\frac{\partial w}{\partial t} = D\frac{\partial^2 w}{\partial q^2}, \tag{5.114}$$

with the delta-functional initial condition

$$w(q, 0) = \delta(q - q_0). \tag{5.115}$$

The simple and important *equation of diffusion* (5.114) may be naturally generalized to the 3D motion[55]:

$$\frac{\partial w}{\partial t} = D\nabla^2 w. \tag{5.116}$$

[53] See, e.g. either [8], or chapter 1 in the monograph by W Coffey *et al*, cited above.

[54] By the way, the goal of the traditional definition (5.78) of the diffusion coefficient, leading to the coefficient 2 in Eq. (5.77), is exactly to have the fundamental equation (5.114) free of numerical coefficients.

[55] As will be discussed in chapter 6, the equation of diffusion also describes several other physical phenomena—in particular, the heat propagation in a uniform, isotropic solid, and in this context is called the *heat conduction equation* or (rather inappropriately) just the 'heat equation'.

Now let us compare this equation with the probability conservation law[56],

$$\frac{\partial w}{\partial t} + \nabla \cdot \mathbf{j}_w = 0, \tag{5.117a}$$

where the vector $\mathbf{j}_w$ has the physical sense of the probability current density. (The validity of this relation is evident from its integral form,

$$\frac{d}{dt}\int_V w d^3r + \oint_S \mathbf{j}_w \cdot d^2\mathbf{q} = 0, \tag{5.117b}$$

which results from the integration of Eq. (5.117*a*) over an arbitrary time-independent volume V limited by surface S, and applying the divergence theorem[57] to the second term.) The continuity relation (5.117*a*) coincides with Eq. (5.116), with D given by Eq. (5.78), only if we take

$$\mathbf{j}_w = -D\nabla w = -\frac{T}{\eta}\nabla w. \tag{5.118}$$

The first form of this relation allows a simple interpretation: the probability flow is proportional to the spatial gradient of the probability density (i.e. in application to $N \gg 1$ similar and independent particles, just to the gradient of their concentration $n = Nw$), with the sign corresponding to the flow from the higher to lower concentrations. This flow is the very essence of the effect of diffusion. The second form of Eq. (5.118) is also not very surprising: the diffusion speed scales as temperature, and is inversely proportional to the viscous drag.

The fundamental Eq. (5.117) has to be satisfied also in the case of a force-driven particle at negligible diffusion ($D \to 0$); in this case

$$\mathbf{j}_w = w\mathbf{v}, \tag{5.119}$$

where $\mathbf{v}$ is the deterministic velocity of the particle. In the high-damping limit we are considering right now, $\mathbf{v}$ has to be just the *drift velocity*:

$$\mathbf{v} = \frac{1}{\eta}\mathscr{F}_{\text{det}} = -\frac{1}{\eta}\nabla U(\mathbf{q}), \tag{5.120}$$

where $\mathscr{F}_{\text{det}}$ is the deterministic force described by the potential energy $U(\mathbf{q})$.

Now that we have descriptions of $\mathbf{j}_w$ due to both the drift *and* the diffusion separately, we may rationally assume that in the general case when both effects are present, the corresponding components (5.118) and (5.119) of the probability current just add up, so that

$$\mathbf{j}_w = \frac{1}{\eta}[w(-\nabla U) - T\nabla w], \tag{5.121}$$

[56] Both forms of Eq. (5.117) are similar to the mass conservation law in classical dynamics (see, e.g. *Part CM* section 8.2), the electric charge conservation law in electrodynamics (see, e.g. *Part EM* section 4.1), and the probability conservation law in quantum mechanics (see, e.g. *Part QM* section 1.4).

[57] See, e.g. Eq. (A.78),

so that Eq. (5.117*a*) takes the form

$$\eta\frac{\partial w}{\partial t} = \nabla(w\nabla U) + T\nabla^2 w. \tag{5.122}$$

This is the *Smoluchowski equation*[58], which is closely related to the *drift–diffusion equation* in multi-particle kinetics—to be discussed in the next chapter.

As a sanity check, let us see what the Smoluchowski equation gives in the stationary limit, $\partial w/\partial t \to 0$ (which evidently may be eventually achieved only if the deterministic potential U is time-independent). Then Eq. (5.117*a*) yields $\mathbf{j}_w = \text{const}$, where the constant describes the deterministic motion of the system as the whole. If such a motion is absent, $\mathbf{j}_w = 0$, then according to Eq. (5.121),

$$w\nabla U + T\nabla w = 0, \quad \text{i.e.} \quad \frac{\nabla w}{w} = -\frac{\nabla U}{T}. \tag{5.123}$$

Since the left-hand side of the last relation is just $\nabla(\ln w)$, it may be easily integrated over $\mathbf{q}$, giving

$$\ln w = -\frac{U}{T} + \ln C, \quad \text{i.e. } w(\mathbf{r}) = C\exp\left\{-\frac{U(\mathbf{q})}{T}\right\}, \tag{5.124}$$

where C is a normalization constant. With both sides multiplied by the number N of similar, independent systems, with the spatial density $n(\mathbf{q}) = Nw(\mathbf{q})$, this equality becomes the Boltzmann distribution (3.26).

Next, as a less trivial example of the Smoluchowski equation's applications, let us use it to solve the 1D Kramers problem (figure 5.10) in the corresponding high-damping limit, $m \ll \eta\tau_A$, where τ_A (still to be calculated) is some time scale of the particle's motion inside the well. It is straightforward to verify that the 1D version of Eq. (5.121),

$$I_w = \frac{1}{\eta}\left[w\left(-\frac{\partial U}{\partial q}\right) - T\frac{\partial w}{\partial q}\right], \tag{5.125a}$$

(where I_w is the probability current at a certain point q, rather than its density) is mathematically equivalent to

$$I_w = -\frac{T}{\eta}\exp\left\{-\frac{U(q)}{T}\right\}\frac{\partial}{\partial q}\left(w\exp\left\{\frac{U(q)}{T}\right\}\right), \tag{5.125b}$$

so that we may write

$$I_w\exp\left\{\frac{U(q)}{T}\right\} = -\frac{T}{\eta}\frac{\partial}{\partial q}\left(w\exp\left\{\frac{U(q)}{T}\right\}\right). \tag{5.126}$$

[58] It is named after M Smoluchowski, who developed this formalism in 1906, apparently independently from the slightly earlier Einstein's work, but in much more detail. This equation has important applications in many fields of science—including such surprising topics as statistics of spikes in neural networks. (Note, however, that in some non-physical fields, Eq. (5.122) is referred to as the *Fokker–Planck equation*, while actually the latter equation is much more general—see the next section.)

As was discussed above, the notion of metastable state's lifetime is well defined only for sufficiently low temperatures

$$T \ll U_0. \tag{5.127}$$

when the lifetime is relatively long, $\tau \gg \tau_A$. Since according to Eq. (5.111*a*), the first term of the continuity equation (5.117*b*) has to be of the order of W/τ, in this limit the term, and hence the gradient of I_w, are exponentially small, so the probability current virtually does not depend on q in the potential barrier region. Let us use this fact at the integration of both sides of Eq. (5.126) over that region:

$$I_w \int_{q'}^{q''} \exp\left\{\frac{U(q)}{T}\right\} dq = -\frac{T}{\eta}\left(w \exp\left\{\frac{U(q)}{T}\right\}\right)\Bigg|_{q'}^{q''}, \tag{5.128}$$

where the integration limits q' and q'' (see figure 5.10) are selected so that so that

$$T \ll U(q') - U(q_1),\ U(q_2) - U(q'') \ll U_0. \tag{5.129}$$

(Evidently, such selection is only possible if the condition (5.127) is satisfied.) In this limit the contribution from the point q'' to the right-hand side of Eq. (5.129) is negligible, because the probability density behind the barrier is exponentially small. On the other hand, the probability at the point q' is close to the value given by its quasi-stationary Boltzmann distribution (5.124), so that

$$w(q')\exp\left\{\frac{U(q')}{T}\right\} = w(q_1)\exp\left\{\frac{U(q_1)}{T}\right\}, \tag{5.130}$$

and Eq. (5.128) yields

$$I_w = \frac{T}{\eta} w(q_1) \Big/ \int_{q'}^{q''} \exp\left\{\frac{U(q) - U(q_1)}{T}\right\} dq. \tag{5.131}$$

Patience, my reader, we are almost done. The probability density $w(q_1)$ at the well's bottom may be expressed in terms of the total probability W of the particle being in the well by using the normalization condition

$$W = \int_{\substack{\text{well's}\\ \text{bottom}}} w(q_1)\exp\left\{\frac{U(q_1) - U(q)}{T}\right\} dq; \tag{5.132}$$

the integration here may be limited to the region where the difference $U(q) - U(q_1)$ is much larger then T, but still much smaller than U_0—cf. Eq. (5.129). According to the Taylor expansion, the shape of virtually any smooth potential $U(q)$ near the point q_1 of its minimum may be well approximated with a quadratic parabola:

$$U(q \approx q_1) - U(q_1) \approx \frac{\kappa_1}{2}(q - q_1)^2, \quad \text{where } \kappa_1 \equiv \left.\frac{d^2U}{dq^2}\right|_{q=q_1} > 0. \tag{5.133}$$

With this approximation, Eq. (5.132) is reduced to the standard Gaussian integral[59]:

$$\begin{aligned} W &= w(q_1)\int_{\substack{\text{well's}\\ \text{bottom}}} \exp\left\{-\frac{\kappa_1(q-q_1)^2}{2T}\right\}dq \approx w(q_1)\int_{-\infty}^{+\infty} \exp\left\{\frac{\kappa_1\tilde{q}^2}{2T}\right\}d\tilde{q} \\ &= w(q_1)\left(\frac{2\pi T}{\kappa_1}\right)^{1/2}. \end{aligned} \quad (5.134)$$

To complete the calculation, we may use the similar approximation,

$$\begin{aligned} U(q\approx q_2) - U(q_1) &\approx \left[U(q_2) - \frac{\kappa_2}{2}(q-q_2)^2\right] - U(q_1) = U_0 - \frac{\kappa_2}{2}(q-q_2)^2, \\ &\text{where } \kappa_2 \equiv -\frac{d^2U}{dq^2}\bigg|_{q=q_2} > 0, \end{aligned} \quad (5.135)$$

to work out the remaining integral in Eq. (5.131), because in the limit (5.129) this integral is dominated by the contribution from a region very close to the barrier top, where the approximation (5.135) is asymptotically exact. As a result, we get

$$\int_{q'}^{q''} \exp\left\{\frac{U(q)-U(q_1)}{T}\right\}dq \approx \exp\left\{\frac{U_0}{T}\right\}\left(\frac{2\pi T}{\kappa_2}\right)^{1/2}. \quad (5.136)$$

Plugging Eq. (5.136), and the $w(q_1)$ expressed from Eq. (5.134), into Eq. (5.131), we finally get

$$I_w = W\frac{(\kappa_1\kappa_2)^{1/2}}{2\pi\eta}\exp\left\{-\frac{U_0}{T}\right\}. \quad (5.137)$$

This expression should be compared with the 1D version of Eq. (5.117*b*) for the segment $[-\infty, q']$. Since this interval covers the region near q_1 where most of the probability density resides, and $I_q(-\infty) = 0$, this equation is merely

$$\frac{dW}{dt} + I_w(q') = 0. \quad (5.138)$$

In our approximation, $I_w(q')$ does not depend on the exact position of the point q', and is given by Eq. (5.137), so that plugging it into Eq. (5.138), we recover the exponential decay law (5.111*a*), with the lifetime τ obeying the Arrhenius law (5.111*b*), and the following attempt time:

$$\tau_A = \frac{2\pi\eta}{(\kappa_1\kappa_2)^{1/2}} \equiv 2\pi(\tau_1\tau_2)^{1/2}, \quad \text{where} \quad \tau_{1,2} \equiv \frac{\eta}{\kappa_{1,2}}. \quad (5.139)$$

Thus the metastable state lifetime is indeed described by the Arrhenius law, with the attempt time scaling as the geometric mean of system's 'relaxation times' near

[59] If necessary, see Eq. (A.36*b*) again.

the potential well bottom (τ_1) and the potential barrier top (τ_2).[60] Let me leave for the reader's exercise to prove that if the potential profile near the well's bottom and/or top is sharp, the attempt time should be modified, but the Arrhenius decay law (5.111) is not affected.

5.7 The Fokker–Planck equation

Eq. (5.139) is just a particular, high-damping limit of a more general result obtained by Kramers. In order to recover all of it, we need to generalize the Smoluchowski equation to arbitrary values of the damping η. In this case, the probability density w is a function of not only the particle's position $\mathbf{q}$ (and time t), but also its momentum $\mathbf{p}$—see Eq. (2.11). Thus the continuity equation (5.117a) needs to be generalized to the 6D phase space $\{\mathbf{q}, \mathbf{p}\}$. Such generalization is natural:

$$\frac{\partial w}{\partial t} + \nabla_q \cdot \mathbf{j}_q + \nabla_p \cdot \mathbf{j}_p = 0, \tag{5.140}$$

where $\mathbf{j}_q$ (which was called $\mathbf{j}_w$ in the last section) is the probability current density in the coordinate space, and ∇_q (which was denoted as ∇ in that section) is the usual gradient operator in that space, while $\mathbf{j}_p$ is the current density in the momentum space, and ∇_p is the similar gradient operator in that space:

$$\nabla_q \equiv \sum_{j=1}^{3} \mathbf{n}_j \frac{\partial}{\partial q_j}, \quad \nabla_p \equiv \sum_{j=1}^{3} \mathbf{n}_j \frac{\partial}{\partial p_j}. \tag{5.141}$$

At negligible fluctuations ($T \to 0$), $\mathbf{j}_p$ may be evaluated using the natural analogy with $\mathbf{j}_q$—see Eq. (5.119). In our new notation, that relation takes the following form,

$$\mathbf{j}_q = w\dot{\mathbf{q}} = w\frac{\mathbf{p}}{m}, \tag{5.142}$$

so it is natural to take

$$\mathbf{j}_p = w\dot{\mathbf{p}} = w\langle \mathscr{F} \rangle, \tag{5.143a}$$

where the (statistical-ensemble) averaged force $\langle \mathscr{F} \rangle$ includes not only the contribution due to potential's gradient, but also the drag force $-\eta\mathbf{v}$ provided by the environment—see Eq. (5.64) and its discussion:

$$\mathbf{j}_p = w\left(-\nabla_q U - \eta\mathbf{v}\right) = -w\left(\nabla_q U + \eta\frac{\mathbf{p}}{m}\right). \tag{5.143b}$$

As a sanity check, it is straightforward to verify that the diffusion-free equation resulting from the combination of Eqs. (5.140), (5.142) and (5.143),

[60] Actually, τ_2 describes the characteristic time of the exponential *growth* of small deviations from the unstable fixed point q_2 at the barrier top, rather than their *decay*, as near the stable point q_1.

$$\left.\frac{\partial w}{\partial t}\right|_{\text{drift}} = -\nabla_q \cdot \left(w\frac{\mathbf{p}}{m}\right) + \nabla_p \cdot \left[w\left(\nabla_q U + \eta\frac{\mathbf{p}}{m}\right)\right], \tag{5.144}$$

allows the following particular solution:

$$w(\mathbf{q}, \mathbf{p}, t) = \delta[\mathbf{q} - \langle\mathbf{q}\rangle(t)]\ \delta[\mathbf{p} - \langle\mathbf{p}\rangle(t)], \tag{5.145}$$

where the statistical-averaged coordinate and momentum satisfy the deterministic equations of motion,

$$\langle\dot{\mathbf{q}}\rangle = \frac{\langle\mathbf{p}\rangle}{m}, \quad \langle\dot{\mathbf{p}}\rangle = -\nabla_q U - \eta\frac{\langle\mathbf{p}\rangle}{m}, \tag{5.146}$$

describing the particle's drift, with the usual deterministic initial conditions.

In order to understand how the diffusion should be accounted for, let us consider a statistical ensemble of free ($\nabla_q U = 0$, $\eta \to 0$) particles that are uniformly distributed in the direct space $\mathbf{q}$ (so that $\nabla_q w = 0$), but possibly localized in the momentum space. For this case, the right-hand side of Eq. (5.144) vanishes, i.e. the time evolution of the probability density w may be only due to diffusion. In the corresponding limit $\langle\mathscr{F}\rangle \to 0$, the Langevin equation (5.107) for each Cartesian coordinate is reduced to

$$m\ddot{q}_j = \tilde{\mathscr{F}}_j(t), \quad \text{i.e.} \quad \dot{p}_j = \tilde{\mathscr{F}}_j(t). \tag{5.147}$$

The last equation is similar to the high-damping 1D equation (5.74) (with $\mathscr{F}_{\text{det}} = 0$), with the replacement $q \to p_j/\eta$, and hence the corresponding contribution to $\partial w/\partial t$ may be described by the last term of equation (5.122), with that replacement:

$$\left.\frac{\partial w}{\partial t}\right|_{\text{diffusion}} = D\nabla^2_{p/\eta} w = \frac{T}{\eta}\eta^2\nabla^2_p w \equiv \eta T\ \nabla^2_p w. \tag{5.148}$$

Now the reasonable assumption that in the arbitrary case the drift and diffusion contributions to $\partial w/\partial t$ just add up, immediately leads us to the full *Fokker–Planck equation*[61]:

$$\frac{\partial w}{\partial t} = -\nabla_q \cdot \left(w\frac{\mathbf{p}}{m}\right) + \nabla_p \cdot \left[w\left(\nabla_q U + \eta\frac{\mathbf{p}}{m}\right)\right] + \eta T\ \nabla^2_p w. \tag{5.149}$$

As a sanity check, let us use this equation to calculate the stationary probability distribution of the momentum of particles with an arbitrary damping η, but otherwise free, in the momentum space, assuming (just for simplicity) their uniform distribution in the direct space, $\nabla_q = 0$. In this case, Eq. (5.149) is reduced to

$$\nabla_p \cdot \left[w\left(\eta\frac{\mathbf{p}}{m}\right)\right] + \eta T\nabla^2_p w = 0, \quad \text{i.e.} \quad \nabla_p \cdot \left(\frac{\mathbf{p}}{m}w + T\nabla_p w\right) = 0. \tag{5.150}$$

[61] It was derived in 1913 in A Fokker's PhD thesis; M Planck was his thesis adviser.

The first integration over the momentum space yields

$$\frac{\mathbf{p}}{m}w + T\nabla_p w = \mathbf{j}_w, \quad \text{or} \quad w\nabla_p\left(\frac{p^2}{2m}\right) + T\nabla_p w = \mathbf{j}_w, \tag{5.151}$$

where $\mathbf{j}_w$ is a vector constant describing a possible general probability flow in the system. In the absence of such flow, $\mathbf{j}_w = 0$, we get

$$\nabla_p\left(\frac{p^2}{2m}\right) + T\frac{\nabla_p w}{w} \equiv \nabla_p\left(\frac{p^2}{2m} + T\ln w\right) = 0, \quad \text{giving}$$
$$w = \text{const} \times \exp\left\{-\frac{p^2}{2mT}\right\}, \tag{5.152}$$

i.e. the Maxwell distribution (3.5). However, the result (5.152) is more general than that obtained in section 3.1, because it shows that the distribution stays the same even at a nonvanishing damping. It is easy to verify that in the more general case of an arbitrary stationary potential $U(\mathbf{q})$, Eq. (5.149) is satisfied with the stationary solution (3.24), also giving $\mathbf{j}_w = 0$.

It is also straightforward to show that if the damping is large (in the sense assumed in the last section), the solution of the Fokker–Planck equation tends to the following product

$$w(\mathbf{q}, \mathbf{p}, t) \to \text{const} \times \exp\left\{-\frac{p^2}{2mT}\right\} \times \mathcal{w}(\mathbf{q}, t), \tag{5.153}$$

where the direct-space distribution $\mathcal{w}(\mathbf{q}, t)$ obeys the Smoluchowski equation (5.122).

Another important particular case is that of the quasi-periodic motion of a particle, with low damping, in a soft potential well. In this case, the Fokker–Planck equation describes both the diffusion of the effective phase Θ of such (generally nonlinear, 'anharmonic') oscillator, and the slow relaxation of its energy. If we are only interested in the latter process, Eq. (5.149) may be reduced to the so-called *energy diffusion equation*[62], which is easier to solve.

However, in most cases, solutions of Eq. (5.149) are rather complicated. (Indeed, the reader should remember that these solutions embody, in the particular case $T = 0$, all classical dynamics of a particle.) Because of this, I will present (rather than derive) only one more of them: the solution of the Kramers problem (figure 5.10). Acting almost exactly as in section 5.6, one can show[63] that at virtually arbitrary damping (but still in the limit $T \ll U_0$), the metastable state's lifetime is again given by the Arrhenius formula (5.111*b*), with the attempt time again expressed by the first of Eqs. (5.139), but with the reciprocal time constants $1/\tau_{1,2}$ replaced with

[62] An example of such an equation, for the particular case of a harmonic oscillator, in given by *Part QM* Eq. (7.214). The Fokker–Planck equation, of course, can give only its classical limit, with $n, n_e \gg 1$.

[63] A detailed description of this calculation (first performed by H Kramers in 1940) may be found in section III.7 of the review paper by S Chandrasekhar [9].

$$\omega'_{1,2} \equiv \left[\omega_{1,2}^2 + \left(\frac{\eta}{2m}\right)^2\right]^{1/2} - \frac{\eta}{2m} \to \begin{cases} \omega_{1,2}, & \text{for } \eta \ll m\omega_{1,2}, \\ 1/\tau_{1,2}, & \text{for } m\omega_{1,2} \ll \eta, \end{cases} \tag{5.154}$$

where $\omega_{1,2} \equiv (\kappa_{1,2}/m)^{1/2}$, and $\kappa_{1,2}$ are the effective spring constants defined by Eqs. (5.133) and (5.135). Thus, in the important particular limit of low damping, Eqs. (5.111*b*) and (5.154) give the famous formula

$$\tau = \frac{2\pi}{(\omega_1\omega_2)^{1/2}} \exp\left\{\frac{U_0}{T}\right\}. \tag{5.155}$$

This Kramers' result for the classical thermal activation of the essentially-Hamiltonian system *over* the potential barrier may be compared with that for its quantum-mechanical tunneling *through* the barrier[64]. Even the crudest, WKB approximation for the latter effect gives the expression

$$\tau_Q = \tau_A \exp\left\{-2\int_{\kappa^2(q)>0} \kappa(q)dq\right\}, \quad \text{with} \quad \frac{\hbar^2\kappa^2(q)}{2m} \equiv U(q) - E, \tag{5.156}$$

showing that generally those two lifetimes have different dependences on the barrier shape. For example, for a nearly-rectangular potential barrier, the exponent that determines the classical lifetime (5.155) depends (linearly) only on the barrier *height* U_0, while that defining the quantum lifetime (5.156) is proportional to the barrier *width*, while scaling as a square root of U_0. However, in the case of 'soft' potential profiles, which are typical for the case of barely emerging (or nearly disappearing) quantum wells (figure 5.11) the classical and quantum results may be simply related.

Indeed, such potential profile $U(q)$ may be well approximated by four leading terms of its Taylor expansion, with the highest term proportional to $(q - q_0)^3$, near some point q_0 in the vicinity of the well. In this approximation, the second derivative d^2U/dq^2 vanishes at the inflection point $q_0 = (q_1 + q_2)/2$, exactly between the well's bottom and the barrier's top (in figure 5.11, q_1 and q_2). Selecting the origin at this point, as is done in figure 5.11, we may reduce the approximation to just two terms[65]:

$$U(q) = aq - \frac{b}{3}q^3. \tag{5.157}$$

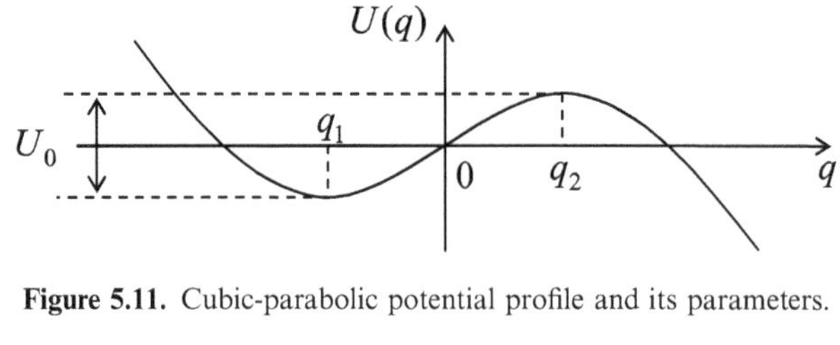

Figure 5.11. Cubic-parabolic potential profile and its parameters.

[64] See, e.g. *Part QM* sections 2.4–2.6.

[65] As a reminder, an absolutely similar approximation arises for the $P(V)$ function, at the analysis of the van der Waals model near the critical temperature in the solution of problem 4.6.

(For the particle's escape into the positive direction of the q-axis, we should have $a, b > 0$.) A straightforward differentiation gives all essential parameters of this cubic parabola: the positions of its minimum and maximum:

$$q_2 = -q_1 = (a/b)^{1/2}, \tag{5.158}$$

the barrier height over the well's bottom:

$$U_0 \equiv U(q_2) - U(q_1) = \frac{4}{3}\left(\frac{a^3}{b}\right)^{1/2}, \tag{5.159}$$

and the effective spring constants at these points:

$$\kappa_1 = \kappa_2 \equiv \left|\frac{d^2U}{dq^2}\right|_{q_{1,2}} = 2(ab)^{1/2}. \tag{5.160}$$

The last expression shows that for this potential profile, the frequencies $\omega_{1,2}$ participating in Eq. (5.155) are equal to each other, so that this result may be rewritten as

$$\tau = \frac{2\pi}{\omega_0}\exp\left\{\frac{U_0}{T}\right\}, \quad \text{with} \quad \omega_0^2 \equiv \frac{2(ab)^{1/2}}{m}. \tag{5.161}$$

On the other hand, for the same profile, the WKB approximation (5.156) (which is accurate when the height of the metastable state energy over the well's bottom, $E - U(q_1) \approx \hbar\omega_0/2$, is much lower than the barrier height U_0) yields[66]

$$\tau_Q = \frac{2\pi}{\omega_0}\left(\frac{\hbar\omega_0}{864\ U_0}\right)^{1/2}\exp\left\{\frac{36}{5}\frac{U_0}{\hbar\omega_0}\right\}. \tag{5.162}$$

The comparison of the dominating, exponential factors in these two results shows that the thermal activation yields a lower lifetime (i.e. dominates the metastable state decay) if temperature is above the crossover value

$$T_c = \frac{36}{5}\ \hbar\omega_0 \equiv 7.2\ \hbar\omega_0. \tag{5.163}$$

This expression for the *cubic*-parabolic barrier may be compared with the similar crossover for a *quadratic*-parabolic barrier[67], for which $T_c = 2\pi\ \hbar\omega_0 \approx 6.28\ \hbar\omega_0$. We see that the numerical factors for the quantum-to-classical crossover temperature for these two different soft potential profiles are close to each other—and much larger than 1, which could result from a naïve estimate.

[66] The main, exponential factor in this result may be obtained simply by ignoring the difference between E and $U(q_1)$, but the correct calculation of the pre-exponential factor requires taking this difference, $\hbar\omega_0/2$, into account—see, e.g. the solution of *Part QM* problem 2.43.

[67] See, e.g. *Part QM* section 2.4.

5.8 Back to the correlation function

Unfortunately, I will not have time to review solutions of other problems using the Smoluchowski and Fokker–Planck equations, but have to mention one conceptual issue. Since it is intuitively clear that these equations provide the complete statistical information about the system under analysis, one may wonder whether they may be used to find the temporal characteristics of the system that were discussed in sections 5.4–5.5 using the Langevin formalism. For any statistical average of a function taken at the same time instant, the answer is evidently *yes*—cf. Eq. (2.11):

$$\langle f[\mathbf{q}(t), \mathbf{p}(t)]\rangle = \int f(\mathbf{q}, \mathbf{p})w(\mathbf{q}, \mathbf{p}, t)d^3q d^3p, \tag{5.164}$$

but what if the function depends on variables taken at *different* times, for example as in the correlation function $K_f(\tau)$ defined by Eq. (5.48)?

To answer this question, let us start from the discrete-variable case, when Eq. (5.164) takes the form (2.7), which, for our current purposes, may be rewritten as

$$\langle f(t)\rangle = \sum_m f_m W_m(t). \tag{5.165}$$

In plain English, this is a sum of all possible values of the function, each multiplied by its probability as a function of time. But this means that the average $\langle f(t)f(t')\rangle$ may be calculated as the sum of all possible products $f_m f_{m'}$, multiplied by the *joint probability* for measurement outcome m at moment t, *and* outcome m' at moment t'. The joint probability may be represented as a product of $W_m(t)$ by the *conditional probability* $W(m', t'| m, t)$. Since the correlation function is well defined only for stationary systems, in the last expression we can take $t = 0$, i.e. find the conditional probability as the solution, $W_{m'}(\tau)$, of the equation describing the system's probability evolution, at time $\tau = t' - t$ (rather than t'), with the special initial condition

$$W_{m'}(0) = \delta_{m',m}. \tag{5.166}$$

On the other hand, since the average $\langle f(t)f(t+\tau)\rangle$ of a stationary process should not depend on t, instead of $W_m(t)$ we may take the stationary probability distribution $W_m(\infty)$, independent of the initial conditions, and may be found as the same special solution, but at time $\tau \to \infty$. As a result, we may write

$$\langle f(t)f(t+\tau)\rangle = \sum_{m,m'} f_m W_m(\infty) f_{m'} W_{m'}(\tau). \tag{5.167}$$

This expression looks simple, but note that this recipe requires one to solve the time evolution equations for each $W_{m'}(\tau)$ for all possible initial conditions (5.166). To see how this recipe works in practice, let us revisit the simplest two-level system (see, e.g. figure 4.13 reproduced in figure 5.12 below in a notation more convenient for our current purposes), and calculate the correlation function of its energy fluctuations.

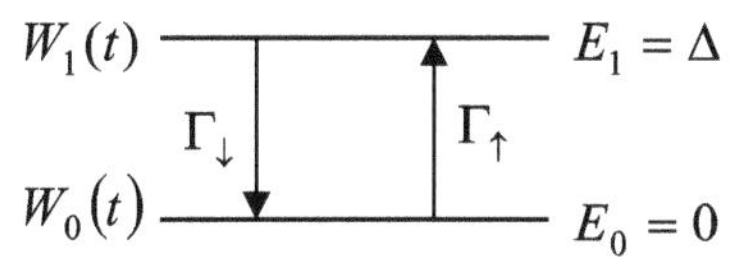

Figure 5.12. Dynamics of a two-level system.

The stationary probabilities of the system states (i.e. their probabilities for $\tau \to \infty$) have been calculated in chapter 2, and then again in section 4.4—see Eq. (4.68). In our current notation (figure 5.12),

$$W_0(\infty) = \frac{1}{1 + e^{-\Delta/T}}, \quad W_1(\infty) = \frac{1}{e^{\Delta/T} + 1},$$
$$\text{so that} \quad \langle E \rangle = W_0(\infty) \times 0 + W_1(\infty) \times \Delta = \frac{\Delta}{e^{\Delta/T} + 1}. \tag{5.168}$$

In order to calculate the conditional probabilities $W_{m'}(\tau)$ with the initial conditions (5.166) (according to Eq. (5.167), we need all four of them, for $\{m, m'\} = \{0, 1\}$), we may use the master equations (4.100), in our current notation reading

$$\frac{dW_1}{d\tau} = -\frac{dW_0}{d\tau} = \Gamma_\uparrow W_0 - \Gamma_\downarrow W_1. \tag{5.169}$$

Since Eq. (5.169) conserves the total probability, $W_0 + W_1 = 1$, only one probability (say, W_1) is an independent variable, and for it, Eq. (5.169) gives a simple, linear differential equation

$$\frac{dW_1}{d\tau} = \Gamma_\uparrow - \Gamma_\Sigma W_1, \quad \text{where} \quad \Gamma_\Sigma \equiv \Gamma_\uparrow + \Gamma_\downarrow, \tag{5.170}$$

which may be readily integrated for an arbitrary initial condition:

$$W_1(\tau) = W_1(0)e^{-\Gamma_\Sigma \tau} + W_1(\infty)(1 - e^{-\Gamma_\Sigma \tau}), \tag{5.171}$$

where $W_1(\infty)$ is given by the second of Eqs. (5.168). (It is straightforward to verify that the solution for $W_0(\tau)$ may be represented in the similar form, with the corresponding change of the state index.)

Now everything is ready to calculate the average $\langle E(t)E(t + \tau)\rangle$ using Eq. (5.167), with $f_{m,m'} = E_{0,1}$. Thanks to our (smart :-) choice of energy origin, of the four terms in the double sum (5.167), all three terms that include at least one factor $E_0 = 0$ vanish, and we have only one term left to calculate:

$$\begin{aligned}
\langle E(t)E(t + \tau)\rangle &= E_1 W_1(\infty) E_1 W_1(\tau)|_{W_1(0)=1} \\
&= E_1^2 W_1(\infty)[W_1(0)e^{-\Gamma_\Sigma \tau} + W_1(\infty)(1 - e^{-\Gamma_\Sigma \tau})]_{W_1(0)=1} \\
&= \frac{\Delta^2}{e^{\Delta/T} + 1}\left[e^{-\Gamma_\Sigma \tau} + \frac{1}{e^{\Delta/T} + 1}(1 - e^{-\Gamma_\Sigma \tau})\right] \\
&\equiv \frac{\Delta^2}{(e^{\Delta/T} + 1)^2}(1 + e^{\Delta/T} e^{-\Gamma_\Sigma \tau}).
\end{aligned} \tag{5.172}$$

From here and the last of Eqs. (5.168), the correlation function of energy fluctuations is[68]

$$\begin{aligned} K_E(\tau) &\equiv \langle \tilde{E}(t)\tilde{E}(t+\tau)\rangle = \langle (E(t) - \langle E(t)\rangle)(E(t+\tau) - \langle E(t)\rangle)\rangle \\ &= \langle E(t)E(t+\tau)\rangle - \langle E\rangle^2 = \Delta^2 \frac{e^{\Delta/T}}{(e^{\Delta/T}+1)^2} e^{-\Gamma_\Sigma \tau}, \end{aligned} \tag{5.173}$$

so that its variance, equal to $K_E(0)$, does not depend on the transition rates $\Gamma_\uparrow$ and $\Gamma_\downarrow$. However, since the rates have to obey the detailed balance relation (4.103), $\Gamma_\downarrow/\Gamma_\uparrow = \exp\{\Delta/T\}$, for this variance we may formally write

$$\frac{K_E(0)}{\Delta^2} = \frac{e^{\Delta/T}}{(e^{\Delta/T}+1)^2} = \frac{\Gamma_\downarrow/\Gamma_\uparrow}{(\Gamma_\downarrow/\Gamma_\uparrow + 1)^2} \equiv \frac{\Gamma_\uparrow\Gamma_\downarrow}{(\Gamma_\uparrow + \Gamma_\downarrow)^2} \equiv \frac{\Gamma_\uparrow\Gamma_\downarrow}{\Gamma_\Sigma^2}, \tag{5.174}$$

so that Eq. (5.173) may be represented in a simpler form:

$$K_E(\tau) = \Delta^2 \frac{\Gamma_\uparrow\Gamma_\downarrow}{\Gamma_\Sigma^2} e^{-\Gamma_\Sigma \tau}. \tag{5.175}$$

We see that the correlation function of the energy decays exponentially with time, with the net rate Γ_Σ. Now using the Wiener–Khinchin theorem (5.58) to calculate its spectral density, we get

$$S_E(\omega) = \frac{1}{\pi}\int_0^\infty \Delta^2 \frac{\Gamma_\uparrow\Gamma_\downarrow}{\Gamma_\Sigma^2} e^{-\Gamma_\Sigma \tau} \cos\omega\tau \, d\tau = \frac{\Delta^2}{\pi}\frac{\Gamma_\uparrow\Gamma_\downarrow}{\Gamma_\Sigma \Gamma_\Sigma^2 + \omega^2}. \tag{5.176}$$

Such Lorentzian dependence on frequency is very typical for discrete-state systems described by master equations. It is interesting that the most widely accepted explanation of the *1/f noise* (also called the 'flicker' or 'excess' noise), which was mentioned in section 5.5, is that it is a result of thermally-activated jumps between metastable states of a statistical ensemble of such two-level systems, with an exponentially-broad statistical distribution of the transition rates $\Gamma_{\uparrow\downarrow}$. Such a broad distribution follows from the Kramers formula (5.155), which is approximately valid for the lifetimes of both states of systems with double-well potential profiles (figure 5.13), for a statistical ensemble with smooth statistical distributions of energy barriers U_0. Such profiles are typical, in particular, for electrons in disordered (amorphous) solid-state materials, which indeed feature high $1/f$ noise.

Returning to the Fokker–Planck equation, we may use the evident generalization of Eq. (5.167) to the continuous-variable case:

$$\begin{aligned} \langle f(t)f(t+\tau)\rangle = &\int d^3q d^3p \int d^3q' d^3p' \\ &\times f(\mathbf{q}, \mathbf{p}) w(\mathbf{q}, \mathbf{p}, \infty) f(\mathbf{q}', \mathbf{p}') w(\mathbf{q}', \mathbf{p}', \tau), \end{aligned} \tag{5.177}$$

[68] The step from the first line of Eq. (5.173) to the second one uses the fact that our system is stationary, so that $\langle E(t+\tau)\rangle = \langle E(t)\rangle = \langle E\rangle = \text{const}$.

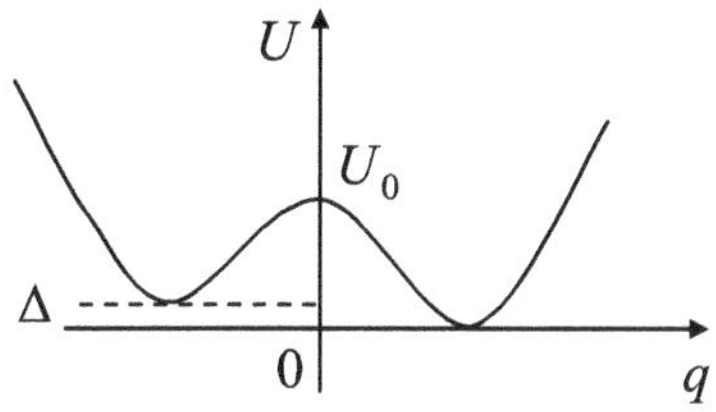

Figure 5.13. Typical double-well potential profile.

where both probability density distributions are the solutions of the equation with the delta-functional initial condition

$$w(\mathbf{q}', \mathbf{p}', 0) = \delta(\mathbf{q}'-\mathbf{q})\delta(\mathbf{p}'-\mathbf{p}). \tag{5.178}$$

For the Smoluchowski equation, valid in the high-damping limit, the expressions are similar, albeit with a lower dimensionality:

$$\langle f(t)f(t+\tau)\rangle = \int d^3q \int d^3q' \; f(\mathbf{q})w(\mathbf{q}, \infty)f(\mathbf{q}')w(\mathbf{q}', \tau), \tag{5.179}$$

$$w(\mathbf{q}', 0) = \delta(\mathbf{q}'-\mathbf{q}). \tag{5.180}$$

To see this formalism in action, let us use it to calculate the correlation function $K_q(\tau)$ of a *linear relaxator*, i.e. an overdamped 1D harmonic oscillator with $m\omega_0 \ll \eta$. In this limit, the coordinate, averaged over the ensemble of environments, obeys a linear equation,

$$\eta\langle\dot{q}\rangle + \kappa\langle q\rangle = 0, \tag{5.181}$$

which describes its exponential relaxation from a certain initial condition q_0 to the equilibrium position $q = 0$, with the reciprocal time constant $\Gamma = \kappa/\eta$:

$$\langle q\rangle(t) = q_0 e^{-\Gamma t}. \tag{5.182}$$

The deterministic equation (5.181) evidently corresponds to the quadratic potential energy $U(q) = \kappa q^2/2$, so that the 1D version of the corresponding Smoluchowski equation (5.122) takes the form

$$\eta\frac{\partial w}{\partial t} = \kappa\frac{\partial}{\partial q}(wq) + T\frac{\partial^2 w}{\partial q^2}. \tag{5.183}$$

It is straightforward to check, by substitution, that this equation, rewritten for the function $w(q',\tau)$, with the delta-functional initial condition (5.180), $w(q', 0) = \delta(q' - q)$, is satisfied with a Gaussian function,

$$w(q', \tau) = \frac{1}{(2\pi)^{1/2}\delta q(\tau)}\exp\left\{-\frac{[q' - \langle q\rangle(\tau)]^2}{2\delta q^2(\tau)}\right\}, \tag{5.184}$$

with its center, $\langle q\rangle(\tau)$, moving in accordance with Eq. (5.182), and a time-dependent variance

$$\delta q^2(\tau) = \delta q^2(\infty)(1 - e^{-2\Gamma\tau}), \quad \text{where} \quad \delta q^2(\infty) = \langle q^2 \rangle = \frac{T}{\kappa}. \tag{5.185}$$

(As a sanity check, the last equality coincides with the equipartition theorem's result.) Finally, the first probability under the integral in Eq. (5.179) may be found from Eq. (5.184) in the limit $\tau \to \infty$ (in which $\langle q \rangle(\tau) \to 0$), by replacing q' with q:

$$w(q, \infty) = \frac{1}{(2\pi)^{1/2}\delta q(\infty)} \exp\left\{-\frac{q^2}{2\delta q^2(\infty)}\right\}. \tag{5.186}$$

Now, all components of the recipe (5.179) are ready, and we can spell it out, for $f(q) = q$, as

$$\begin{aligned}\langle q(t)q(t+\tau)\rangle &= \frac{1}{2\pi\delta q(\tau)\delta q(\infty)} \int_{-\infty}^{+\infty} dq \int_{-\infty}^{+\infty} dq' \; q \exp\left\{-\frac{q^2}{2\delta q^2(\infty)}\right\} \\ &\quad \times q' \exp\left\{-\frac{(q' - qe^{-\Gamma\tau})^2}{2\delta q^2(\tau)}\right\}.\end{aligned} \tag{5.187}$$

The integral over q' may be worked out first, by replacing this integration variable with $(q'' + qe^{-\Gamma\tau})$ and hence dq' with dq'':

$$\begin{aligned}\langle q(t)q(t+\tau)\rangle &= \frac{1}{2\pi\delta q(\tau)\delta q(\infty)} \int_{-\infty}^{+\infty} q \exp\left\{-\frac{q^2}{2\delta q^2(\infty)}\right\} dq \\ &\quad \times \int_{-\infty}^{+\infty} (q'' + qe^{-\Gamma\tau}) \exp\left\{-\frac{q''^2}{2\delta q^2(\tau)}\right\} dq''.\end{aligned} \tag{5.188}$$

The internal integral of the first term in the parentheses equals zero (as that of an odd function in symmetric integration limits), while that with the second term is a standard Gaussian integral, giving

$$\begin{aligned}\langle q(t)q(t+\tau)\rangle &= \frac{1}{(2\pi)^{1/2}\delta q(\infty)} e^{-\Gamma\tau} \int_{-\infty}^{+\infty} q^2 \exp\left\{-\frac{q^2}{2\delta q^2(\infty)}\right\} dq \\ &\equiv \frac{2T}{\pi^{1/2}\kappa} e^{-\Gamma\tau} \int_{-\infty}^{+\infty} \xi^2 \exp\{-\xi^2\} d\xi.\end{aligned} \tag{5.189}$$

The last integral[69] equals $\pi^{1/2}/2$, so taking into account that for this stationary system centered at the coordinate origin, the full ensemble average $\langle q \rangle = 0$, we finally get a very simple result,

$$\begin{aligned}K_q(\tau) &\equiv \langle \tilde{q}(t)\tilde{q}(t+\tau)\rangle = \langle q(t)q(t+\tau)\rangle - \langle q \rangle^2 = \langle q(t)q(t+\tau)\rangle \\ &= \frac{T}{\kappa} e^{-\Gamma\tau}.\end{aligned} \tag{5.190}$$

[69] See, e.g. Eq. (A.36*c*).

As a sanity check, for $\tau = 0$ it yields $K_q(0) \equiv \langle q^2 \rangle = T/\kappa$, in accordance with Eq. (5.185). As τ is increased the correlation function decreases monotonically—see the solid-line sketch in figure 5.8.

So, the solution of this very simple problem has required straightforward but somewhat bulky calculations. On the other hand, the same result may be obtained literally in one line using the Langevin formalism—namely, as the Fourier transform (5.59) of the spectral density (5.68) in the corresponding limit $m\omega \ll \eta$, with $S_{\mathscr{F}}(\omega)$ given by Eq. (5.73a):[70]

$$K_q(\tau) = 2\int_0^\infty S_q(\omega)\cos\omega\tau\ d\omega = 2\int_0^\infty \frac{\eta T}{\pi}\frac{1}{\kappa^2 + (\eta\omega)^2}\cos\omega\tau\ d\omega$$
$$\equiv 2\frac{T\Gamma}{\pi}\int_0^\infty \frac{\cos\xi}{(\Gamma\tau)^2 + \xi^2}\ d\xi = \frac{T}{\kappa}e^{-\Gamma\tau}. \qquad (5.191)$$

This example illustrates the fact that for linear systems (and small fluctuations in nonlinear systems) the Langevin approach is usually much simpler than the one based on the Fokker–Planck or Smoluchowski equations. However, again, the latter approach is indispensable for the analysis of fluctuations of arbitrary intensity in nonlinear systems.

To conclude this chapter, I have to emphasize again that the Fokker–Planck and Smoluchowski equations give a quantitative description of time evolution of nonlinear Brownian systems with finite dissipation in the *classical* limit. The description of quantum properties of such dissipative ('open') and nonlinear *quantum* systems is more complex[71], and only a few simple problems of such theory have been solved so far[72], typically using a particular model of the environment, e.g. as a large set of harmonic oscillators with different statistical distributions of their parameters, leading to different frequency dependences of the generalized susceptibility $\chi(\omega)$.

5.9 Problems

Problem 5.1. Treating the first 30 digits of number $\pi = 3.1415...$ as a statistical ensemble of integers k (equal to 3, 1, 4, 1, 5,...), calculate the average $\langle k \rangle$, and the rms fluctuation δk. Compare the results with those for the ensemble of completely random decimal integers 0, 1, 2,..,9, and comment.

Problem 5.2. Calculate the variance of fluctuations of a magnetic moment $\mathscr{m}$ placed into an external magnetic field $\mathscr{H}$, within the same two models as in problem 2.4:[73]

[70] The involved table integral may be found, e.g. in Eq. (A.38).

[71] See, e.g. *Part QM* section 7.6.

[72] See, e.g. the solutions of the 1D Kramers problem for quantum systems with low damping by A Caldeira and A Leggett [10], and with high damping by A Larkin and Yu Ovchinnikov [11].

[73] Note that these two cases may be considered as the non-interacting limits of, respectively, the Ising model (4.23) and the classical limit of the Heisenberg model (4.21), whose analysis within the Weiss approximation was the subject of problem 4.18.

(i) a spin-1/2 with a gyromagnetic ratio γ, and
(ii) a classical magnetic moment $\mathscr{m}$, of a fixed magnitude $\mathscr{m}_0$, but with free orientation,

both in thermal equilibrium at temperature T. Discuss and compare the results.

Hint: Mind all three Cartesian components of the vector $\mathscr{m}$.

Problem 5.3. For a field-free, two-site Ising system with energy values $E_m = -Js_1s_2$, in thermal equilibrium at temperature T, calculate the variance of energy fluctuations. Explore the low-temperature and high-temperature limits of the result.

Problem 5.4. For a uniform three-site Ising ring with ferromagnetic coupling (and no external field), calculate the correlation coefficients $K_s \equiv \langle s_k s_{k'} \rangle$ for both $k = k'$ and $k \neq k'$.

Problem 5.5.* For a field-free 1D Ising system of $N \gg 1$ 'spins', in thermal equilibrium at temperature T, calculate the correlation coefficient $K_s \equiv \langle s_l s_{l+n} \rangle$, where l and $(l + n)$ are the numbers of two specific spins in the chain.

Hint: You may like to start with the calculation of the statistical sum for an open-ended chain with arbitrary $N > 1$ and arbitrary coupling coefficients J_k, and then consider its mixed partial derivative over a part of these parameters.

Problem 5.6. Within the framework of Weiss' molecular-field theory, calculate the variance of spin fluctuations in the d-dimensional Ising model. Use the result to derive the conditions of quantitative validity of this theory.

Problem 5.7. Calculate the variance of fluctuations of the energy of a quantum harmonic oscillator with frequency ω, in thermal equilibrium at temperature T, and express it via the average value of the energy.

Problem 5.8. The spontaneous electromagnetic field in a closed volume V is in thermal equilibrium with temperature T. Assuming that V is sufficiently large, calculate the variance of fluctuations of the total energy of the field, and express the result via its average energy and temperature. How large should the volume V be for your results to be qualitatively valid? Evaluate this limitation for room temperature.

Problem 5.9. Express the rms uncertainty of the occupancy N_k of a certain energy level ε_k by:

(i) a classical particle,
(ii) a fermion, and
(iii) a boson,

in thermodynamic equilibrium, via its average occupancy $\langle N_k \rangle$, and compare the results.

Problem 5.10. Express the variance of the number of particles, $\langle \tilde{N}^2 \rangle_{V,T,\mu}$, via the isothermal compressibility $\kappa_T \equiv -(1/V)(\partial V/\partial P)_{T,N}$ of the same single-phase system, in thermal equilibrium at temperature T.

Problem 5.11.* Starting from the Maxwell distribution of velocities, calculate the low-frequency spectral density of fluctuations of the pressure $P(t)$ of an ideal gas of N classical particles, in thermal equilibrium at temperature T, and estimate their variance. Compare the former result with the solution of problem 3.2.

Hints: You may consider a cylindrically-shaped container of volume $V = LA$ (see the figure below), calculate fluctuations of the force $\mathscr{F}(t)$ exerted by the confined particles on its plane lid of area A, approximating it as a delta-correlated process (5.62), and then recalculate the fluctuations into those of the pressure $P \equiv \mathscr{F}/A$.

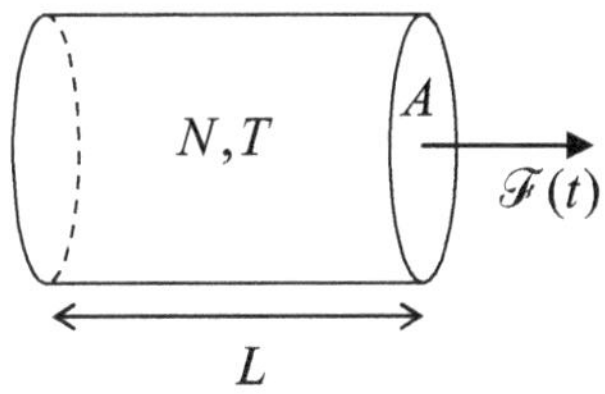

Problem 5.12. Calculate the low-frequency spectral density of fluctuations of the electric current $I(t)$ due to random passage of charged particles between two conducting electrodes—see the figure below. Assume that the particles are emitted, at random times, by one of the electrodes, and are fully absorbed by the counterpart electrode. Can your result be mapped on some aspect of the electromagnetic blackbody radiation?

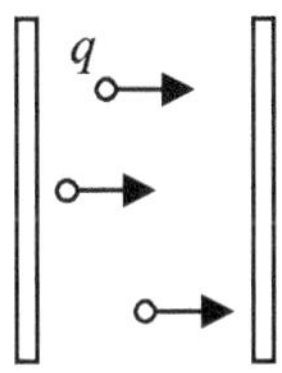

Hint: For the current $I(t)$, use the same delta-correlated-process approximation as for the force $\mathscr{F}(t)$ in the previous problem.

Problem 5.13.[74] A very long, uniform string, of mass μ per unit length, is attached to a firm support, and stretched with a constant force ('tension') $\mathscr{T}$—see the figure below. Calculate the spectral density of the random force $\mathscr{F}(t)$ exerted by the string

[74] This problem, conceptually important for the quantum mechanics of open systems, was offered in chapter 7 of *Part QM* of this series, and is repeated here for the benefit of readers who, for any reason, skipped that course.

on the support point, within the plane normal to its length, in thermal equilibrium at temperature T.

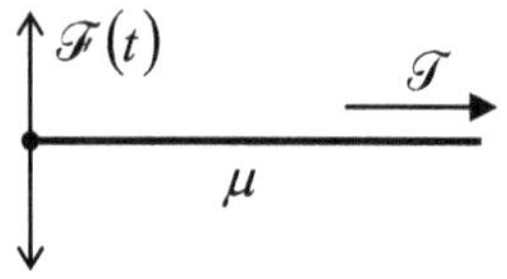

Hint: You may assume that the string is so long that a transverse wave, propagating along it from the support point, never comes back.

Problem 5.14.[75] Each of two 3D harmonic oscillators, with mass m and resonance frequency ω_0, has the electric dipole moment $\mathbf{d} = q\mathbf{s}$, where $\mathbf{s}$ is the oscillator's displacement from its equilibrium position. Use the Langevin formalism to calculate the average potential of electrostatic interaction of these two oscillators (a particular case of the so-called *London dispersion force*), separated by distance $r \gg (T/m\omega_0^2)^{1/2}$, in thermal equilibrium at temperature $T \gg \hbar\omega_0$. Also, explain why the approach used in the solution of the very similar problem 2.15 is not directly applicable to this case.

Hint: You may like to use the following integral:

$$\int_0^\infty \frac{1 - \xi^2}{\left[\left(1 - \xi^2\right)^2 + (a\xi)^2\right]^2} d\xi = \frac{\pi}{4a}.$$

Problem 5.15.* Within the van der Pol approximation[76], calculate major statistical properties of fluctuations of classical self-oscillations, at:

(i) the free ('autonomous') run of the oscillator, and
(ii) its phase locking by an external sinusoidal force,

assuming that the fluctuations are caused by a weak external noise with a smooth spectral density $S_f(\omega)$. In particular, calculate the self-oscillation linewidth.

Problem 5.16. Calculate the correlation function of the coordinate of a 1D harmonic oscillator with small Ohmic damping at thermal equilibrium. Compare the result with that for the autonomous self-oscillator (the subject of the previous problem).

---

[75] This problem, for the case of arbitrary temperature, was the subject of problem 7.6, with problem 5.15 of that part serving as the background. However, the method used in the model solutions of those problems requires one to prescribe, to the oscillators, different frequencies ω_1 and ω_2 at first, and only after this more general problem has been solved, pursue the limit $\omega_1 \to \omega_2$.The goal of this problem is to demonstrate that the Langevin formalism enables a solution taking $\omega_1 = \omega_2 \equiv \omega_0$ from the very beginning.

[76] See, e.g. *Part CM* sections 5.2–5.5. Note that in quantum mechanics, a similar approach is called the *rotating-wave approximation* (RWA)—see, e.g. *Part QM* sections 6.5, 7.6, 9.2, and 9.4.

Problem 5.17. Consider a very long, uniform, two-wire transmission line (see the figure below) with a wave impedance $\mathscr{Z}$, which allows propagation of TEM electromagnetic waves with negligible attenuation, in thermal equilibrium at temperature T. Calculate the variance $\langle \mathscr{V}^2 \rangle_{\Delta\nu}$ of the voltage $\mathscr{V}$ between the wires within a small interval $\Delta\nu$ of cyclic frequencies.

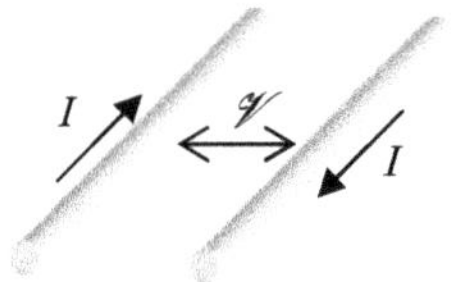

Hint: As an E&M reminder[77], in the absence of dispersive materials, TEM waves propagate with a frequency-independent velocity (equal to the speed c of light, if the wires are in vacuum), with the voltage $\mathscr{V}$ and the current I (see figure above) related as $\mathscr{V}(x, t)/I(x, t) = \pm\mathscr{Z}$, where $\mathscr{Z}$ is the line's wave impedance.

Problem 5.18. Now consider a similar long transmission line but terminated, at one end, with an impedance-matching Ohmic resistor $R = \mathscr{Z}$. Calculate the variance $\langle \mathscr{V}^2 \rangle_{\Delta\nu}$ of the voltage across the resistor, and discuss the relation between the result and the Nyquist formula (5.81), including numerical factors.

Hint: Take into account that a load with resistance $R = \mathscr{Z}$ absorbs incident TEM waves without reflection.

Problem 5.19. An overdamped classical 1D particle escapes from a potential well with a smooth bottom, but a sharp top of the barrier—see the figure below. Perform the necessary modification of the Kramers formula (5.139).

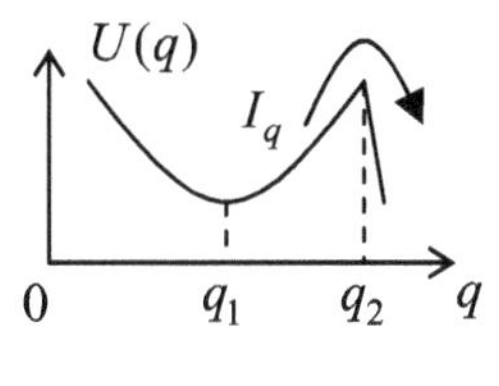

Problem 5.20. Perhaps the simplest mathematical model of the diffusion is the 1D *discrete random walk*: each time interval τ, a particle leaps, with equal probability, to any of two adjacent sites of a 1D lattice with a spatial period a. Prove that the particle's displacement during a time interval $t \gg \tau$ obeys Eq. (5.77), and calculate the corresponding diffusion coefficient D.

Problem 5.21. A classical particle may occupy any of N similar sites. Its weak interaction with the environment induces random, incoherent jumps from the occupied site to any other site, with the same time-independent rate Γ. Calculate the correlation function and the spectral density of fluctuations of the instant occupancy $n(t)$ (equal to either 1 or 0) of a site.

[77] See, e.g. *Part EM* section 7.6.

References

[1] Morgan K 2018 *Phys. Today* **71** 29
[2] Coffey W, Kalmykov Yu and Waldron J 1996 *The Langevin equation* (World Scientific)
[3] Crossno J *et al* 2015 *Appl. Phys. Lett.* **106** 023121
[4] Naveh Y *et al* 1998 *Phys. Rev.* B **58** 15371
[5] Blanter Ya and Büttiker M 2000 *Phys. Rep.* **336** 1
[6] Balbi A 2008 *The Music of the Big Bang* (Springer)
[7] Koch R *et al* 1982 *Phys. Rev.* B **26** 74
[8] Stratonovich R 1963 *Topics in the Theory of Random Noise* vol 1 (Gordon and Breach)
[9] Chandrasekhar S 1943 *Rev. Mod. Phys.* **15** 1
[10] Caldeira A and Leggett A 1981 *Phys. Rev. Lett.* **46** 211
[11] Larkin A and Ovchinnikov Yu 1983 *JETP Lett.* **37** 382

IOP Publishing

Chapter 6

Elements of kinetics

This chapter gives a brief introduction to the basic notions of physical kinetics. Its main focus is on the Boltzmann equation, especially within the simple relaxation-time approximation (RTA), which allows, in particular, an approximate but reasonable and simple description of transport phenomena (such as the electric current and thermoelectric effects) in gases, including electron gases in metals and semiconductors.

6.1 The Liouville theorem and the Boltzmann equation

Physical kinetics (not to be confused with 'kinematics'!) is the branch of statistical physics that deals with systems out of thermodynamic equilibrium. Major effects addressed by kinetics include:

(i) for *autonomous* systems (those out of external fields): the transient processes (*relaxation*), that lead from an arbitrary initial state of a system to its thermodynamic equilibrium;

(ii) for systems in time-dependent (say, sinusoidal) external fields: the field-induced periodic oscillations of the system's parameters; and

(iii) for systems in time-independent ('dc') external fields: dc transport.

In the last case, we are dealing with stationary ($\partial/\partial t = 0$ everywhere), but *non-equilibrium* situations, in which the effect of an external field, continuously driving the system out of equilibrium, is balanced by the simultaneous relaxation—the trend back to equilibrium. Perhaps the most important effect of this class is the dc current in conductors[1], which alone justifies the inclusion of the basic notions of kinetics into any set of core physics courses.

[1] This topic was briefly addressed in *Part EM* chapter 4, carefully avoiding all issues related to the thermal effects.

doi:10.1088/2053-2563/aaf503ch6

Actually, the reader who has reached this point of the notes, already has some taste of physical kinetics, because the subject of the last part of chapter 5 *was* the kinetics of a 'Brownian particle', i.e. of a 'heavy' system interacting with environment consisting of many 'lighter' components. Indeed, the equations discussed in that part—whether the Smoluchowski equation (5.122) or the Fokker–Planck equation (5.149)—are valid if the environment is in thermodynamic equilibrium, but the system of our interest is not necessarily so. As a result, we could use those equations to discuss such non-equilibrium phenomena as the Kramers problem of the metastable state's lifetime.

In contrast, this chapter is devoted to the more traditional subject of kinetics: systems of many *similar* particles—generally, interacting with each other, but not too strongly, so that the energy of the system still may be partitioned into a sum of the components, with the interparticle interactions considered as a weak perturbation. Actually, we have already started the job of describing such a system in the beginning of section 5.7. Indeed, in the absence of particle interactions (i.e. when it is unimportant whether the particle of our interest is 'light' or 'heavy'), the probability current densities in the coordinate and momentum spaces are given, respectively, by Eq. (5.142) and the first form of Eq. (5.143*a*), so that the continuity equation (5.140) takes the form

$$\frac{\partial w}{\partial t} + \nabla_q \cdot (w\dot{\mathbf{q}}) + \nabla_p \cdot (w\dot{\mathbf{p}}) = 0. \tag{6.1}$$

If similar particles do *not* interact, this equation for the single-particle probability density $w(\mathbf{q}, \mathbf{p}, t)$ is valid for each of them, and the result of its solution may be used to calculate any ensemble-average characteristic of the system as a whole.

Let us rewrite Eq. (6.1) in the Cartesian component form,

$$\frac{\partial w}{\partial t} + \sum_j \left[\frac{\partial}{\partial q_j}\left(w\dot{q}_j\right) + \frac{\partial}{\partial p_j}\left(w\dot{p}_j\right)\right] = 0, \tag{6.2}$$

where the index j lists all degrees of freedom of the particle, and assume that its motion (perhaps in an external, time-dependent field) may be described by a Hamiltonian function $\mathscr{H}(q_j, p_j, t)$. Plugging into Eq. (6.2) the Hamiltonian equations of motion[2]:

$$\dot{q}_j = \frac{\partial \mathscr{H}}{\partial p_j}, \quad \dot{p}_j = -\frac{\partial \mathscr{H}}{\partial q_j}, \tag{6.3}$$

we get

$$\frac{\partial w}{\partial t} + \sum_j \left[\frac{\partial}{\partial q_j}\left(w\frac{\partial \mathscr{H}}{\partial p_j}\right) - \frac{\partial}{\partial p_j}\left(w\frac{\partial \mathscr{H}}{\partial q_j}\right)\right] = 0. \tag{6.4}$$

[2] See, e.g. *Part CM* section 10.1.

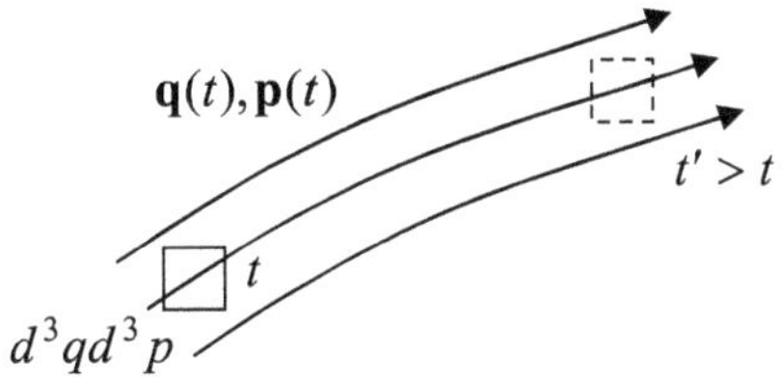

Figure 6.1. The Liouville theorem's interpretation: probability's conservation at its flow in the [**q**, **p**] space.

At the parentheses' differentiation, the equal mixed terms $w\partial^2\mathscr{H}/\partial q_j\partial p_j$ and $w\partial^2\mathscr{H}/\partial p_j\partial q_j$ cancel, and using Eq. (6.3) again, we get the so-called *Liouville theorem*[3]

$$\frac{\partial w}{\partial t} + \sum_j \left(\frac{\partial w}{\partial q_j}\dot{q}_j + \frac{\partial w}{\partial p_j}\dot{p}_j \right) = 0. \tag{6.5}$$

Since the left-hand part of this equation is just the full derivative of the probability density w considered as a function of the generalized coordinates $q_j(t)$ of a particle, its generalized momenta components $p_j(t)$, and (possibly) time t,[4] the Liouville theorem (6.5) may be represented in a surprisingly simple form:

$$\frac{dw(\mathbf{q}, \mathbf{p}, t)}{dt} = 0. \tag{6.6}$$

Physically this means that the elementary probability $dW = wd^3qd^3p$ to find a Hamiltonian particle in a small volume of the coordinate-momentum space [**q**, **p**], with its center moving in accordance to the deterministic law (6.3), does not change with time—see figure 6.1.

At first glance, this may not look surprising, because according to the fundamental Einstein relation (5.78), one needs non-Hamiltonian forces (such as the kinematic friction) to have diffusion. On the other hand, it is striking that the Liouville theorem is valid even for (Hamiltonian) systems with deterministic chaos[5], in which the deterministic trajectories corresponding to slightly different initial conditions become increasingly mixed with time.

For an ideal gas of 3D particles, we may use the ordinary Cartesian coordinates r_j (with $j = 1, 2, 3$) for the generalized coordinates q_j, so that p_j become the Cartesian components mv_j of the usual (linear) momentum, and the elementary volume is just d^3rd^3p—see figure 6.1. In this case Eqs. (6.3) are just

$$\dot{r}_j = \frac{p_j}{m} \equiv v_j, \quad \dot{p}_j = \mathscr{F}_j, \tag{6.7}$$

where $\mathscr{F}$ is the force exerted on the particle, so that the Liouville theorem may be rewritten as

---

[3] Actually, this is just one of several theorems bearing the name of J Liouville (1809–82).

[4] See, e.g. Eq. (A.23).

[5] See, e.g. *Part CM* section 9.3.

$$\frac{\partial w}{\partial t} + \sum_{j=1}^{3}\left(v_j\frac{\partial w}{\partial r_j} + \mathscr{F}_j\frac{\partial w}{\partial p_j}\right) = 0, \tag{6.8}$$

and conveniently represented in the vector form

$$\frac{\partial w}{\partial t} + \mathbf{v}\cdot\nabla_r w + \mathscr{F}\cdot\nabla_p w = 0. \tag{6.9}$$

Of course the situation becomes much more complex if the particles interact. Generally, a system of N similar particles in 3D space has to be described by probability density being a function of $6N + 1$ arguments ($3N$ Cartesian coordinates, plus $3N$ momentum components, plus time). An analytical or numerical solution of any equation describing time evolution of such a function for a typical system of $N \sim 10^{23}$ particles is evidently a hopeless task. Hence, kinetics of realistic ensembles has to rely on making reasonable approximations that would simplify the situation.

One of the most useful approximations (sometimes called *Stosszahlansatz*, German for the 'collision-number assumption') was suggested by L Boltzmann for a gas of particles that move freely most of the time, but interact during short time intervals, when a particle comes close to either an immobile scattering center (say, an impurity in a conductor's crystal lattice) or to another particle of the gas. Such brief *scattering events* may change the particle's momentum. Boltzmann argued that they may be still approximately described by Eq. (6.9), with the addition of a special term (called the *scattering integral*) to its right-hand side:

$$\frac{\partial w}{\partial t} + \mathbf{v}\cdot\nabla_r w + \mathscr{F}\cdot\nabla_p w = \left.\frac{\partial w}{\partial t}\right|_{\text{scattering}}. \tag{6.10}$$

This is the *Boltzmann transport equation*. As will be discussed below, it may give a very reasonable description of not only classical, but also quantum particles, though it evidently neglects the quantum-mechanical coherence effects[6]—besides those that may be hidden inside the scattering integral.

The concrete form of the scattering integral depends on the scatterers. If the scattering centers do not belong to the ensemble under consideration (an example is given, again, by impurity atoms in a conductor), then the scattering integral may be obtained by an evident generalization of the master equation (4.100):

$$\left.\frac{\partial w}{\partial t}\right|_{\text{scatteering}} = \int d^3p' \left[\Gamma_{\mathbf{p}'\to\mathbf{p}} w(\mathbf{r}, \mathbf{p}', t) - \Gamma_{\mathbf{p}\to\mathbf{p}'} w(\mathbf{r}, \mathbf{p}, t)\right], \tag{6.11}$$

where the physical sense of $\Gamma_{\mathbf{p}\to\mathbf{p}'}$ is the rate (i.e. the probability per unit time) for the particle to be scattered from the state with the momentum $\mathbf{p}$ into the state with the momentum $\mathbf{p}'$—see figure 6.2.

[6] Indeed the quantum state coherence is described by off-diagonal elements of the density matrix, while the classical probability w represents only the diagonal elements of the matrix. However, at least for the ensembles close to thermal equilibrium, this is a reasonable approximation—see the discussion in section 2.1.

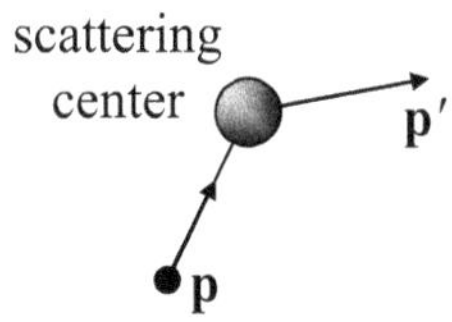

Figure 6.2. A particle scattering event.

Most elastic interactions are *reciprocal*, i.e. obey the following relation (closely related to the reversibility of time in Hamiltonian systems): $\Gamma_{\mathbf{p}\to\mathbf{p}'} = \Gamma_{\mathbf{p}'\to\mathbf{p}}$, so that Eq. (6.11) may be rewritten as[7]

$$\left.\frac{\partial w}{\partial t}\right|_{\text{scatteering}} = \int d^3p' \ \Gamma_{\mathbf{p}\to\mathbf{p}'}\left[w(\mathbf{r}, \mathbf{p}', t) - w(\mathbf{r}, \mathbf{p}, t)\right]. \tag{6.12}$$

With such scattering integral, Eq. (6.10) stays linear in w, but becomes an *integro-differential equation*, typically harder to solve than differential equations.

The equation becomes even more complex if the scattering is due to mutual interaction of the particle members of the system—see figure 6.3.

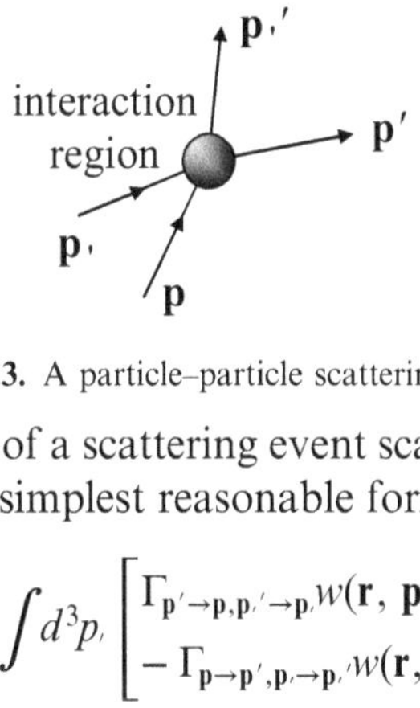

Figure 6.3. A particle–particle scattering event.

In this case, the probability of a scattering event scales as a product of two single-particle probabilities, and the simplest reasonable form of the scattering integral is[8]

$$\left.\frac{\partial w}{\partial t}\right|_{\text{scatteering}} = \int d^3p' \int d^3p_{,} \begin{bmatrix} \Gamma_{\mathbf{p}'\to\mathbf{p},\mathbf{p}_{,}'\to\mathbf{p}_{,}} w(\mathbf{r}, \mathbf{p}', t) w(\mathbf{r}, \mathbf{p}_{,}', t) \\ -\Gamma_{\mathbf{p}\to\mathbf{p}',\mathbf{p}_{,}\to\mathbf{p}_{,}'} w(\mathbf{r}, \mathbf{p}, t) w(\mathbf{r}, \mathbf{p}_{,}, t) \end{bmatrix}. \tag{6.13}$$

The integration dimensionality in Eq. (6.13) takes into account the fact that due to the conservation of the total momentum at scattering,

[7] One may wonder whether this approximation may work for Fermi particles, for whom the Pauli principle forbids scattering into the already occupied state, so that for the scattering $\mathbf{p} \to \mathbf{p}'$, the factor $w(\mathbf{r}, \mathbf{p}, t)$ in Eq. (6.12) has to be multiplied by the probability $[1—w(\mathbf{r}, \mathbf{p}', t)]$ that the final state is available. This is a valid argument, but one should notice that if this modification has been done with both terms of Eq. (6.12), it becomes

$$\left.\frac{\partial w}{\partial t}\right|_{\text{scatteering}} = \int d^3p' \ \Gamma_{\mathbf{p}\to\mathbf{p}'}\{w(\mathbf{r}, \mathbf{p}', t)[1 - w(\mathbf{r}, \mathbf{p}, t)] - w(\mathbf{r}, \mathbf{p}, t)[1 - w(\mathbf{r}, \mathbf{p}', t)]\}.$$

Opening both square brackets, we see that the probability density products cancel, bringing us back to Eq. (6.12).

[8] This was the approximation used by L Boltzmann to prove the famous *H-theorem*, stating that entropy of the gas described by Eq. (6.13) may only grow (or stay constant) in time, $dS/dt \geqslant 0$. Since the model is very approximate, that result does not seem too fundamental nowadays, despite all its historic significance.

$$\mathbf{p} + \mathbf{p}_{\prime} = \mathbf{p}' + \mathbf{p}_{\prime}', \tag{6.14}$$

one of the momenta is not an independent argument, so that the integration in Eq. (6.13) may be restricted to a 6D p-space rather than the 9D one. For the reciprocal interaction, Eq. (6.13) may also be a bit simplified, but it still keeps Eq. (6.10) a *nonlinear* integro-differential transport equation, excluding such powerful solution methods as the Fourier expansion—which hinges on the linear superposition principle.

This is why most useful results based on the Boltzmann transport equation depend on its further simplifications, most notably the *relaxation-time approximation*—RTA for short[9]. This approximation is based on noticing that in the absence of spatial gradients ($\nabla = 0$), and external forces ($\mathscr{F} = 0$), Eq. (6.10) yields

$$\frac{\partial w}{\partial t} = \left.\frac{\partial w}{\partial t}\right|_{\text{scattering}}, \tag{6.15}$$

so that the thermally-equilibrium probability distribution $w_0(\mathbf{r}, \mathbf{p}, t)$ has to turn any scattering integral into zero. Hence at *small* deviations from the equilibrium,

$$\tilde{w}(\mathbf{r}, \mathbf{p}, t) \equiv w(\mathbf{r}, \mathbf{p}, t) - w_0(\mathbf{r}, \mathbf{p}, t) \to 0, \tag{6.16}$$

the scattering integral should be proportional to the deviation $\tilde{w}$, and its simplest reasonable model is

$$\left.\frac{\partial w}{\partial t}\right|_{\text{scatteering}} = -\frac{\tilde{w}}{\tau}, \tag{6.17}$$

where τ is a phenomenological constant (which, according to Eq. (6.15), has to be positive for the system's stability) called the *relaxation time*. Its physical meaning will be more clear in the next section.

The relaxation-time approximation is quite reasonable if the angular distribution of the scattering rate is dominated by small angles between vectors $\mathbf{p}$ and $\mathbf{p}'$—as it is, for example, for the Rutherford scattering by a Coulomb center[10]. Indeed, in this case the two values of the function w, participating in Eq. (6.12), are close to each other for most scattering events, so that the loss of the second momentum argument ($\mathbf{p}'$) is not too essential. However, using the Boltzmann-RTA equation that results from combining Eqs. (6.10) and (6.17),

$$\frac{\partial w}{\partial t} + \mathbf{v} \cdot \nabla_r w + \mathscr{F} \cdot \nabla_p w = -\frac{\tilde{w}}{\tau}, \tag{6.18}$$

we should always remember this is just an approximation, sometimes giving completely wrong results. For example, it prescribes the same time scale (τ) to the relaxation of the net *momentum* of the system, and to its *energy* relaxation, while in many real systems the latter process (that results from inelastic collisions) may be substantially longer.

[9] Sometimes this approximation is called the 'BGK model', after P Bhatnager, E Gross, and M Krook who suggested it in 1954. (The same year, a similar model was considered by P Welander.)
[10] See, e.g. *Part CM* section 3.7.

Naturally, in the following sections I will describe only those applications of the Boltzmann-RTA equation that give a reasonable description of physical reality.

6.2 The Ohm law and the Drude formula

Despite its shortcomings, Eq. (6.18) is adequate for quite a few applications. Perhaps the most important of them is deriving the Ohm law for dc current in a 'nearly-ideal' gas of charged particles, whose only important deviation from ideality is the rare scattering effects described by Eq. (6.17). As a result, in equilibrium it is described by the stationary probability w_0 of an ideal gas (see section 3.1):

$$w_0(\mathbf{r}, \mathbf{p}, t) = \frac{g}{(2\pi\hbar)^3}\langle N(\varepsilon)\rangle, \tag{6.19}$$

where g is the degeneracy factor (say, $g = 2$ for electrons due to their spin), and $\langle N(\varepsilon)\rangle$ is the average occupancy of a quantum state with momentum $\mathbf{p}$, that obeys either the Fermi–Dirac or the Bose–Einstein distribution:

$$\langle N(\varepsilon)\rangle = \frac{1}{e^{(\varepsilon-\mu)/T} \pm 1}, \qquad \varepsilon = \varepsilon(\mathbf{p}). \tag{6.20}$$

(Up to a point, the following calculations will be valid for both statistics, and hence, in the limit $\mu/T \to -\infty$, for a classical gas as well.)

Now let a uniform dc electric field $\mathscr{E}$ be applied to the gas of particles with electric charge q, exerting the force $\mathscr{F} = q\mathscr{E}$ on each of them. Then the stationary solution to Eq. (6.18), with $\partial/\partial t = 0$, should also be stationary and spatially-uniform ($\nabla_r = 0$), so that this equation is reduced to

$$q\mathscr{E} \cdot \nabla_p w = -\frac{\tilde{w}}{\tau}. \tag{6.21}$$

Let us assume the electric field to be relatively low as well, so that the perturbation $\tilde{w}$ it produces is relatively small[11]. Then on the left-hand side of Eq. (6.21) we can neglect that perturbation, by replacing w with w_0, because that side already has a small factor ($\mathscr{E}$). As a result, this equation yields

$$\tilde{w} = -\tau q\mathscr{E} \cdot \nabla_p w_0 \equiv -\tau q\mathscr{E} \cdot \left(\nabla_p \varepsilon\right)\frac{\partial w_0}{\partial \varepsilon}, \tag{6.22}$$

where the second step implies isotropy of the parameters μ and T, i.e. their independence of the direction of the particle's momentum $\mathbf{p}$. But the gradient $\nabla_p \varepsilon$ is nothing other than the particle's velocity $\mathbf{v}$—for a quantum particle, its group

[11] Since the scale of the fastest change of w_0 in the momentum space is of the order of $\partial w_0/\partial p = (\partial w_0/\partial \varepsilon)(d\varepsilon/dp) \sim (1/T)v$, where v is the scale of particle's speed, the necessary condition of the linear approximation (6.22) is $e\mathscr{E}\tau \ll T/v$, i.e. if $e\mathscr{E}l \ll T$, where $l \equiv v\tau$ has the meaning of the effective mean free path. Since the left-hand part of the last inequality is just the average energy given to the particle by the electric field between two scattering events, the condition may be interpreted as the smallness of the gas' 'overheating' by the applied field. However, another condition is also necessary—see the last paragraph of this section.

velocity[12]. (This fact is easy to verify for the isotropic and parabolic dispersion law, pertinent to classical particles moving in free space,

$$\varepsilon(\mathbf{p}) = \frac{p^2}{2m} \equiv \frac{p_1^2 + p_2^2 + p_3^2}{2m}. \tag{6.23}$$

Indeed, in this case the *j*th Cartesian components of the vector $\nabla_p \varepsilon$ is

$$\left(\nabla_p \varepsilon\right)_j \equiv \frac{\partial \varepsilon}{\partial p_j} = \frac{p_j}{m} = v_j, \tag{6.24}$$

so that $\nabla_p \varepsilon = \mathbf{v}$.) Hence, Eq. (6.22) may be rewritten as

$$\tilde{w} = -\tau q \mathscr{E} \cdot \mathbf{v} \frac{\partial w_0}{\partial \varepsilon}. \tag{6.25}$$

Let us use this result to calculate the electric current density $\mathbf{j}$. The contribution of each particle to the current density is $q\mathbf{v}$, so that the total density is

$$\mathbf{j} = \int q\mathbf{v} w d^3p \equiv q \int \mathbf{v}(w_0 + \tilde{w})\, d^3p. \tag{6.26}$$

Since in the equilibrium state (with $w = w_0$), the current has to be zero, the integral of the first term in the parentheses has to vanish. For the integral of the second term, plugging in Eq. (6.25), and then using Eq. (6.19), we get

$$\mathbf{j} = q^2\tau \int \mathbf{v}(\mathscr{E} \cdot \mathbf{v}) \left(-\frac{\partial w_0}{\partial \varepsilon}\right) d^3p = \frac{g q^2 \tau}{(2\pi\hbar)^3} \int \mathbf{v}\,(\mathscr{E} \cdot \mathbf{v}) \left[-\frac{\partial \langle N(\varepsilon)\rangle}{\partial \varepsilon}\right] d^2p_\perp\, dp_{||}, \tag{6.27}$$

where $d^2p_\perp$ is the elementary area of the constant energy surface in the momentum space, while $dp_{||}$ is the momentum differential's component normal to that surface. The real power of this result[13] is that it is valid even for particles with an arbitrary dispersion law $\varepsilon(\mathbf{p})$ (which may be rather complicated, for example, for particles moving in space-periodic potentials[14]), and gives, in particular, a fair description of conductivity's anisotropy in crystals.

For free particles whose dispersion law is isotropic and parabolic, as in Eq. (6.23), the constant energy surface is a sphere of radius p, so that $d^2p_\perp = p^2 d\Omega = p^2 \sin\theta d\theta d\varphi$, while $dp_{||} = dp$. In the spherical coordinates, with the polar axis directed along the electric field vector $\mathscr{E}$, we get $(\mathscr{E} \cdot \mathbf{v}) = \mathscr{E} v \cos\theta$. Now separating the vector $\mathbf{v}$ outside the parentheses into the component $v\cos\theta$ directed along the vector $\mathscr{E}$, and two perpendicular components, $v\sin\theta\cos\varphi$ and $v\sin\theta\sin\varphi$, we see that the integrals of the last two components over the angle φ give zero. Hence, as we

[12] See, e.g. *Part QM* section 2.1.

[13] It was obtained by A Sommerfeld in 1927.

[14] See, e.g. *Part QM* sections 2.7, 2.8, and 3.4. (In this case, $\mathbf{p}$ should be understood as the quasi-momentum rather than genuine momentum.)

could expect, in the isotropic case the net current is directed along the electric field and obeys the linear *Ohm law*,

$$\mathbf{j} = \sigma \mathscr{E}, \tag{6.28}$$

with a field-independent, scalar[15] *electric conductivity*

$$\sigma = \frac{gq^2\tau}{(2\pi\hbar)^3}\int_0^{2\pi} d\varphi \int_0^{\pi} \sin\theta d\theta \cos^2\theta \int_0^{\infty} p^2 dp\, v^2 \left[-\frac{\partial\langle N(\varepsilon)\rangle}{\partial\varepsilon}\right]. \tag{6.29}$$

(Note that σ is proportional to q^2 and hence does not depend on the particle charge sign[16].)

Since $\sin\theta d\theta$ is just $-d(\cos\theta)$, the integral over θ equals (2/3). The integral over $d\varphi$ is of course just 2π, while that over p may be readily transformed to one over the particle's energy $\varepsilon(\mathbf{p}) = p^2/2m$: $p^2 = 2m\varepsilon$, $v^2 = 2\varepsilon/m$, $p = (2m\varepsilon)^{1/2}$, so that $dp = (m/2\varepsilon)^{1/2}d\varepsilon$, and $p^2dpv^2 = (2m\varepsilon)(m/2\varepsilon)^{1/2}d\varepsilon\,(2\varepsilon/m) \equiv (8m\varepsilon^3)^{1/2}d\varepsilon$. As a result, the conductivity equals

$$\sigma = \frac{gq^2\tau}{(2\pi\hbar)^3}\frac{4\pi}{3}\int_0^{\infty}(8m\varepsilon^3)^{1/2}\left[-\frac{\partial\langle N(\varepsilon)\rangle}{\partial\varepsilon}\right]d\varepsilon\,. \tag{6.30}$$

Now we may work out the integral in Eq. (6.30) by parts, first rewriting $[-\partial\langle N(\varepsilon)\rangle/\partial\varepsilon]d\varepsilon$ as $-d[\langle N(\varepsilon)\rangle]$. Due to the fast (exponential) decay of the factor $\langle N(\varepsilon)\rangle$ at $\varepsilon \to \infty$, its product by the factor $(8m\varepsilon^3)^{1/2}$ vanishes at both integration limits, and we get

$$\begin{aligned}\sigma &= \frac{gq^2\tau}{(2\pi\hbar)^3}\frac{4\pi}{3}\int_0^{\infty}\langle N(\varepsilon)\rangle\, d[(8m\varepsilon^3)^{1/2}]\\ &\equiv \frac{gq^2\tau}{(2\pi\hbar)^3}\frac{4\pi}{3}(8m)^{1/2}\int_0^{\infty}\langle N(\varepsilon)\rangle\frac{3}{2}\varepsilon^{1/2}d\varepsilon\\ &\equiv \frac{q^2\tau}{m}\times\frac{gm^{3/2}}{\sqrt{2}\pi^2\hbar^3}\int_0^{\infty}\langle N(\varepsilon)\rangle\varepsilon^{1/2}d\varepsilon.\end{aligned} \tag{6.31}$$

But according to Eq. (3.40), the last factor in this expression (after the $\times$ sign) is just the particle density $n \equiv N/V$, so that the Sommerfeld's result is reduced, for arbitrary temperature, and any particle statistics, to the very simple *Drude formula*[17],

$$\sigma = \frac{q^2\tau}{m}n, \tag{6.32}$$

[15] As Eq. (6.27) shows, if the dispersion law $\varepsilon(\mathbf{p})$ is anisotropic, the current density direction may be different from that of the electric field. In this case, conductivity should be described by a tensor $\sigma_{jj'}$, rather than a scalar.

[16] This is why in order to determine the dominating type of charge carriers in semiconductors (electrons or holes, see section 6.4 below), the Hall effect, which lacks such ambivalence (see, e.g. *Part QM* section 3.2), needs to be used.

[17] It was derived in 1900 by P Drude. Note that Drude also used the same arguments to derive a very simple (and very reasonable) approximation for the complex electric conductivity in the ac field of frequency ω: $\sigma(\omega) = \sigma(0)/(1 - i\omega\tau)$, with $\sigma(0)$ given by Eq. (6.32); sometimes the name 'Drude formula' is used for this expression rather than for Eq. (6.32). Let me leave its derivation, from the Boltzmann-RTA equation, for the reader's exercise.

which should be well familiar to the reader from an undergraduate physics course. As a reminder, here is its simple classical derivation[18].

Let τ be the average time after which a scattering events causes a particle to loose all the deterministic component of its velocity, $\mathbf{v}_{\text{drift}}$, provided by the electric field $\mathscr{E}$ on the top of particle's random thermal motion (which does not contribute to the net current). Using the 2nd Newton law to describe particle's acceleration by the field, $d\mathbf{v}_{\text{drift}}/dt = q\mathscr{E}/m$, we get $\langle \mathbf{v}_{\text{drift}} \rangle = \tau q\mathscr{E}/m$. Multiplying this result by the particle charge q and density $n \equiv N/V$, we get the Ohm law $\mathbf{j} = \sigma\mathscr{E}$, with σ given by Eq. (6.32).

The Sommerfeld's derivation of the Drude formula poses an important conceptual question. The structure of Eq. (6.30) implies that the only quantum states contributing to the electric conductivity are those whose derivative $[-\partial\langle N(\varepsilon)\rangle/\partial\varepsilon]$ is significant. For the Fermi particles such as electrons, in the limit $T \ll \varepsilon_F$, these are the states at the very surface of the Fermi sphere. On the other hand, Eq. (6.32) involves the density n of *all* electrons. So, exactly which electrons are responsible for the conductivity: all of them, or only those on the Fermi surface? For the resolution of this paradox, let us return to Eq. (6.22) and analyze the physical meaning of that result. For that, let us compare it with the following model distribution:

$$w_{\text{model}} \equiv w_0(\mathbf{r}, \mathbf{p} - \tilde{\mathbf{p}}, t), \tag{6.33}$$

where $\tilde{\mathbf{p}}$ is some constant, small vector, which describes a small shift of the unperturbed distribution w_0 as a whole, in the momentum space. Performing the Taylor expansion of Eq. (6.33) in this small parameter, and keeping only two leading terms, we get

$$w_{\text{model}} \approx w_0(\mathbf{r}, \mathbf{p}, t) + \tilde{w}_{\text{model}}, \quad \text{with} \quad \tilde{w}_{\text{model}} = -\tilde{\mathbf{p}} \cdot \nabla_p w_0(\mathbf{r}, \mathbf{p}, t). \tag{6.34}$$

Comparing the last expression with the first form of Eq. (6.22), we see that they coincide if

$$\tilde{\mathbf{p}} = q\mathscr{E}\tau \equiv \mathscr{F}\tau. \tag{6.35}$$

This means that Eq. (6.22) describes a small shift of the equilibrium distribution of all particles (in the momentum space) by $q\mathscr{E}\tau$ along the direction of electric field, justifying the cartoon shown in figure 6.4.

At $\mathscr{E} = 0$, the system is in equilibrium, so that the quantum states inside the Fermi sphere ($p < p_F$), are occupied, while those outside of it are empty (figure 6.4a). Electron scattering events may happen only between states within a very thin layer ($|p^2/2m - \varepsilon_F| \sim T$) at the Fermi surface, because only in this layer are the states partially occupied, so that both components of the product $w(\mathbf{r}, \mathbf{p}, t)[1 - w(\mathbf{r}, \mathbf{p}', t)]$, mentioned in section 6.1, do not vanish. These scattering events, on the average, do not change the equilibrium probability distribution, because they are uniformly spread over the Fermi surface.

Now let the electric field be turned on instantly. Immediately it starts accelerating all electrons in its direction, i.e. the whole Fermi sphere starts moving in the

[18] See also *Part EM* section 4.2.

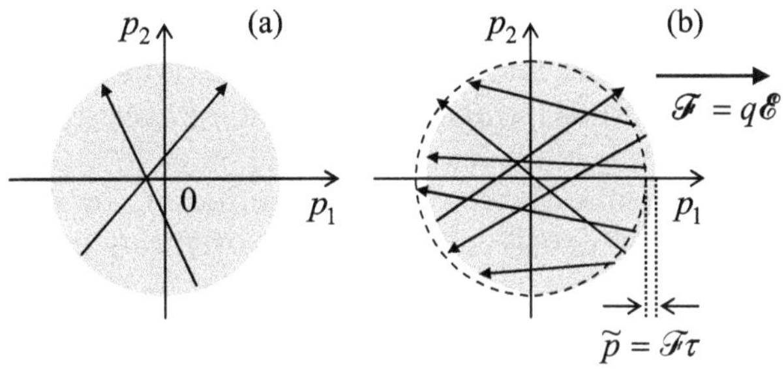

Figure 6.4. Filling of momentum states in a degenerate electron gas: (a) in the absence and (b) in the presence of an external electric field $\mathscr{E}$. Arrows show representative scattering events.

momentum space, along the field's direction in the real space. For elastic scattering events (with $|p'| = |p|$), this creates an addition of occupied states at the leading front of the accelerating sphere, and an addition of free states on its trailing edge (figure 6.4b). As a result, now there are more scattering events bringing electrons from the leading edge to the trailing edge of the sphere than in the opposite direction. This creates the average backflow of the state occupancy in the momentum space. These two trends eventually cancel each other, and the Fermi sphere approaches a stationary (though not a thermal-equilibrium!) state, with the shift (6.35) relatively to its thermal-equilibrium position.

Now figure 6.4b may be used to answer to the question which of the two different interpretations of the Drude formula is correct, and the answer is: *either*. On one hand, we can look at the electric current at a result of the shift (6.35) of *all* electrons in the momentum space. On the other hand, each filled quantum state deep inside the sphere gives exactly the same contribution into the net current density as it did without the field. All these internal contributions to the net current cancel each other, so that the applied field changes the situation only at the Fermi surface. Thus it is equally legitimate to say that only the surface states are responsible for the nonvanishing net current[19].

Let me also mention another paradox related to the Drude formula, which is often misunderstood (not only by students :-). As was emphasized above, τ is nonvanishing even at *elastic* scattering—that by itself does not change the total energy of the electron gas. The question is how can such scattering be responsible for the Ohmic resistivity $\rho \equiv 1/\sigma$, and hence for the Joule heat production, with the power density $\mathscr{p} = \mathbf{j} \cdot \mathscr{E} = \rho j^2$?[20] The answer is that the Drude/Sommerfeld formulas describe just the 'bottleneck' of the Joule heat formation. In the scattering picture (figure 6.4b) the states filled by elastically scattered electrons are located above the (shifted) Fermi surface, and these electrons eventually need to relax onto it via some inelastic process,

[19] So here, as it frequently happens in physics, formulas (or graphical sketches, such as figure 6.4b) give a more clear and unambiguous description of the reality than words—the privilege lacked by many 'scientific' disciplines, rich with unending, shallow verbal debates. Note also that, as frequently happens in physics, the dual interpretation of σ is expressed by two different but equal integrals (6.30) and (6.31), related by the integration by parts.

[20] This formula is probably self-evident, but if you need, you may revisit *Part EM* section 4.4.

which releases their excessive energy in the form of heat (in solid state, described by phonons—see section 2.6). The rate and other features of these inelastic phenomena do not participate in the Drude formula directly, but for keeping the theory valid (in particular, keeping the probability distribution w close to its equilibrium value w_0), their intensity has to be sufficient to avoid gas overheating by the applied field. In some poorly conducting materials, charge carrier overheating effects, resulting in deviations from the Ohm law, i.e. from the linear relation (6.28) between $\mathbf{j}$ and $\mathscr{E}$, may be readily observed already at rather practicable electric fields.

One final comment is that the Sommerfeld theory of the Ohmic conductivity works very well for the electron gas in most conductors. The scheme shown in figure 6.4 helps to understand why: for degenerate Fermi gases the energies of all particles whose scattering contributes to transport properties, are close ($\varepsilon \approx \varepsilon_F$), and prescribing them all the same relaxation time τ is very reasonable. In contrast, in classical gases, with their relatively broad distribution of ε, some results given by the Boltzmann-RTA equation (6.18) are valid only by the order of magnitude.

6.3 Electrochemical potential and the drift–diffusion equation

Now let us generalize our calculation to the case when the particle transport takes place in the presence of a time-independent spatial gradient of the probability distribution, $\nabla_r w \neq 0$, caused for example by that of the particle concentration $n = N/V$ (and hence, according to Eq. (3.40), of the chemical potential μ), while still considering temperature T constant. For this generalization, we should just keep the second term in the left-hand part of Eq. (6.18). If the gradient of w is sufficiently small, we can repeat the arguments of the last section and replace w with w_0 in this term as well. With the applied electric field $\mathscr{E}$ represented as $(-\nabla\phi)$,[21] where ϕ is the electrostatic potential, Eq. (6.25) now becomes

$$\tilde{w} = \tau\, \mathbf{v} \cdot \left(\frac{\partial w_0}{\partial \varepsilon} q \nabla \phi - \nabla w_0 \right). \qquad (6.36)$$

Since in any of the equilibrium distributions (6.20), $\langle N(\varepsilon)\rangle$ is a function of ε and μ only in the combination $(\varepsilon - \mu)$, it obeys the following relation,

$$\frac{\partial \langle N(\varepsilon)\rangle}{\partial \mu} = -\frac{\partial \langle N(\varepsilon)\rangle}{\partial \varepsilon}. \qquad (6.37)$$

Using this relation, the gradient of $w_0 \propto \langle N(\varepsilon)\rangle$ may be represented as[22]

$$\nabla w_0 = -\frac{\partial w_0}{\partial \varepsilon} \nabla \mu, \quad \text{for } T = \text{const}, \qquad (6.38)$$

[21] Since we will not encounter ∇_p in the balance of this chapter, from this point on, the subscript r of the operator ∇ is dropped for the notation brevity.

[22] Since we consider w_0 as a function of two *independent* arguments $\mathbf{r}$ and $\mathbf{p}$, taking its gradient, i.e. the differentiation of this function over $\mathbf{r}$, does not involve its differentiation over the kinetic energy ε—which is a function of $\mathbf{p}$ only.

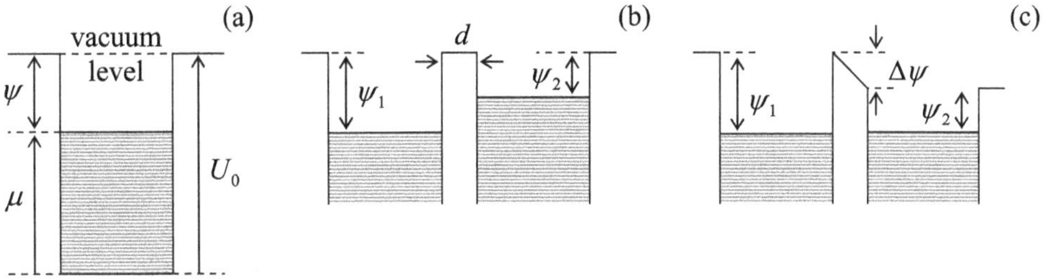

Figure 6.5. Potential profiles of (a) a single conductor and (b, c) a system of two closely located conductors, for two different biasing situations: (b) zero electrostatic field (the 'flat-band condition'), and (c) zero voltage $\Delta\mu'$.

so that Eq. (6.36) becomes

$$\tilde{w} = \tau \frac{\partial w_0}{\partial \varepsilon} \mathbf{v} \cdot (q\nabla\varphi + \nabla\mu) \equiv \tau \frac{\partial w_0}{\partial \varepsilon} \mathbf{v} \cdot \nabla\mu', \tag{6.39}$$

where the following sum,

$$\mu' \equiv \mu + q\phi, \tag{6.40}$$

is called the *electrochemical potential.* Now repeating the calculation of the electric current, carried out in the last section, we get the following generalization of the Ohm law (6.28):

$$\mathbf{j} = \sigma(-\nabla\mu'/q) \equiv \sigma\,\mathcal{E}, \tag{6.41}$$

where the *effective electric field* $\mathcal{E}$ is proportional to the gradient of the electrochemical potential, rather of the electrostatic potential:

$$\mathcal{E} \equiv -\frac{\nabla\mu'}{q} = \mathscr{E} - \frac{\nabla\mu}{q}. \tag{6.42}$$

The physics of this extremely important and general result[23] may be explained in two ways. First, let us have a look at the energy spectrum of a degenerate Fermi-gas confined in a volume of finite size, but otherwise free. In order to ensure such a confinement, we need a piecewise-constant potential $U(\mathbf{r})$—a 'hard-wall, flat-bottom potential well'—see figure 6.5a. (In a solid-state conductor, such a profile is provided by the crystal lattice of positively charged ions of the crystal lattice.) The well should be of a sufficient depth $U_0 > \varepsilon_F \equiv \mu|_{T=0}$ in order to provide the confinement of the overwhelming majority of the particles, with energies below and somewhat above the Fermi level ε_F. This means that there should be a substantial energy gap,

$$\psi \equiv U_0 - \mu \gg T, \tag{6.43}$$

between the Fermi energy of a particle inside the well, and its potential energy U_0 outside the well. (The latter value of energy is usually called the *vacuum level.*)

[23] Note that Eq. (6.42) does not include the phenomenological parameter τ of the relaxation-time approximation, signaling that it is more general than the RTA. Indeed, it is based entirely on the relation between the second and third terms in the left-hand part of the general Boltzmann equation (6.10), rather than on any details of the scattering integral on its right-hand side.

The difference defined by Eq. (6.43) is called the *workfunction*[24]; for most metals, it is between 4 and 5 eV, so that the relation $\psi \gg T$ is well fulfilled for the room temperatures ($T \sim 0.025$ eV)—and actually for all temperatures up to the metal's evaporation point.

Now let us consider two conductors, with different values of ψ, separated by a small gap d—see figure 6.5b,c. Panel b shows the case when the electric field $\mathscr{E} = -\nabla\phi$ in the free-space gap between the conductors equals zero, i.e. their *electrostatic* potentials ϕ are equal[25]. If there is an opportunity for particles to cross the gap (e.g. by either the thermally-activated hopping *over* the potential barrier, discussed in sections 5.6–5.7, or the quantum-mechanical tunneling *through* it), there will be an average flux of particles from the conductor with the higher Fermi level to that with the lower Fermi level[26], because the chemical equilibrium requires their equality—see sections 1.5 and 2.7. If the particles have an electric charge (as electrons do), the equilibrium will be automatically achieved by them recharging the effective capacitor formed by the conductors, until the electrostatic energy difference $q\Delta\phi$ reaches the value reproducing that of the workfunctions (figure 6.5c). So for the equilibrium potential difference[27] we may write

$$q\Delta\phi = \Delta\psi = -\Delta\mu. \tag{6.44}$$

At this equilibrium, the electric field in the gap between the conductors is

$$\mathscr{E} \equiv -\frac{\Delta\phi}{d}\mathbf{n} = \frac{\Delta\mu}{qd}\mathbf{n} = \frac{\nabla\mu}{q}; \tag{6.45}$$

in figure 6.5c this field is clearly visible as the tilt of the electric potential profile. Comparing Eq. (6.45) with the definition (6.42) of the effective electric field $\mathcal{E}$, we see that the equilibrium, i.e. the absence of current through the potential barrier, is achieved exactly when $\mathcal{E} = 0$, in accordance with Eq. (6.41).

The electric field dichotomy, $\mathscr{E} \leftrightarrow \mathcal{E}$, raises a natural question: which of these fields we are speaking about in the everyday and laboratory practice? Upon some contemplation, the reader should agree that most of our electric field measurements are done indirectly, by measuring corresponding voltages—with voltmeters. A vast majority of these instruments belong to the *electrodynamic* variety, which is based on the measurement of a small current flowing through the voltmeter. As Eq. (6.41) shows, electrodynamic voltmeters measure the *electrochemical* potential difference $\Delta\mu'/q$. However, there exist a rare breed of *electrostatic* voltmeters (also called 'electrometers') that measure the *electrostatic* potential difference $\Delta\phi$ between two

[24] Sometimes it is also called the 'electron affinity', though this term is mostly used for atoms and molecules.

[25] In semiconductor physics and engineering, the situation shown in figure 6.5b is called the *flat-band condition*, because any electric field applied normally to a surface of a semiconductor leads to the so-called energy *band bending*—see the next section.

[26] As measured from a common reference value, for example from the vacuum level—rather from the bottom of an individual potential well as in figure 6.5a.

[27] In physics literature, it is usually called the *contact potential difference*, while in electrochemistry (for which it is one of the key notions), the term *Volta potential* is more common.

conductors. One way to implement such an instrument is to use an ordinary, electrodynamic voltmeter, but with the reference point set at the flat-band condition (figure 6.5b) between the conductors. This condition may be detected by vanishing electric charge on the adjacent surfaces of the conductors, and hence by the absence of its modulation in time, caused by a specially arranged periodic variation of the distance between the surfaces.

Now let me return to Eq. (6.41) to make two very important remarks. First, it says that in the presence of electric field, the current vanishes only if $\nabla\mu' = 0$, i.e. the electrochemical potential μ', rather than the chemical potential μ, has to be position-independent in a system in thermodynamic (thermal, chemical, and electric) equilibrium of a conducting system. This result by no means contradicts the fundamental thermodynamic relations for μ discussed in section 1.5, or the statistical relations involving μ, which were discussed in section 2.7 and beyond; indeed, according to Eq. (6.40), $\mu'(\mathbf{r})$ is 'merely' the chemical potential referred to the local value of the electrostatic energy $q\phi(\mathbf{r})$, and in all previous parts of the course this energy was implied to be constant through the system.

Second, note another interpretation of Eq. (6.41), which may be achieved by modifying Eq. (6.38) for the particular case of a classical gas. Indeed, the gas' local density $n \equiv N/V$ obeys Eq. (3.32), which may be represented as

$$n(\mathbf{r}) = \text{const} \times \exp\left\{\frac{\mu(\mathbf{r})}{T}\right\}. \tag{6.46}$$

Taking the spatial gradient of the both sides of this relation (still at constant T), we get

$$\nabla n = \text{const} \times \frac{1}{T}\exp\left\{\frac{\mu}{T}\right\}\nabla\mu = \frac{n}{T}\nabla\mu, \tag{6.47}$$

so that $\nabla\mu = (T/n)\nabla n$, and Eq. (6.41), with σ given by Eq. (6.32), may be recast as

$$\mathbf{j} = \sigma\left(-\frac{\nabla\mu'}{q}\right) \equiv \frac{q^2\tau}{m}n\left(-\nabla\phi - \frac{1}{q}\nabla\mu\right) \equiv q\frac{\tau}{m}(nq\mathscr{E} - T\nabla n). \tag{6.48}$$

Hence the current density may be viewed as consisting of two independent parts: one due to the 'usual' electric field $\mathscr{E} = -\nabla\phi$, and another due to the particle diffusion—see Eq. (5.118) and its discussion. This is exactly the physics of the 'mysterious' term $\nabla\mu$ in Eq. (6.42), though it may be represented in the simple form (6.48) only in the classical limit.

Besides being very useful for applications, Eq. (6.48) gives us a pleasant surprise. Namely, plugging it into the continuity equation for electric charge[28],

$$\frac{\partial(qn)}{\partial t} + \nabla\cdot\mathbf{j} = 0, \tag{6.49}$$

[28] If this relation is not evident, please revisit *Part EM* section 4.1.

we get (after the division of all terms by $q\tau/m$) the so-called *drift–diffusion equation*[29]:

$$\frac{m}{\tau}\frac{\partial n}{\partial t} = \nabla(n\nabla U) + T\nabla^2 n, \quad \text{with } U \equiv q\phi. \tag{6.50}$$

Comparing it with Eq. (5.122), we see that the drift–diffusion equation is identical to the Smoluchowski equation[30], provided that we parallel the ratio τ/m with the mobility $\mu_m = 1/\eta$ of the Brownian particle. Now using the Einstein relation (5.78), we see that the effective diffusion constant D of the classical gas of similar particles is

$$D = \frac{\tau T}{m}. \tag{6.51a}$$

This important relation is more frequently represented in either of two other forms. First, since the rare scattering events we are considering do not change the statistics of the gas in thermal equilibrium, we may still use the Maxwell-distribution result (3.9) for the average-square velocity $\langle v^2 \rangle$, to recast Eq. (6.51*a*) as

$$D = \frac{1}{3}\langle v^2 \rangle\, \tau. \tag{6.51b}$$

One more popular form of the same relation uses the notion of the *mean free path l*, which may be defined as the average distance passed by the particle between two sequential scattering events:

$$D = \frac{1}{3} l\, \langle v^2 \rangle^{1/2}, \quad \text{with} \quad l \equiv \langle v^2 \rangle^{1/2} \tau. \tag{6.51c}$$

In the forms (6.51*b*)–(6.51*c*), the result for D makes more physical sense, because it may be readily derived (admittedly, with some uncertainty of the numerical coefficient) from simple kinetic arguments—a task left for the reader. Note that since the definition of τ in Eq. (6.17) is phenomenological, so is the above definition of l; this is why different definitions of this parameter, which may differ by a numerical factor of the order of 1, are possible.

Note also that using Eq. (6.51*a*), Eq. (6.48) is frequently rewritten as an expression for the *particle flow density* $\mathbf{j}_n \equiv n\mathbf{j}_w = \mathbf{j}/q$:

$$\mathbf{j}_n = n\mu_m q\mathscr{E} - D\nabla n, \tag{6.52}$$

with the first term on the right-hand side describing particles' drift, and the second one, their diffusion. I will discuss the application of this equation, to the most important case of non-degenerate (quasi-classical) gases of electrons and holes in semiconductors, in the next section.

To complete this section, let me emphasize again that the mathematically-similar drift–diffusion equation (6.50) and the Smoluchowski equation (5.122) describe

[29] Sometimes this term is associated with Eq. (6.52). One may also run into the term 'convection–diffusion equation' for Eq. (6.50) with the replacement (6.51*a*).

[30] And hence, at negligible ∇U, identical to the diffusion equation (5.116).

different physical situations. Indeed, our (or rather Einstein and Smoluchowski's) treatment of the Brownian motion in chapter 5 was based on a strong hierarchy of the total system, consisting of a large 'Brownian particle' in an environment of many smaller particles—'molecules'. On the other hand, in this chapter we are considering a gas of similar particles. Nevertheless, the equations describing the dynamics of their probability distribution, are the same—at least within the framework of the Boltzmann transport equation within the relaxation-time approximation (6.17) of the scattering integral. The origin of this similarity is that Eq. (6.12) is clearly applicable to a Brownian particle as well, with each 'scattering' event being the particle's hit by a random molecule of its environment. Since, due to the mass hierarchy, the particle momentum change at each such event is very small, the scattering integral has to be local, i.e. depend only on w at the same momentum $\mathbf{p}$ as the left-hand part of the Boltzmann equation, so that the relaxation time approximation (6.17) is absolutely natural—indeed, more natural than for our current case of similar particles.

6.4 Charge carriers in semiconductors

Now let me make a detour to demonstrate the application of the concepts discussed in the last section to understanding the basic kinetic properties of semiconductors and key semiconductor structures—which are the basis of most modern electronic and optoelectronic devices, and hence of all our IT civilization. For that, we will need to start with their equilibrium properties.

I will use an approximate but reasonable picture in which the energy of the electron subsystem in a solid may be partitioned into the sum of effective energies ε of independent electrons. Quantum mechanics says[31] that in such periodic structures as crystals, the stationary state energy ε of a particle interacting with the atomic lattice follows one of periodic functions $\varepsilon_n(\mathbf{q})$ of the *quasi-momentum* $\mathbf{q}$, oscillating between two extreme values $\varepsilon_n|_{\min}$ and $\varepsilon_n|_{\max}$. These *allowed energy bands* are separated by *bandgaps*, of widths $\Delta_n \equiv \varepsilon_n|_{\min} - \varepsilon_{n-1}|_{\max}$, with no allowed states inside them. Semiconductors and insulators (dielectrics) are defined as such crystals that in equilibrium at $T = 0$, all electron states in several energy bands (with the highest of them called the *valence band*) are completely filled, $\langle N(\varepsilon_v)\rangle = 1$, while those in the upper bands, starting from the lowest, *conduction band*, are completely empty, $\langle N(\varepsilon_c)\rangle = 0$.[32] Since the electrons follow the Fermi–Dirac statistics (2.115), this means that at $T \to 0$, the Fermi energy $\varepsilon_F \equiv \mu(0)$ is located somewhere between

[31] See, e.g. *Part QM* sections 2.7 and 3.4, though this material is not necessary for following discussions of this section. If the reader is not familiar with the notion of quasi-momentum (alternatively called the 'crystal momentum'), the following semi-qualitative interpretation may be useful: $\mathbf{q}$ is the result of quantum averaging of the genuine electron momentum $\mathbf{p}$ over the crystal lattice period. In contrast to $\mathbf{p}$, which is not conserved because of the electron interaction with the atomic lattice, $\mathbf{q}$ is an integral of motion—in the absence of other forces.

[32] In insulators, the bandgap Δ is so large (e.g. ~9 eV in SiO_2) that the conduction band remain unpopulated in all practical situations, so that the following discussion is only relevant for semiconductors, with their moderate bandgaps—such as 1.14 eV in the most important case of silicon at room temperature.

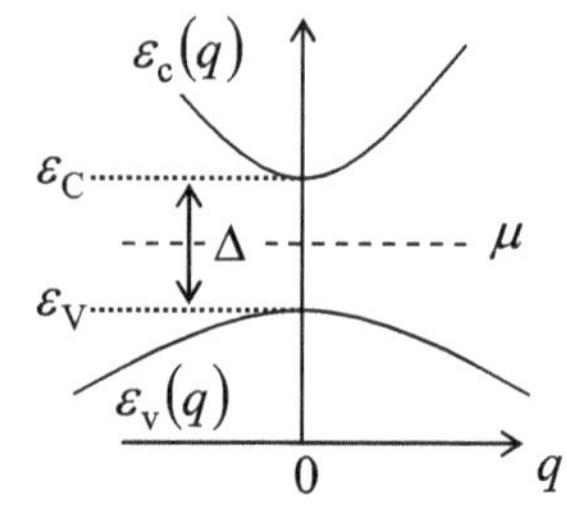

Figure 6.6. Calculating μ in an intrinsic semiconductor.

the valence band's maximum $\varepsilon_v|_{max}$ (usually called simply ε_V), and the conduction band's minimum $\varepsilon_c|_{min}$ (called ε_C)—see figure 6.6.

Let us calculate the population of both branches $\varepsilon_n(\mathbf{q})$, and the chemical potential μ in equilibrium at $T > 0$. Since the functions $\varepsilon_n(\mathbf{q})$ are typically smooth, near the bandgap edges the dispersion laws $\varepsilon_c(\mathbf{q})$ and $\varepsilon_v(\mathbf{q})$ may be well approximated with quadratic parabolas. For our analysis, let us take the parabolas the simplest, isotropic form, with origins at the same quasi-momentum, taking it for the reference point[33]:

$$\varepsilon = \begin{cases} \varepsilon_C + q^2/2m_C, & \text{for } \varepsilon \geqslant \varepsilon_C, \\ \varepsilon_V - q^2/2m_V, & \text{for } \varepsilon \leqslant \varepsilon_V, \end{cases} \quad \text{with} \quad \varepsilon_C - \varepsilon_V \equiv \Delta. \tag{6.53}$$

The positive constants m_C and m_V are usually called, respectively, the electron and hole effective masses. (In typical semiconductors, m_C is a few times smaller than the free electron mass m_e, while m_V is closer to m_e.)

Due to the similarity between the top line of Eq. (6.53) and the dispersion law (3.3) of free particles, we may re-use Eq. (3.40), with the appropriate particle mass m, the degeneracy factor g, and the energy origin, to calculate the density of populated states (in semiconductor physics, called *electrons* in the narrow sense of the word):

$$n \equiv \frac{N_e}{V} = \int_{\varepsilon_C}^{\infty} \langle N(\varepsilon)\rangle g_3(\varepsilon) d\varepsilon \equiv \frac{g_C m_C^{3/2}}{\sqrt{2}\pi^2\hbar^3} \int_0^{\infty} \langle N(\tilde{\varepsilon}+\varepsilon_C)\rangle\, \tilde{\varepsilon}^{1/2} d\tilde{\varepsilon}, \tag{6.54}$$

where $\tilde{\varepsilon} \equiv \varepsilon - \varepsilon_C \geqslant 0$. Similarly, the density p of 'no-electron' excitations (called *holes*) in the valence band is the number of *unfilled* states in the band, and hence may be calculated as

$$\begin{aligned} p \equiv \frac{N_h}{V} &= \int_{-\infty}^{\varepsilon_V} [1 - \langle N(\varepsilon)\rangle]\, g_3(\varepsilon) d\varepsilon \\ &\equiv \frac{g_V m_V^{3/2}}{\sqrt{2}\pi^2\hbar^3} \int_0^{\infty} [1 - \langle N(\varepsilon_V - \tilde{\varepsilon})\rangle]\, \tilde{\varepsilon}^{1/2} d\tilde{\varepsilon}, \end{aligned} \tag{6.55}$$

[33] It is easy (and hence is left for the reader's exercise) to verify that all equilibrium properties of charge carriers remain the same (with some effective values of m_C and m_V) if $\varepsilon_c(\mathbf{q})$ and $\varepsilon_v(\mathbf{q})$ are arbitrary quadratic forms of the Cartesian components of the quasi-momentum. A displacement of the branches $\varepsilon_c(\mathbf{q})$ and $\varepsilon_v(\mathbf{q})$ in the quasi-momentum space is also unimportant for statistical and most transport properties of the semiconductors, though it is very important for their optical properties—which I will not have time to discuss in any detail.

where in this case, $\tilde{\varepsilon} \geqslant 0$ is defined as $(\varepsilon_V - \varepsilon)$. If the electrons and holes[34] are in the thermal and chemical equilibrium, the functions $\langle N(\varepsilon)\rangle$ in these two relations should follow the Fermi–Dirac distribution (2.115) with the same temperature T and the same chemical potential μ. Moreover, in our current case of an undoped (*intrinsic*) semiconductor, these densities have to be equal,

$$n = p \equiv n_i, \tag{6.56}$$

because if this *electroneutrality condition* was violated, the volume would acquire a nonvanishing electric charge density $\rho = e(p - n)$, which would result, for a bulk sample, in an extremely high electric field energy. From this condition, we get a system of two equations,

$$\begin{aligned} n_i &= \frac{g_C m_C^{3/2}}{\sqrt{2}\pi^2\hbar^3}\int_0^\infty \frac{\tilde{\varepsilon}^{1/2} d\tilde{\varepsilon}}{\exp\{(\tilde{\varepsilon} + \varepsilon_C - \mu)/T\} + 1} \\ &= \frac{g_V m_V^{3/2}}{\sqrt{2}\pi^2\hbar^3}\int_0^\infty \frac{\tilde{\varepsilon}^{1/2} d\tilde{\varepsilon}}{\exp\{(\tilde{\varepsilon} - \varepsilon_V + \mu)/T\} + 1}, \end{aligned} \tag{6.57}$$

whose solution gives both the requested charge carrier density n_i and the Fermi level μ.

For an arbitrary ratio Δ/T, this solution may be carried out only numerically, but in most practical cases, this ratio is very large. (Again, for Si at room temperature, $\Delta \approx 1.14$ eV, while $T \approx 0.025$ eV.) In this case, we may use the same classical approximation as in Eq. (3.45), to reduce Eqs. (6.54) and (6.55) to simple expressions

$$n = n_C \exp\left\{\frac{\mu - \varepsilon_C}{T}\right\}, \quad p = n_V \exp\left\{\frac{\varepsilon_V - \mu}{T}\right\}, \quad \text{for } T \ll \Delta, \tag{6.58}$$

where the temperature-dependent parameters

$$n_C \equiv \frac{g_C}{\hbar^3}\left(\frac{m_C T}{2\pi}\right)^{3/2} \quad \text{and} \quad n_V \equiv \frac{g_V}{\hbar^3}\left(\frac{m_V T}{2\pi}\right)^{3/2} \tag{6.59}$$

are, physically, the effective numbers of states (per unit volume) available for the thermal excitation in, respectively, the conduction and valence bands. For usual semiconductors (with $g_C \sim g_V \sim 1$, and $m_C \sim m_V \sim m_e$), at room temperature, these numbers are of the order of 3×10^{25} m^{-3} $\equiv 3 \times 10^{19}$ cm^{-3}. (Note that all results based on Eqs. (6.58) are only valid if both n and p are much lower than, respectively, n_C and n_V.)

With the substitution of Eqs. (6.58), the system of equations (6.56) allows a straightforward solution:

$$\mu = \frac{\varepsilon_V + \varepsilon_C}{2} + \frac{T}{2}\left(\ln\frac{g_V}{g_C} + \frac{3}{2}\ln\frac{m_V}{m_C}\right), \quad n_i = (n_C n_V)^{1/2}\exp\left\{-\frac{\Delta}{2T}\right\}. \tag{6.60}$$

[34] The collective name for them in semiconductor physics is *charge carriers*—or just 'carriers'.

Since in all practical materials the logarithms in the first of these expressions are never much larger than 1,[35] it shows that the Fermi level in intrinsic semiconductors never deviates substantially from the so-called *midgap value* $(\varepsilon_V + \varepsilon_C)/2$—see the dashed line in the (schematic) figure 6.6. In the result for n_i, the last (exponential) factor is very small, so that the equilibrium number of charge carriers is much lower than that of the atoms—for the most important case of silicon at room temperature, $n_i \sim 10^{10}$ cm^{-3}. The exponential temperature dependence of n_i (and hence of the electric conductivity $\sigma \propto n_i$) of intrinsic semiconductors is the basis of several applications, for example simple *germanium resistance thermometers*, efficient in the whole range from ~0.5 K to ~100 K. Another useful application of the same fact is the extraction of the bandgap of a semiconductor from the experimental measurement of the temperature dependence of σ—frequently, in just two well-separated temperature points.

However, most applications require much higher concentration of carriers. It may be increased dramatically by placing into a semiconductor a relatively small number of slightly different atoms—either *donors* (e.g. phosphorus atoms for Si) or *acceptors* (e.g. boron atoms for Si). Let us analyze the first opportunity, called *n-doping*, using the same simple energy band model (6.53). If the donor atom is only slightly different from those in the crystal lattice, it may be easily ionized—giving an additional electron to the conduction band, and hence becoming a positive ion. This means that the effective ground state energy ε_D of the additional electrons is just slightly below the conduction band edge ε_C—see figure 6.7a.[36]

Reviewing the arguments that have led us to Eqs. (6.58), we see that at relatively low doping, when the strong inequalities $n \ll n_C$ and $p \ll n_V$ still hold, these relations are not affected by the doping, so that the concentrations of electrons and holes given by these equalities still obey a universal (doping-independent) relation[37]

$$np = n_i^2. \tag{6.61}$$

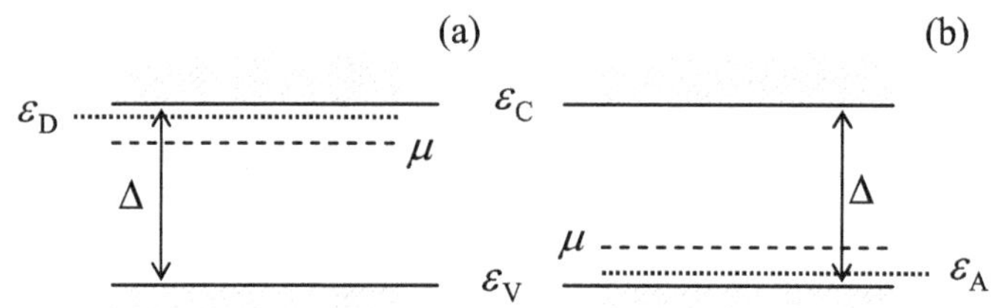

Figure 6.7. The Fermi levels μ in (a) n-doped and (b) p-doped semiconductors. Hatching shows the regions of available electron state energies.

[35] Note that in the case of simple electron spin degeneracy ($g_V = g_C = 2$), the first logarithm vanishes altogether. However, in many semiconductors, the degeneracy is factored by the number of similar energy bands (e.g. silicon has six similar conduction bands), and the factor $\ln(g_V/g_C)$ may slightly affect quantitative results.

[36] Note that in comparison with figure 6.6, here the (for most purposes, redundant) information on the q-dependence of the energies is collapsed, leaving the horizontal axis of such a *band-edge diagram* free for showing their possible spatial dependences—see figures 6.8, 6.10, and 6.11 below.

[37] Very similar relations may be met in the theory of chemical reactions (where it is called the *law of mass action*), and other disciplines—including such exotic examples as the theoretical ecology.

However, for a doped semiconductor the electroneutrality condition looks different from Eq. (6.56), because the total density of positive charges in a unit volume is not p, but rather $(p + n_+)$, where n_+ is the density of positively-ionized ('activated') donor atoms, so that the electroneutrality condition becomes

$$n = p + n_+. \tag{6.62}$$

If virtually all dopants are activated, as in most practical cases[38], then we may take $n_+ = n_D$, where n_D is the total concentration of donor atoms, i.e. their number per unit volume, and Eq. (6.62) becomes

$$n = p + n_D. \tag{6.63}$$

Plugging in the expression $p = n_i^2/n$, following from Eq. (6.61), we get a simple quadratic equation for n, with the following physically acceptable (positive) solution:

$$n = \frac{n_D}{2} + \left(\frac{n_D^2}{4} + n_i^2\right)^{1/2}. \tag{6.64}$$

This result shows that the doping affects n (and hence $\mu = \varepsilon_C - T\ln[n_C/n]$, and $p = n_i^2/n$) only if the dopant concentration n_D is comparable with, or higher than the intrinsic carrier density n_i given by Eq. (6.60). For most applications, n_D is made much higher than n_i; in this case Eq. (6.64) yields

$$n \approx n_D \gg n_i, \quad p = \frac{n_i^2}{n} \approx \frac{n_i^2}{n_D} \ll n, \quad \mu \approx \mu_p \equiv \varepsilon_C - T\ln\frac{n_C}{n_D}. \tag{6.65}$$

Because of the reasons to be discussed very soon, modern electron devices require doping densities above 10^{18} cm^{-3}, so that the logarithm in Eq. (6.65) is not much larger than 1. This means that the Fermi level rises from the midgap (6.60) to a position only slightly below the conduction band edge ε_C—see figure 6.7a.

The opposite case of purely p-doping, with n_A acceptor atoms per unit volume, and a small activation (negative ionization) energy $\varepsilon_A - \varepsilon_V \ll \Delta$,[39] may be considered absolutely similarly, using the electroneutrality condition in the form

$$n + n_- = p, \tag{6.66}$$

where n_- is the number of activated (and hence negatively charged) acceptors. For the relatively high concentration ($n_i \ll n_A \ll n_V$), virtually all acceptors are activated, so that $n_- \approx n_A$, Eq. (6.66) may be approximated as $n + n_A = p$, and the analysis gives the results dual to Eq. (6.65):

$$p \approx n_A \gg n_i, \quad n = \frac{n_i^2}{p} \approx \frac{n_i^2}{n_A} \ll p, \quad \mu \approx \mu_n \equiv \varepsilon_V + T\ln\frac{n_V}{n_A}, \tag{6.67}$$

[38] Let me leave it for the reader's exercise to prove that this assumption is always valid unless the doping density n_D becomes comparable to n_C, and as a result, the Fermi energy μ moves into a $\sim T$-wide vicinity of ε_D.

[39] For the typical donors (P) and acceptors (B) in silicon, both ionization energies, $(\varepsilon_C - \varepsilon_D)$ and $(\varepsilon_A - \varepsilon_V)$, are close to 45 meV, i.e. are indeed much smaller than $\Delta \approx 1.14$ eV.

so that in this case, the Fermi level is just slightly above the valence band edge (figure 6.7b), and the number of holes far exceeds that of electrons—again, in the narrow sense of the word. Let me leave the analysis of the simultaneous n- and p-doping (which enables, in particular, so-called *compensated semiconductors* with the sign-variable difference $n - p \approx n_D - n_A$) for the reader's exercise.

Now let us consider how does a sample of a doped semiconductor (say, a p-doped one) respond to a static external electrostatic field $\mathscr{E}$ applied normally to its surface[40]. (In semiconductor integrated circuits, such a field is usually created by a voltage applied to a special highly-conducting *gate electrode*, separated from the semiconductor surface by a thin insulating layer.) Assuming that the field penetrates into the sample by a distance λ much larger than the crystal lattice period a (the assumption to be verified *a posteriori*), we may calculate the distribution of the electrostatic potential ϕ using the macroscopic version of the Poisson equation[41]. Assuming that the semiconductor occupies the semi-space $x > 0$, and that $\mathscr{E} = \mathbf{n}_x\mathscr{E}$, the equation reduces to the following 1D form[42]

$$\frac{d^2\phi}{dx^2} = -\frac{\rho(x)}{\kappa\varepsilon_0}. \tag{6.68}$$

Here κ is the dielectric constant of the semiconductor matrix—excluding the dopants and charge carriers, which in this approach are treated as 'stand-alone' charges, with the volumic density

$$\rho = e\,(p - n_- - n). \tag{6.69}$$

(As a sanity check, Eqs. (6.68) and (6.69) show that if $\mathscr{E} \equiv -d\phi/dx = 0$, then $\rho = 0$, bringing us back to the electroneutrality condition (6.66), and hence the 'flat' band-edge diagrams shown in 6.7b and 6.8a.)

In order to get a closed system of equations for the case $\mathscr{E} \neq 0$, we should take into account that the electrostatic potential $\phi \neq 0$, penetrating into the sample with the field[43], adds the potential component $q\phi(x) = -e\phi(x)$ to the energy of each electron, and hence shifts the whole local system of single-electron energy levels 'vertically' by this amount—down for $\phi > 0$, and up for $\phi < 0$. As a result, the field penetration leads to what is called *band bending*—see the band-edge diagrams schematically shown in figure 6.8b,c for two possible polarities of the applied field, which affects the distribution $\phi(x)$ via the boundary condition[44]

---

[40] A simplified version of this analysis is discussed in *Part EM* section 2.1.

[41] See, e.g. *Part EM* section 3.4.

[42] I am sorry for using, for the SI electric constant ε_0, the same Greek letter as for single-particle energies, but both notations are traditional, and the difference between these uses will be clear from the context.

[43] It is common (though not necessary) to select the reference point so that deep in the semiconductor, $\phi = 0$; in what follows I will use this convention.

[44] Here $\mathscr{E}$ is the field *just inside* the semiconductor. The free-space field necessary to create it is κ times larger—see, e.g. the same *Part EM* section 3.4, in particular Eq. (3.56).

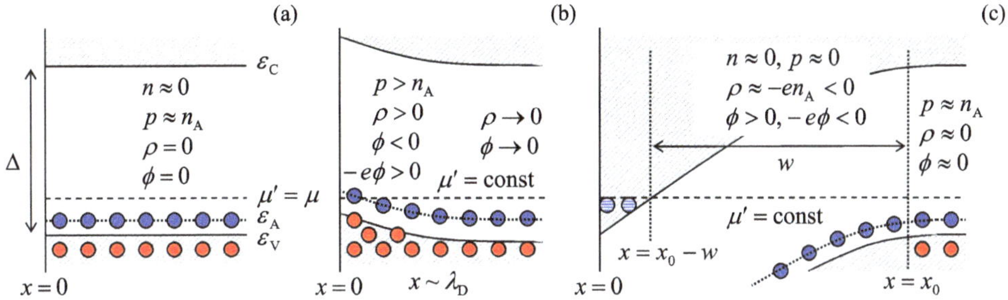

Figure 6.8. The band-edge diagrams of the electric field penetration into a uniform p-doped semiconductor: (a) $\mathscr{E} = 0$, (b) $\mathscr{E} < 0$, and (c) $\mathscr{E} > \mathscr{E}_c > 0$. Solid red points depict positive charges, solid blue points, negative charges, and hatched blue points, possible electrons in the inversion layer—all very schematically.

$$\frac{d\phi}{dx}(0) = -\mathscr{E}. \tag{6.70}$$

Note that the electrochemical potential μ' (which, in accordance with the discussion in section 6.3, replaces the chemical potential in presence of the electric field)[45], has to stay constant through the system in equilibrium, keeping the electric current vanishing—see Eq. (6.41). For arbitrary doping parameters, the system of equations (6.58) (with the replacements $\varepsilon_V \to \varepsilon_V - e\phi$, and $\mu \to \mu'$), (6.68)–(6.70), plus the relation between n_- and n_A (describing the acceptor activation), does not allow an analytical solution. However, as was discussed above, in the most practical cases $n_A \gg n_i$, we may use the approximate relations $n_- \approx n_A$ and $n \approx 0$ at virtually any values of μ' within the locally shifted bandgap $[\varepsilon_V - e\phi(x), \varepsilon_C - e\phi(x)]$, so that the substitution of these relations, and the second of Eqs. (6.58), with the mentioned replacements, into Eq. (6.69) yields

$$\begin{aligned}\rho &\approx en_V \exp\left\{\frac{\varepsilon_V - e\phi - \mu'}{T}\right\} - en_A \\ &\equiv en_A\left[\left(\frac{n_V}{n_A}\exp\left\{\frac{\varepsilon_V - \mu'}{T}\right\}\right)\exp\left\{-\frac{e\phi}{T}\right\} - 1\right].\end{aligned} \tag{6.71}$$

The x-independent electrochemical potential (a.k.a. Fermi level) μ' in this relation should be equal to the value of the chemical potential $\mu(x)$ in the semiconductor's bulk ($x \to \infty$), given by the last of Eqs. (6.67), which turns the expression in the parentheses into 1. With these substitutions, Eq. (6.68) becomes

$$\frac{d^2\phi}{dx^2} = -\frac{en_A}{\kappa\varepsilon_0}\left[\exp\left\{-\frac{e\phi}{T}\right\} - 1\right], \quad \text{for } \varepsilon_V - e\phi(x) < \mu' < \varepsilon_C - e\phi(x). \tag{6.72}$$

This nonlinear differential equation may be solved analytically, but in order to avoid being distracted by this, rather bulky solution, let me first consider the case

[45] In semiconductor physics literature, the value of μ' is usually called the *Fermi level*, even in the absence of the degenerate Fermi sea typical for metals—cf. section 3.3. In this section, I will follow this common terminology.

when the electrostatic potential is sufficiently small—either because the external field is small, or we focus on the distances sufficiently far from the surface—see figure 6.8 again. In this case, in the Taylor expansion of the exponent in Eq. (6.72), with respect to small ϕ, we may keep only two leading terms, turning it into a linear equation,

$$\frac{d^2\phi}{dx^2} = \frac{e^2 n_A}{\kappa\varepsilon_0 T}\phi, \quad \text{i.e.} \quad \frac{d^2\phi}{dx^2} = \frac{\phi}{\lambda_D^2}, \quad \text{where} \quad \lambda_D \equiv \left(\frac{\kappa\varepsilon_0 T}{e^2 n_A}\right)^{1/2}, \tag{6.73}$$

with the well-known exponential solution, satisfying also the boundary condition $\phi \to 0$ at $x \to \infty$:

$$\phi = C\exp\left\{-\frac{x}{\lambda_D}\right\}, \quad \text{at} \quad e|\phi| \ll T. \tag{6.74}$$

The constant λ_D given by last of Eqs. (6.73) is called the *Debye screening length*. It may be rather substantial; for example, at $T_K = 300$ K, even for the relatively high doping, $n_A \approx 10^{18}$ cm^{-3}, typical for modern silicon ($\kappa \approx 12$) integrated circuits, it is close to 4 nm—still much larger than the lattice constant $a \sim 0.3$ nm, so that the above analysis is indeed qualitatively valid. Note also that λ_D does not depend on the charge's sign; hence it should be no large surprise that repeating our analysis for an n-doped semiconductor, we may find out that Eqs. (6.73) and (6.74) are valid for that case as well, with the only replacement $n_A \to n_D$.

If the applied field $\mathscr{E}$ is weak, Eq. (6.74) is valid in the whole sample, and the constant C in it may be readily calculated using the boundary condition (6.70), giving

$$\phi|_{x=0} \equiv C = \lambda_D\mathscr{E} \equiv \left(\frac{\kappa\varepsilon_0 T}{e^2 n_A}\right)^{1/2}\mathscr{E}. \tag{6.75}$$

This formula allows us to express the condition of validity of the linear approximation leading to Eq. (6.74), $e|\phi| \ll T$, in terms of the applied field:

$$|\mathscr{E}| \ll \mathscr{E}_{\max}, \quad \text{with} \quad \mathscr{E}_{\max} \equiv \frac{T}{e\lambda_D} \equiv \left(\frac{T n_A}{\kappa\varepsilon_0}\right)^{1/2}; \tag{6.76}$$

in the above example, $\mathscr{E}_{\max} \sim 60$ kV cm^{-1}. On the lab scale, such a field is not low at all (it is twice higher than the threshold of electric breakdown in air at ambient conditions), but may be readily reached in solid-state structures, which are much less prone to the breakdown[46]. This is why we should be interested what happens if the applied field is higher than this value.

The semi-qualitative answer is relatively simple if the field is directed out of the p-doped semiconductor (in our nomenclature, $\mathscr{E} < 0$—see figure 6.8b). As the valence

[46] Even some amorphous thin-film insulators, such as properly formed silicon and aluminum oxides, can withstand fields up to ~10 MV cm^{-1}.

band bends up by a few T, the local hole concentration $p(x)$, and hence the charge density $\rho(x)$, grow exponentially—see Eq. (6.71). Hence the effective local length of the nonlinear field penetration, $\lambda_{ef}(x) \propto \rho^{-1/2}(x)$, shrinks exponentially. A quantitative analysis of this effect using Eq. (6.72) does not make much sense, because as soon as $\lambda_{ef}(0)$ decreases to $\sim a$, the macroscopic Poisson equation (6.68) is no longer valid quantitatively. For typical semiconductors, this happens at fields where the edge $\varepsilon_V - e\phi(0)$ of the bent valence band at the surface is raised above the Fermi level μ'. In this case, the valence-band electrons near the sample's surface form a degenerate Fermi gas, with an 'open' Fermi surface—essentially a metal, which a very small (atomic-scale) *Thomas–Fermi screening depth*[47]:

$$\lambda_{ef}(0) \sim \lambda_{TF} \equiv \left[\frac{\kappa\varepsilon_0}{e^2 g_3(\varepsilon_F)}\right]^{1/2}. \tag{6.77}$$

The effects taking place at the opposite polarity of the field, $\mathscr{E} > 0$, are much more interesting—and more useful for applications. Indeed, in this case, the band bending down leads to an exponential decrease of $\rho(x)$ as soon as the valence band edge $\varepsilon_V - e\phi(x)$ drops down by just a few T below its unperturbed value ε_V. If the applied field is large enough, $\mathscr{E} > \mathscr{E}_{max}$ (as it is in the situation shown in figure 6.8c), on the left of such point x_0 the so-called *depletion layer*, of a certain width w, is formed. Within this layer, not only is the electron density n negligible, but so is the hole density p as well, so that the only substantial contribution to the charge density ρ is given by the fully ionized acceptors: $\rho \approx -en_- \approx -en_A$, and Eq. (6.72) becomes very simple:

$$\frac{d^2\phi}{dx^2} = \frac{en_A}{\kappa\varepsilon_0} = \text{const}, \quad \text{for} \quad x_0 - w < x < x_0. \tag{6.78}$$

Let us use this equation to calculate the largest possible width w of the depletion layer, and the critical value, $\mathscr{E}_c$, of the applied field necessary for this. By definition, at $\mathscr{E} = \mathscr{E}_c$ the left boundary of the layer, where $\varepsilon_V - e\phi(x) = \varepsilon_C$, i.e. $e\phi(x) = \varepsilon_V - \varepsilon_A \equiv \Delta$, just touches the semiconductor surface: $x_0 - w = 0$, i.e. $x_0 = w$. (Figure 6.8c shows the case when $\mathscr{E}$ is slightly larger than $\mathscr{E}_c$.) For this, Eq. (6.78) has to be solved with the following boundary conditions:

$$\phi(0) = \frac{\Delta}{e}, \quad \frac{d\phi}{dx}(0) = -\mathscr{E}_c, \quad \phi(w) = 0, \quad \frac{d\phi}{dx}(w) = 0. \tag{6.79}$$

Note that the first of these conditions is strictly valid only if $T \ll \Delta$, i.e. the assumption we have made from the very beginning, while the last two conditions are asymptotically correct only if $\lambda_D \ll w$—the assumption we should not forget to check after the solution.

After all the undergraduate experience with projective motion problems, the reader certainly knows by heart that the solution of Eq. (6.78) is a quadratic parabola, so that let me immediately write its final form satisfying the boundary conditions (6.79):

[47] As a reminder, the derivation of this formula was the task of problem 3.14

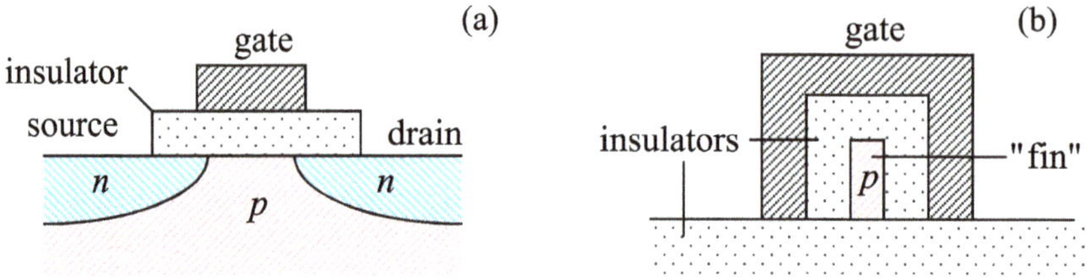

Figure 6.9. Two main species of the n-FET: (a) the *bulk FET*, and (b) the *FinFET*. While on panel (a) the current flow from the source to the drain is parallel to the plane of drawing, on panel (b) it is normal to the plane, and the n-doped source and drain contact the thin 'fin' from two sides off this plane.

$$\phi(x) = \frac{en_A}{\kappa\varepsilon_0}\frac{(w-x)^2}{2}, \quad \text{with} \quad w = \left(\frac{2\kappa\varepsilon_0\Delta}{e^2 n_A}\right)^{1/2}, \quad \text{at} \quad \mathscr{E}_c = \frac{2\Delta}{e\varepsilon_0 w}. \qquad (6.80)$$

Comparing the result for w with Eq. (6.73), we see that if our basic condition $T \ll \Delta$ is fulfilled, then $\lambda_D \ll w$, confirming the qualitative validity of the whole solution (6.80). For the same particular parameters as in the example before ($n_A \approx 10^{18}$ cm^{-3}, $\kappa \approx 10$), and $\Delta \approx 1$ eV, Eqs. (6.80) give $w \approx 40$ nm and $\mathscr{E}_c \approx 600$ kV cm^{-1}—still a practicable field. (As figure 6.8c shows, in order to create it, we need a gate voltage only slightly larger than Δ/e, i.e. close to 1 V for typical semiconductors.)

Figure 6.8c also shows that if the applied field exceeds this critical value, near the surface of the semiconductor the conduction band edge drops below the Fermi level. This is the so-called *inversion layer*, in which electrons with energies below μ' form a highly conductive degenerate Fermi gas. However, typical rates of electron tunneling from the bulk through the depletion layer are very low, so that after the inversion layer has been created (say, by the gate voltage application), it may be only populated from another source—hence the hatched blue points in figure 6.8c. This is exactly the fact used in the workhorse device of semiconductor integrated circuits—the *field effect transistor* (FET)—see figure 6.9.

In the 'bulk' variety of this structure (figure 6.9a), a gate electrode overlaps a gap between two similar highly-n-doped regions near the surface, called *source* and *drain*, formed inside a p-doped semiconductor. It is more or less obvious (and will be shown in a moment) that in the absence of gate voltage, the electrons cannot pass through the p-doped region, so that virtually no current flows between the source and the drain, even if a modest voltage is applied between these electrodes. However, if the gate voltage is positive and large enough to induce the electric field $\mathscr{E} > \mathscr{E}_c$ at the surface of the p-doped semiconductor, it creates the inversion layer, as shown in figure 6.8c, and the electron current between the source and drain electrodes may readily flow through this *surface channel*. (Very unfortunately, in this course I do not have time for a detailed analysis of transport properties of this cornerstone electron device, and have to refer the reader to special literature[48].)

[48] The classical monograph in this field is [1]. (The newer 3rd edition, circa 2006, is more tilted toward technical details.) I can also recommend a detailed textbook by R Pierret [2].

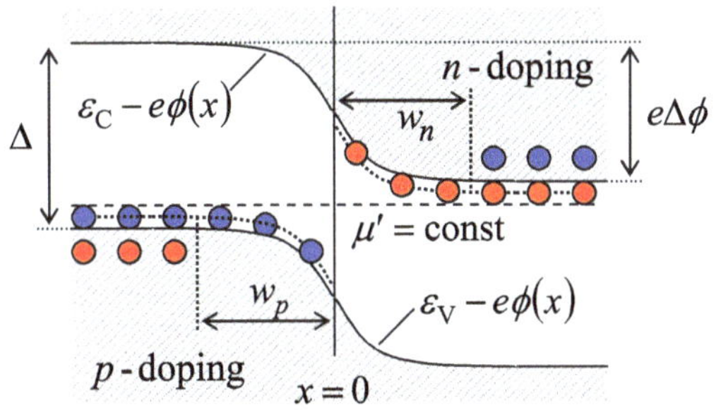

Figure 6.10. The band-edge diagram of a p–n junction in thermodynamic equilibrium (T = const, μ' = const). The notation is the same as in figures 6.7 and 6.8.

Figure 6.9a makes it obvious that another major (and virtually unavoidable) structure of semiconductor integrated circuits is the famous *p–n junction*—an interface between p- and n-doped regions. Let us analyze its simple model, in which the interface is in the plane $x = 0$, and the doping profiles $n_D(x)$ and $n_A(x)$ are step-like, making an abrupt jump at the interface:

$$n_A(x) = \begin{cases} n_A = \text{const}, & \text{at } x < 0, \\ 0, & \text{at } x > 0, \end{cases} \qquad n_D(x) = \begin{cases} 0, & \text{at } x < 0, \\ n_D = \text{const}, & \text{at } x > 0. \end{cases} \qquad (6.81)$$

(This model is very reasonable for modern integrated circuits, where the doping is performed by *implantation*, using high-energy ion beams.)

To start with, let us assume that no voltage is applied between the p- and n-regions, so that the system may be in thermodynamic equilibrium. In the equilibrium, the Fermi level μ' should be flat through the structure, and at $x \to -\infty$ and $x \to +\infty$, where $\phi \to 0$, the level structure has to approach the levels of those shown, respectively, on panels a and b of figure 6.7. In addition, the distribution of the electric potential $\phi(x)$, shifting the level structure vertically by $-e\phi(x)$, has to be continuous to avoid unphysical infinite electric fields. With that, we inevitably arrive at the band-edge diagram that is (schematically) shown in 6.10.

The diagram shows that the contact of differently doped semiconductors gives rise to a built-in electric potential difference $\Delta\phi$, equal to the difference of their values of μ in the absence of the contact—see Eqs. (6.65) and (6.67):

$$e\Delta\phi \equiv e\phi(+\infty) - e\phi(-\infty) = \mu_n - \mu_p = \Delta - T \ln \frac{n_C n_V}{n_D n_A}, \qquad (6.82)$$

which is usually just slightly smaller than the bandgap[49]. (Physically, this is essentially the same contact potential difference as was discussed, for the case of metals, in section 6.3—see figure 6.5.) The arising internal electrostatic field $\mathscr{E} = -d\phi/dx$ induces, in both semiconductors, depletion layers similar to that induced by an external field (figure 6.8c). Their widths w_p and w_n may also calculated

[49] Frequently, Eq. (6.82) is also rewritten in the form $e\Delta\varphi = T \ln(n_D n_A / n_i^2)$. In the view of Eq. (6.60), this equality is formally correct, but may be misleading, because the intrinsic carrier density n_i is an exponential function of temperature, and is physically irrelevant for this particular problem.

similarly, by solving the following boundary problem of electrostatics, generally similar to that given by Eqs. (6.78) and (6.79):

$$\frac{d^2\phi}{dx^2} = \frac{e}{\kappa\varepsilon_0} \times \begin{cases} n_A, & \text{for } -w_p < x < 0, \\ n_D, & \text{for } 0 < x < +w_n, \end{cases} \tag{6.83}$$

$$\phi(w_n) = \phi(-w_p) + \Delta\phi, \quad \frac{d\phi}{dx}(w_n) = \frac{d\phi}{dx}(-w_p) = 0,$$
$$\phi(-0) = \phi(+0), \quad \frac{d\phi}{dx}(-0) = \frac{d\phi}{dx}(+0), \tag{6.84}$$

also exact only in the limit $T \ll \Delta$, $n_i \ll n_D$, n_A. Its (easy) solution gives a result similar to Eq. (6.80):

$$\phi = \text{const} + \begin{cases} en_A(w_p + x)^2/2\kappa\varepsilon_0, & \text{for } -w_p < x < 0, \\ \Delta\phi - en_D(w_n - x)^2/2\kappa\varepsilon_0, & \text{for } 0 < x < +w_n, \end{cases} \tag{6.85}$$

with expressions for w_p and w_n giving the following formula for the full depletion layer width:

$$w \equiv w_p + w_n = \left(\frac{2\kappa\varepsilon_0\Delta\phi}{en_{ef}}\right)^{1/2}, \quad \text{with } n_{ef} \equiv \frac{n_A n_D}{n_A + n_D}, \quad \text{i.e. } \frac{1}{n_{ef}} = \frac{1}{n_A} + \frac{1}{n_D}. \tag{6.86}$$

This expression is similar to that given by Eq. (6.80), so that for typical highly doped semiconductors ($n_{ef} \sim 10^{18}$ cm^{-3}) it gives for w a similar estimate of a few tens nm[50]. Returning to figure 6.9a, we see that this scale imposes an essential limit on the reduction of bulk FETs (whose scaling down is at the heart of the well-known *Moore's law*)[51], explaining why such high doping is necessary. In the early 2010s, the problems with implementing even higher doping, plus dissipated power issues, have motivated the transition of advanced silicon integrated circuit technology from the bulk FETs to the *FinFET* (also called 'double-gate', or 'tri-gate', or 'wrap-around-gate') variety of these devices, schematically shown in figure 6.9b, despite its essentially 3D structure and hence a more complex fabrication technology. In the FinFETs, the role of *p–n* junctions is reduced, but these structures remain an important feature of semiconductor integrated circuits.

Now let us have a look at the *p–n* junction in equilibrium from the point of view of Eq. (6.52). In the simple model we are considering now (in particular, at $T \ll \Delta$), this equation is applicable separately to the electron and hole subsystems, because in this model the gases of these charge carriers are classical in all parts of the system, and the *generation–recombination* processes[52] coupling these subsystems have relatively small rates—see below. Hence, for the electron subsystem, we may rewrite Eq. (6.52) as

---

[50] Note that such w is again much larger than λ_D—the fact that justifies the first two boundary conditions (6.84).

[51] Another important limit is quantum-mechanical tunneling through the gate insulator, whose thickness has to be scaled down in parallel with lateral dimensions of a FET, including the channel length.

[52] In the semiconductor physics lingo, the 'generation' event is the thermal excitation of an electron from the valence band to the conduction band, leaving a hole behind, while the reciprocal event of filling such a hole by a conduction-band electron is called the 'recombination'.

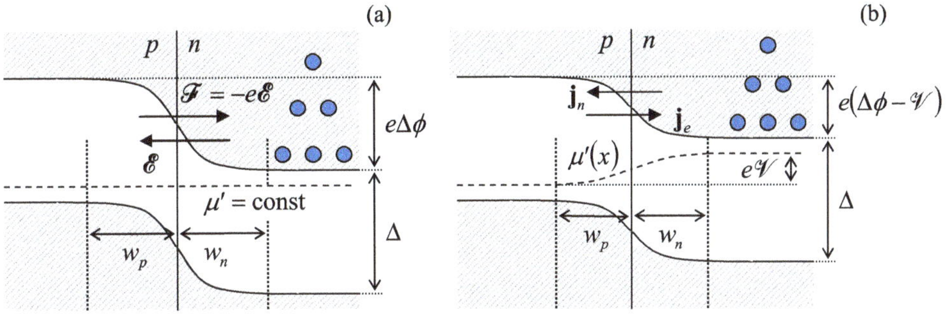

Figure 6.11. Electrons in the conduction band of a *p*–*n* junction at: (a) $\mathscr{V} = 0$, and (b) $\mathscr{V} > 0$. For clarity, other charges (of the holes and all ionized dopant atoms) are not shown.

$$j_n = n\mu_\text{m} q\mathscr{E} - D_n \frac{\partial n}{\partial x}, \tag{6.87}$$

where $q = -e$. Let us discuss how each term of the right-hand of this equality depends on system's parameters. Because of the *n*-doping at $x > 0$, there are many more electrons in this part of the system. According to the Boltzmann distribution (6.58), some number of them,

$$n_> \propto \exp\left\{-\frac{e\Delta\phi}{T}\right\}, \tag{6.88}$$

have energies above the conduction band edge in the *p*-doped part (see 6.11*a*) and try to diffuse into this part through the depletion layer; the second term on the right-hand side of Eq. (6.87) describes this diffusion flow of electrons from the *n*-side to the *p*-side of the structure (in figure 6.11, from the right to the left). On the other hand, the intrinsic electric field $\mathscr{E} = -\partial\phi/\partial x$ inside the depletion layer, directed as figure 6.11a shows, exerts on the electrons the force $\mathscr{F} = q\mathscr{E} \equiv -e\mathscr{E}$ in the opposite direction (from *p* to *n*), creating a counter-flow of electrons, described by the first, 'drift' term on the right-hand side of Eq. (6.87).[53]

The explicit calculation of these two flows[54] shows, unsurprisingly, that in the equilibrium, they are exactly equal and opposite, so that $j_n = 0$, and such analysis does not give us any new information. However, the prospective of two electron

[53] Note that if an external photon with energy $\hbar\omega > \Delta$ generates an electron–hole pair somewhere inside the depletion layer, this electric field would immediately drive its electron component to the right, and the hole component to the left, thus generating a pulse of electric current through the junction. This is the physical basis of the whole vast technological field of *photovoltaics*, currently strongly driven by the demand for renewable electric power. Due to the progress of this technology, the cost of solar power systems has dropped from ~\$300 per watt in the mid-1950s to the current ~\$1 per watt, and its global generation has increased to ~4×10^{14} watt hours per year—though this is still only ~1.4% of the whole generated electric power.

[54] I will not try to reproduce this calculation (which may be found in any of the semiconductor physics books mentioned above), because getting all its scaling factors right requires using some model of the recombination process, and in this course, there is just no time for their quantitative discussion. However, see Eq. (6.93) below.

counter-flows, given by Eq. (6.87), enables us to predict the functional dependence of j_n on a modest external voltage $\mathscr{V}$, with $|\mathscr{V}| < \Delta\phi$, applied to the junction. Indeed, since the semiconductor regions outside the depletion region are much more conductive that that layer, virtually all applied voltage (i.e. the difference of values of the electrochemical potential μ') drops across this layer, changing the total band edge shift—see figure 6.11b:[55]

$$e\Delta\phi \to e\Delta\phi + \Delta\mu' \equiv e\Delta\phi + q\mathscr{V} \equiv e(\Delta\phi - \mathscr{V}). \tag{6.89}$$

This change results in an exponential change of the number of electrons able to diffuse into the p-side of the junction—cf. Eq. (6.88):

$$n_>(\mathscr{V}) \approx n_>(0) \exp\left\{\frac{e\mathscr{V}}{T}\right\}, \tag{6.90}$$

and hence in a proportional change of the diffusion flow j_n of electrons from the n-side to the p-side of the system, i.e. of the oppositely directed density of the electron current $j_e = -ej_n$—see figure 6.11b.

On the other hand, the drift counter-flow of electrons is not altered too much by the applied voltage: though it does change the electrostatic field $\mathscr{E} = -\nabla\phi$ inside the depletion layer, and also the depletion layer width[56], these changes are incremental, not exponential. As a result, the net density of the current carried by electrons may be approximately expressed as

$$j_e(\mathscr{V}) = j_{\text{diffusion}} - j_{\text{drift}} \approx j_e(0) \exp\left\{\frac{e\mathscr{V}}{T}\right\} - \text{const.} \tag{6.91a}$$

As was discussed above, at $\mathscr{V} = 0$, this current has to vanish, so that the constant in Eq. (6.91a) has to equal $j_e(0)$, and we may rewrite this equality as

$$j_e(\mathscr{V}) = j_e(0)\left(\exp\left\{\frac{e\mathscr{V}}{T}\right\} - 1\right). \tag{6.91b}$$

Now repeating this analysis for the current j_h of the holes (an exercise highly recommended to the reader), we get a similar expression, with the *same* sign before $e\mathscr{V}$,[57] though with a different scaling factor, $j_h(0)$ instead of $j_e(0)$. As a result, the total electric current density obeys the famous *Shokley law*

[55] In our model, a positive sign of $\mathscr{V} \equiv \Delta\mu'/q \equiv -\Delta\mu'/e$ corresponds to the additional electric field, $-\nabla\mu'/q \equiv \nabla\mu'/e$, directed in the positive direction of the x-axis (in figure 6.11, from the left to the right), i.e. to the positive terminal of the voltage source connected to the p-doped semiconductor—which is the common convention.

[56] This change, schematically shown in figure 6.11b, may be readily calculated by making the replacement (6.89) in the first of Eqs. (6.86).

[57] This sign invariance may look strange, due to the opposite (positive) electric charge of the holes. However, this difference in the charge sign is compensated by the opposite direction of the hole diffusion—see figure 6.10. (Note also that the actual charge carriers in the valence band are still electrons, and the positive charge of holes is just a convenient representation of the specific dispersion law in this energy band, with a negative kinetic energy—see figure 6.6, the second line of Eq. (6.53), and a more detailed discussion of this issue in *Part QM* section 2.8.)

$$j(\mathscr{V}) \equiv j_e(\mathscr{V}) + j_h(\mathscr{V}) = j(0)\left(\exp\left\{\frac{e\mathscr{V}}{T}\right\} - 1\right), \quad \text{with} \quad j(0) \equiv j_e(0) + j_h(0), \tag{6.92}$$

describing the main *p–n* junction's property as an *electric diode*—a two-terminal device passing the current more 'readily' in one direction (from the *p*- to the *n*-terminal) than in the opposite one[58]. Besides numerous practical applications in electrical and electronic engineering, such diodes have very interesting statistical properties, in particular performing very non-trivial transformations of the spectra of deterministic and random signals. Very unfortunately, I do not have time for their discussion, and have to refer the interested reader to the special literature[59].

Still, before proceeding to our next (and last!) topic, let me give for the reader's reference, without proof, the expression for the scaling factor $j(0)$ in Eq. (6.92), which follows from a simple, but broadly used model of the recombination process:

$$j(0) = en_{\mathrm{i}}^2\left(\frac{D_e}{l_e n_{\mathrm{A}}} + \frac{D_h}{l_h n_{\mathrm{D}}}\right). \tag{6.93}$$

Here l_e and l_h are the characteristic lengths of diffusion of electrons and holes before their recombination, which may be expressed by Eq. (5.113), $l_e = (D_e\tau_e)^{1/2}$ and $l_h = (D_h\tau_h)^{1/2}$, with τ_e and τ_h being the characteristic times of recombination of the so-called *minority carriers*—of electrons in the *p*-doped part, and of holes in the *n*-doped part of the structure. Since the recombination is an inelastic process, its times are typically rather long—of the order of 10^{-7} s, i.e. much longer that the typical times of elastic scattering of the same carriers, that define their diffusion coefficients—see Eq. (6.51).

6.5 Heat transfer and thermoelectric effects

Now let me return to our analysis of kinetic effects using the Boltzmann-RTA equation, and extend it even further, to the effects of a non-zero (albeit small) temperature gradient. Again, since for any of the statistics (6.20) the average occupancy $\langle N(\varepsilon)\rangle$ is a function of just one combination of all its arguments, $\xi \equiv (\varepsilon - \mu)/T$, its partial derivatives obey not only Eq. (6.37), but also the following relation:

$$\frac{\partial\langle N(\varepsilon)\rangle}{\partial T} = -\frac{\varepsilon - \mu}{T^2}\frac{\partial\langle N(\varepsilon)\rangle}{\partial \xi} = \frac{\varepsilon - \mu}{T}\frac{\partial\langle N(\varepsilon)\rangle}{\partial \mu}. \tag{6.94}$$

As a result, Eq. (6.38) is generalized as

$$\nabla w_0 = -\frac{\partial w_0}{\partial \varepsilon}\left(\nabla\mu + \frac{\varepsilon - \mu}{T}\nabla T\right), \tag{6.95}$$

[58] Some metal-semiconductor junctions, called *Schottky diodes*, have similar rectifying properties (and may be better fitted for high-power applications than silicon *p–n* junctions), but their properties are more complex because of the rather involved chemistry and physics of interfaces between different materials.

[59] See, e.g. the monograph by R Stratonovich, cited in section 4.2.

giving the following generalization of Eq. (6.39):

$$\tilde{w} = \tau \frac{\partial w_0}{\partial \varepsilon} \mathbf{v} \cdot \left(\nabla \mu' + \frac{\varepsilon - \mu}{T} \nabla T \right). \tag{6.96}$$

Now, calculating the current density as in section 6.3, we get a result that is traditionally represented as

$$\mathbf{j} = \sigma \left(-\frac{\nabla \mu'}{q} \right) + \sigma \mathcal{S}(-\nabla T), \tag{6.97}$$

where the constant $\mathcal{S}$, called the *Seebeck coefficient*[60] (or the 'thermoelectric power', or just 'thermopower') is given by the following relation:

$$\sigma \mathcal{S} = \frac{gq\tau}{(2\pi\hbar)^3} \frac{4\pi}{3} \int_0^\infty (8m\varepsilon^3)^{1/2} \frac{(\varepsilon - \mu)}{T} \left[-\frac{\partial \langle N(\varepsilon) \rangle}{\partial \varepsilon} \right] d\varepsilon. \tag{6.98}$$

Working out this integral for the most important case of a degenerate Fermi gas, with $T \ll \varepsilon_F$, we have to be careful, because the center of the sharp peak of the last factor under the integral coincides with the zero point of the previous factor, $(\varepsilon - \mu)/T$. This uncertainty may be resolved using the Sommerfeld expansion formula (3.59). Indeed, for a smooth function $f(\varepsilon)$ obeying Eq. (3.60), so that $f(0) = 0$, we may use Eq. (3.61) to rewrite the expansion formula as

$$\int_0^\infty f(\varepsilon) \left[-\frac{\partial \langle N(\varepsilon) \rangle}{\partial \varepsilon} \right] d\varepsilon = f(\mu) + \frac{\pi^2 T^2}{6} \frac{d^2 f(\varepsilon)}{d\varepsilon^2} \bigg|_{\varepsilon = \mu}. \tag{6.99}$$

In particular, for working out the integral (6.98), we may take $f(\varepsilon) \equiv (8m\varepsilon^3)^{1/2}(\varepsilon - \mu)/T$. (Evidently, for this function, the condition $f(0) = 0$ is satisfied.) Then $f(\mu) = 0$, $d^2f/d\varepsilon^2|_{\varepsilon=\mu} = 3(8m\mu)^{1/2}/T \approx 3(8m\varepsilon_F)^{1/2}/T$, and Eq. (6.98) yields

$$\sigma \mathcal{S} = \frac{gq\tau}{(2\pi\hbar)^3} \frac{4\pi}{3} \frac{\pi^2 T^2}{6} \frac{3(8m\varepsilon_F)^{1/2}}{T}. \tag{6.100}$$

Comparing the result with Eqs. (3.54) and (6.32), for the constant $\mathcal{S}$ we get a simple expression independent of τ:[61]

$$\mathcal{S} = \frac{\pi^2}{2q} \frac{T}{\varepsilon_F} = \frac{c_V}{q}, \quad \text{for } T \ll \varepsilon_F, \tag{6.101}$$

---

[60] Named after T Seebeck who experimentally discovered, in 1821 (independently of J Peltier), the effect expressed by Eq. (6.103)—see below.

[61] Again, such independence infers that Eq. (6.101) has a broader validity than in our simple model of an isotropic gas. This is indeed the case: this result turns out to be valid for any form of the Fermi surface, and for any dispersion law $\varepsilon(\mathbf{p})$. Note, however, that all calculations of this section are valid for the crude RTA model in that τ is an energy-independent parameter; for real metals, more accurate description of experimental results may be obtained by tweaking this model to take this dependence into account—see, e.g. chapter 13 in the monograph by N Ashcroft and N D Mermin, cited in section 3.5.

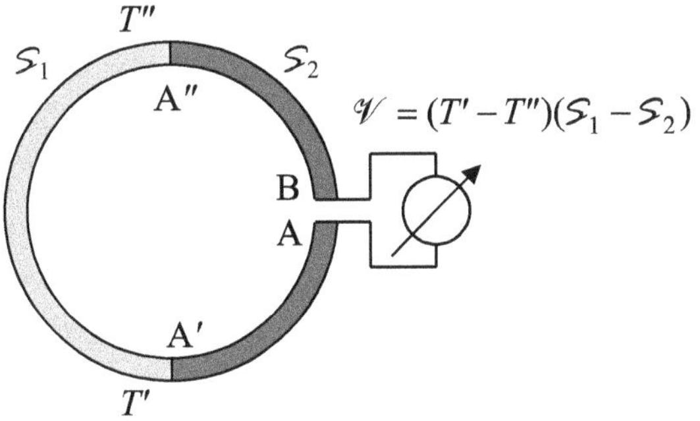

Figure 6.12. The Seebeck effect in a thermocouple.

where $c_V \equiv C_V/N$ is the heat capacity of the gas per unit particle, in this case given by Eq. (3.70).

In order to understand the physical meaning of the Seebeck coefficient, it is sufficient to consider a conductor carrying no current. For this case, Eq. (6.97) yields

$$\nabla(\mu'/q + \mathcal{S}T) = 0. \qquad (6.102)$$

Thus, in these conditions a temperature gradient creates a proportional gradient of the electrochemical potential μ', and hence the effective electric field $\mathcal{E}$ defined by Eq. (6.42). This is the *Seebeck effect.* Figure 6.12 shows the standard way of its measurement, using an ordinary (electrodynamic) voltmeter that measures the difference of μ' at its terminals, and a pair of junctions (in this context, called the *thermocouple*) of two different materials, with different coefficients $\mathcal{S}$. Integrating Eq. (6.102) around the loop from point A to point B, and neglecting the temperature drop across the voltmeter, we get the following simple expression for the thermally-induced difference of the electrochemical potential, usually called either the *thermo-electric power* or the 'thermo emf':

$$\begin{aligned}\mathcal{V} &\equiv \frac{\mu'_B}{q} - \frac{\mu'_A}{q} = \frac{1}{q}\int_A^B \nabla\mu' \cdot d\mathbf{r} = -\int_A^B \mathcal{S}\nabla T \cdot d\mathbf{r} \\ &= -\mathcal{S}_1\int_{A'}^{A''} \nabla T \cdot d\mathbf{r} - \mathcal{S}_2\left(\int_A^{A'} \nabla T \cdot d\mathbf{r} + \int_{A''}^{B} \nabla T \cdot d\mathbf{r}\right) \\ &= -\mathcal{S}_1(T'' - T') - \mathcal{S}_2(T' - T'') \equiv (\mathcal{S}_1 - \mathcal{S}_2)(T' - T'').\end{aligned} \qquad (6.103)$$

(Note that according to Eq. (6.103), any attempt to measure such voltage across any two points of a *uniform* conductor would give results depending on the voltmeter lead materials, due to an unintentional gradient of temperature in them.)

Using thermocouples is a very popular, inexpensive method of temperature measurement—especially in the few-hundred-°C range (where gas- and fluid-based thermometers are not too practicable), if a 1 °C scale accuracy is sufficient. The 'temperature responsivity' ($\mathcal{S}_1 - \mathcal{S}_2$) of a typical popular thermocouple, chromel-constantan[62], is about 70 μV °C^{-1}. In order to understand why typical values of $\mathcal{S}$

[62] Both these materials are *alloys*, i.e. solid solutions: chromel is 10% chromium in 90% nickel, while constantan is 45% nickel and 55% copper.

are so small, let us discuss the Seebeck effect's physics. Superficially, it is very simple: particles, heated by an external source, diffuse from it toward the colder parts of the conductor, carrying electrical current with them if they are electrically charged. However, this naïve argument neglects the fact that at $\mathbf{j} = 0$, there should be no total flow of particles. For a more accurate interpretation, note that the Seebeck effect is described by the factor $(\varepsilon - \mu)/T$ inside the integral (6.98), which changes its sign at the Fermi surface, i.e. at the same energy where the term $[-\partial\langle N(\varepsilon)\rangle/\partial\varepsilon]$, describing the state availability for transport (due to their intermediate occupancy $0 < \langle N(\varepsilon)\rangle < 1$), reaches its peak. The only reason why that integral does not vanish completely, and hence $\mathcal{S} \neq 0$, is the growth of first factor under the integral (which describes the number of available quantum states) with ε, so the hotter particles (with $\varepsilon > \mu$) are more numerous and hence carry more heat than the colder ones.

The Seebeck effect is not the only result of a temperature gradient; the same diffusion of particles also causes the less subtle effect of *heat flow* from the region of higher T to that with lower T, i.e. the effect of *thermal conductivity*, well known from our everyday practice. The density of this flow (i.e. that of thermal energy) may be calculated similarly to that of the electric current—see Eq. (6.26), with the natural replacement of the electric charge q of each particle with its thermal energy $(\varepsilon - \mu)$:

$$\mathbf{j}_{\mathrm{h}} = \int (\varepsilon - \mu)\mathbf{v}w d^3p. \tag{6.104}$$

(Indeed, we may look at this expression as at the difference between the total energy flow density, $\mathbf{j}_\varepsilon = \int \varepsilon \mathbf{v} w d^3p$, and the product of the average energy needed to add a particle to the system (μ) by the particle flow density, $\mathbf{j}_n = \int \mathbf{v} w d^3p \equiv \mathbf{j}/q$.)[63] Again, at equilibrium ($w = w_0$) the heat flow vanishes, so that w in Eq. (6.104) may be replaced with its perturbation $\tilde{w}$, which already has been calculated—see Eq. (6.96). The substitution of that expression into Eq. (6.104), and its transformation exactly similar to the one performed above for the electric current $\mathbf{j}$, yields

$$\mathbf{j}_{\mathrm{h}} = \sigma\Pi\left(-\frac{\nabla\mu'}{q}\right) + \kappa(-\nabla T), \tag{6.105}$$

with the coefficients Π and κ given, in our approximation, by the following equalities:

$$\sigma\Pi = \frac{gq\tau}{(2\pi\hbar)^3}\frac{4\pi}{3}\int_0^\infty (8m\varepsilon^3)^{1/2}(\varepsilon - \mu)\left[-\frac{\partial\langle N(\varepsilon)\rangle}{\partial\varepsilon}\right]d\varepsilon, \tag{6.106}$$

$$\kappa = \frac{g\tau}{(2\pi\hbar)^3}\frac{4\pi}{3}\int_0^\infty (8m\varepsilon^3)^{1/2}\frac{(\varepsilon - \mu)^2}{T}\left[-\frac{\partial\langle N(\varepsilon)\rangle}{\partial\varepsilon}\right]d\varepsilon. \tag{6.107}$$

[63] One more explanation of the factor $(\varepsilon - \mu)$ in Eq. (6.104) is that according to Eqs. (1.37) and (1.56), for a uniform system of N particles this factor is just $(E - G)/N \equiv (TS - PV)/N$. The full differential of the numerator is $TdS + SdT - PdV - VdP$, so that in the absence of the mechanical work $d\mathcal{W} = -PdV$, and changes of temperature and pressure, it is just $TdS \equiv dQ$—see Eq. (1.19).

Besides the missing factor T in the denominator, the integral in Eq. (6.106) is the same as the one in Eq. (6.98), so that the constant Π (called the *Peltier coefficient*), is simply and fundamentally related to the Seebeck coefficient:

$$\Pi = \mathcal{S}T. \tag{6.108}$$

The simplicity of this relation (first discovered experimentally in 1854 by W Thompson, a.k.a. Lord Kelvin) is not occasional. This is one of *Onsager reciprocal relations* between kinetic coefficients (suggested by L Onsager in 1931), which are model-independent, i.e. valid within very general assumptions. Unfortunately, I have no time left for a discussion of this interesting topic (closely related to the fluctuation–dissipation theorem discussed in section 5.5), and have to refer the interested reader to its detailed discussions available in literature[64].

On the other hand, the integral in Eq. (6.107) is different, but may be readily calculated, for the most important case of a degenerate Fermi gas, using the Sommerfeld expansion in the form (6.99), with $f(\varepsilon) \equiv (8m\varepsilon^3)^{1/2}(\varepsilon - \mu)^2/T$, for which $f(\mu) = 0$ and $d^2f/d\varepsilon^2|_{\varepsilon=\mu} = 2(8m\mu^3)^{1/2}/T \approx 2(8m\varepsilon_F{}^3)^{1/2}/T$, so that

$$\kappa = \frac{g\tau}{(2\pi\hbar)^3}\frac{4\pi}{3}\frac{\pi^2}{6}T^2\frac{2(8m\varepsilon_F^3)^{1/2}}{T} \equiv \frac{\pi^2}{3}\frac{n\tau T}{m}. \tag{6.109}$$

Comparing the result with Eq. (6.32), we get the so called *Wiedemann–Franz law*[65]

$$\frac{\kappa}{\sigma} = \frac{\pi^2}{3}\frac{T}{q^2}. \tag{6.110}$$

This relation between the electric conductivity σ and the *thermal conductivity* κ is more general than our formal derivation might imply. Indeed, it is may be shown that the Wiedemann–Franz law is also valid for an arbitrary dispersion law anisotropy (i.e. an arbitrary Fermi surface shape) and, moreover, well beyond the relaxation-time approximation. (For example, it is also valid for the scattering integral (6.12) with an arbitrary angular dependence of rate Γ, provided that the scattering is elastic.) Experiments show that the law is well obeyed by most metals, but only at relatively low temperatures, when the thermal conductance due to electrons is well above the one due to lattice vibrations, i.e. phonons—see section 2.6. Moreover, for a non-degenerate gas, Eq. (6.107) should be treated with utmost care, in the context of the definition (6.105) of this coefficient κ. (Let me leave this issue for the reader's analysis.)

Now let us discuss the effects described by Eq. (6.105), starting from the less obvious, first term on its right-hand side. It describes the so-called *Peltier effect*, which may be measured in the loop geometry similar to that shown in figure 6.12,

---

[64] See, for example, section 15.7 in [3]. Note, however, that the range of validity of the Onsager relations is still debated—see, e.g. [4].

[65] It was named after G Wiedemann and R Franz who noticed the constancy of ratio κ/σ for various materials, at the same temperature, as early as in 1853. The direct proportionality of the ratio to the absolute temperature was noticed by L Lorenz in 1872. Due to this contribution, the Wiedemann–Franz law is frequently represented, in the SI temperature units, as $\kappa/\sigma = LT_K$, where the constant $L \equiv (\pi^2/3)k_B/e^2$, called the *Lorenz number*, is close to 2.45×10^{-8} W·Ω·K^{-2}. Theoretically, Eq. (6.110) was derived in 1928 by A Sommerfeld.

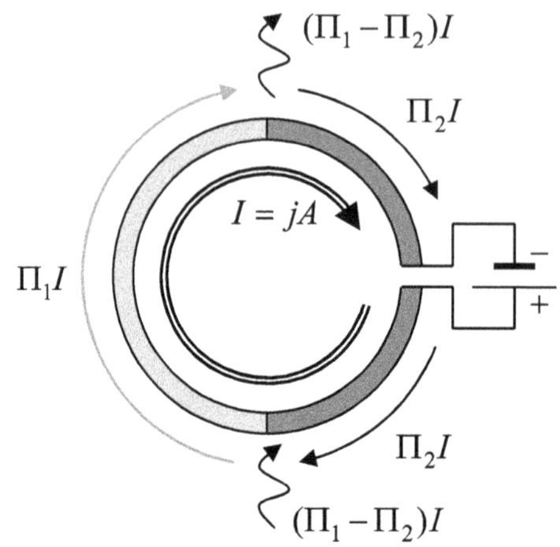

Figure 6.13. The Peltier effect at T = const.

but now driven by an external voltage source—see 6.13. The voltage drives a certain dc current $I = jA$ (where A is conductor's cross-section area), necessarily the same in the whole loop. However, according to Eq. (6.105), if the materials 1 and 2 are different, the power $\mathscr{P} = j_\text{h} A$ of the associated heat flow is different in two parts of the loop[66]. Indeed, if the whole system is kept at the same temperature ($\nabla T = 0$), the integration of that relation over the cross-sections of each part yields

$$\mathscr{P}_{1,2} = \Pi_{1,2}A_{1,2}\sigma_{1,2}\left(-\frac{\nabla\mu'}{q}\right)_{1,2} = \Pi_{1,2}A_{1,2}j_{1,2} = \Pi_{1,2}I_{1,2} = \Pi_{1,2}I, \tag{6.111}$$

where, at the second step, Eq. (6.41) for the electric current density has been used. This equality means that in order to sustain a constant temperature, the following power difference,

$$\Delta\mathscr{P} = (\Pi_1 - \Pi_2)I, \tag{6.112}$$

has to be extracted from one junction of the two materials (in figure 6.13, shown on the top), and inserted into the counterpart junction.

If a constant temperature is not maintained, the former junction is heated (in excess of the bulk, Joule heating), while the latter one is cooled, thus implementing a thermoelectric heat pump/refrigerator. Such *Peltier refrigerators*, which require neither moving parts nor fluids, are very convenient for modest (by a few tens °C) cooling of relatively small components of various systems—from sensitive radiation detectors on mobile platforms (including spacecraft), all the way to cold drinks in vending machines. It is straightforward to use the above formulas to show that the practical efficiency of active materials used in such thermoelectric refrigerators may be characterized by the following dimensionless figure-of-merit,

[66] Let me emphasize that here we are discussing the heat *transferred* through a conductor, not the Joule heat *generated* in it by the current. (The latter effect is quadratic, rather than linear, in the current.)

Table 6.1. Approximate values of two major thermal coefficients of some materials at 20 °C.

Material	κ ($W \cdot m^{-1} \cdot K^{-1}$)	c_V ($J \cdot K^{-1} \cdot m^{-3}$)
Air[(a),(b)]	0.026	1.2×10^3
Teflon ($[C_2F_4]_n$)	0.25	0.6×10^6
Water[(b)]	0.60	4.2×10^6
Amorphous silicon dioxide	1.1-1.4	1.5×10^6
Undoped silicon	150	1.6×10^6
Aluminum[(c)]	235	2.4×10^6
Copper[(c)]	400	3.4×10^6
Diamond	2200	1.8×10^6

[a] At ambient pressure.

[b] In fluids (gases and liquids), heat flow may be much enhanced by temperature-gradient-induced turbulent circulation – *convection*, which is highly dependent on the system's geometry. The given values correspond to conditions preventing the convection.

[c] In the context of the Wiedemann–Franz law (valid for metals only!), the values of κ for Al and Cu correspond to the Lorenz numbers, respectively, 2.22×10^{-8} $W \cdot \Omega \cdot K^{-2}$ and 2.39×10^{-8} $W \cdot \Omega \cdot K^{-2}$, in a pretty impressive comparison with the theoretical value of $2.45 \times 10^{-8} W \cdot \Omega \cdot K^{-2}$ given by Eq. (6.110).

$$\mathrm{ZT} \equiv \frac{\sigma \mathcal{S}^2}{\kappa} T. \tag{6.113}$$

For the best thermoelectric materials found so far, the values of ZT at room temperature are in the range from 2 to 3, providing the $\mathrm{COP}_{\mathrm{cooling}}$, defined by Eq. (1.69), of the order of 0.5—a few times lower than that of traditional, mechanical refrigerators. The search for composite materials (including those with nanoparticles) with higher values of ZT is one of very active fields of applied solid state physics[67].

Now let us discuss the second term of Eq. (6.105), in the absence of $\nabla\mu'$ (and hence of the electric current) giving

$$\mathbf{j}_\mathrm{h} = -\kappa \nabla T, \tag{6.114}$$

This equality should be familiar to the reader, because it describes the very common effect of *thermal conductivity*. Indeed, this linear relation is much more general than the particular expression (6.107) for κ: for sufficiently small temperature gradients it is valid for virtually any medium—for example, in insulators. (The left column in table 6.1 gives typical values of κ for most common and/or representative materials.) Due to its universality and importance, Eq. (6.114) has deserved its own name—the *Fourier law*[68].

[67] See, e.g. [5].

[68] It was suggested (in 1822) by the same universal scientific genius J-B J Fourier, who has not only developed such key mathematical tool as the Fourier series, but also discovered what is now called the greenhouse effect!

Acting absolutely similarly to the derivation of other continuity equations, such as Eqs. (5.117) for the probability, and Eq. (6.49) for the electric charge[69], let us consider the conservation of the aggregate variable corresponding to $\mathbf{j}_h$—the internal energy E within a time-independent volume V. According to the basic Eq. (1.18), in the absence of media's expansion ($dV = 0$ and hence $d\mathscr{W} = 0$), the energy[70] has only the thermal component, so it may change only due to a flow of heat flow through its boundary surface S:

$$\frac{dE}{dt} = -\oint_S \mathbf{j}_h \cdot d^2\mathbf{r}. \tag{6.115}$$

In the simplest case of thermally-independent C_V, we may integrate Eq. (1.22) over temperature to write[71]

$$E = C_V T = \int_V c_V T d^3r, \tag{6.116}$$

where c_V is the volumic specific heat, i.e. the heat capacity per unit volume (see the right column in table 6.1).

Now applying to the right-hand side of Eq. (6.115) the divergence theorem[72], and taking into account that for a time-independent volume the full and partial derivatives over time are equivalent, we get

$$\int_V \left(c_V \frac{\partial T}{\partial t} + \nabla \cdot \mathbf{j}_h \right) d^3r = 0, \tag{6.117}$$

This equality should hold for any time-independent volume V, which is possible only if the function under the integral equals zero at any point. Using Eq. (6.114), we get the following partial differential equation, called the *heat conduction equation* (or, rather inappropriately, the 'heat equation'):

$$c_V(\mathbf{r}) \frac{\partial T}{\partial t} - \nabla \cdot [\kappa(\mathbf{r}) \nabla T] = 0, \tag{6.118}$$

where the spatial arguments of the coefficients c_V and κ are spelled out to emphasize that this equation is valid even for non-uniform media. (Note, however, that Eq. (6.114) and hence Eq. (6.118) are valid only if the medium is *isotropic*.)

In a uniform medium, the thermal conductivity κ may be taken out from the external spatial differentiation, and the heat conduction equation becomes similar to

---

[69] They are all similar to continuity equations for other quantities—e.g. the mass density (see *Part CM* section 8.3) and the quantum-mechanical probability density (see *Part QM* sections 1.4 and 9.6).

[70] According to Eq. (1.25), in the case of negligible thermal expansion it does not matter whether we speak about the internal energy E or the enthalpy H.

[71] If the dependence of c_V on temperature may be ignored only within a limited temperature interval, Eqs. (6.116) and (6.118) may be still used within that interval, for temperature deviations from some reference value.

[72] I hope the reader knows it by heart by now, but if not—see, e.g. Eq. (A.79).

the diffusion equation (5.116), and also to the drift–diffusion equation (6.50) in the absence of drift ($\nabla U = 0$):

$$\frac{\partial T}{\partial t} = D_T \nabla^2 T, \quad \text{with } D_T \equiv \frac{\kappa}{c_V}. \tag{6.119}$$

This means, in particular, that the solutions of these equations, discussed earlier in this course (such as Eqs. (5.112) and (5.113) for the evolution of the delta-functional initial perturbation) are valid for Eq. (6.119) as well, with the only replacement $D \rightarrow D_T$. This is why I will leave other examples of applications of this equation for the reader's exercise.

Let me finish this chapter by emphasizing again that due to time/space restrictions I was able to barely scratch the surface of physical kinetics[73].

6.6 Problems

Problem 6.1. Use the Boltzmann equation in the relaxation-time approximation to derive the Drude formula for the complex ac conductivity $\sigma(\omega)$, and give a physical interpretation of the result's trend at high frequencies.

Problem 6.2. Apply the variable separation method[74] to the drift–diffusion equation (6.50) to calculate the time evolution of the particle density distribution in an unlimited uniform medium, provided that at $t = 0$, the particles are released from their uniform distribution in a plane layer of thickness $2a$:

$$n = \begin{cases} n_0, & \text{for } -a \leqslant x \leqslant +a, \\ 0, & \text{otherwise.} \end{cases}$$

Problem 6.3. Solve the previous problem using an appropriate Green's function for the 1D version of the diffusion equation, and discuss the relative convenience of the results.

Problem 6.4.* Calculate the electric conductance of a narrow, uniform conducting link between two bulk conductors, in the low-voltage and low-temperature limit, neglecting the electron interaction and scattering inside the link.

Problem 6.5. Calculate the effective capacitance (per unit area) of a broad plane sheet of a degenerate 2D electron gas, separated by distance d from a metallic ground plane.

Problem 6.6. Give a quantitative description of the dopant atom ionization, which would be consistent with the conduction and valence band occupation statistics, using the same simple model of an n-doped semiconductor as in section 6.4 (see figure 6.7a),

[73] A much more detailed coverage of this important part of physics may be found, for example, in the textbook by L Pitaevskii and E Lifshitz [6]. A deeper discussion of the Boltzmann equation may be found, e.g. in [7]. For a discussion of applied aspects of kinetics see, e.g. [8].

[74] An introduction to this method may be found, for example, in *Part EM* section 2.5.

and taking into account that the ground state of the dopant atom is typically doubly degenerate, due to two possible spin orientations of the bound electron. Use the results to verify Eq. (6.65), within the displayed limits of its validity.

Problem 6.7. Generalize the solution of the previous problem to the case when the n-doping of a semiconductor is complemented with its simultaneous p-doping by n_A acceptor atoms per unit volume, whose energy $\varepsilon_A - \varepsilon_V$ of activation, i.e. of accepting an additional electron and hence becoming a negative ion, is much less than the bandgap Δ—see the figure below.

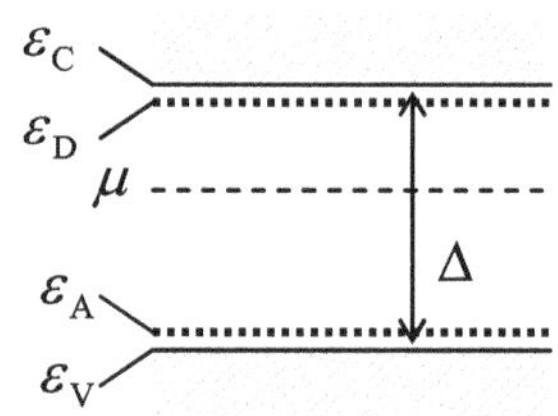

Problem 6.8. A nearly-classical gas of N particles with mass m, is in thermal equilibrium at temperature T, in a closed container of volume V. At some moment, an orifice of a very small area A is open in the container's wall, allowing the particles to escape into the surrounding vacuum[75]. In the limit of very low density $n \equiv N/V$, use simple kinetic arguments to calculate the rms velocity of the escaping particles during the time when the number of escaped particles is still negligible. Formulate the limits of validity of your results in terms of V, A, and the mean free path l.

Problem 6.9. For the system analyzed in the previous problem, calculate the rate of particle flow through the orifice (the so-called *effusion rate*). Discuss the limits of validity of your result.

Problem 6.10. Use simple kinetic arguments to estimate:

(i) the diffusion coefficient D,
(ii) the thermal conductivity κ, and
(iii) the *shear viscosity* η,

of a nearly-ideal classical gas with mean free path l. Compare the result for D with that calculated in section 6.3 from the Boltzmann-RTA equation.

Hint: In fluid dynamics, the shear viscosity (frequently called simply 'viscosity') is defined as the coefficient η in the following relation[76]:

[75] In chemistry-related fields, this process is frequently called *effusion*.
[76] See, e.g. *Part CM* Eq. (8.56). Please note the difference between the shear viscosity coefficient η considered in this problem, and the drag coefficient η whose calculation was the task of problem 3.2. Despite the similar traditional notation, and belonging to the same realm (kinematic friction), these coefficients have different definitions, and even different dimensionalities.

$$\frac{d\mathscr{F}_{j'}}{dA_j} = \eta \frac{\partial v_{j'}}{\partial r_j},$$

where $d\mathscr{F}_{j'}$ is the j'th Cartesian component of the tangential force between two parts of a fluid, separated by an imaginary interface normal to some direction $\mathbf{n}_j$ (with $j \neq j'$, and hence $\mathbf{n}_j \perp \mathbf{n}_{j'}$), exerted over an elementary area dA_j of this surface, and $\mathbf{v}(\mathbf{r})$ is the velocity of the fluid at the interface.

Problem 6.11. Use simple kinetic arguments to relate the mean free path l in a nearly-ideal classical gas, with the full cross-section σ of mutual scattering of its particles[77]. Use the result to evaluate the thermal conductivity and the viscosity coefficient estimates, made in the previous problem, for the molecular nitrogen, with the molecular mass $m \approx 4.7 \times 10^{-26}$ kg and the effective ('van der Waals') diameter $d_{\rm ef} \approx 4.5 \times 10^{-10}$m, at ambient conditions, and compare them with experimental results.

Problem 6.12. Use the Boltzmann-RTA equation to calculate the thermal conductivity of a nearly-ideal classical gas, measured in conditions when the applied thermal gradient does not create a net particle flow. Compare the result with that following from the simple kinetic arguments (problem 6.10), and discuss their relation.

Problem 6.13. Use the heat conduction equation (6.119) to calculate the time evolution of temperature in the center of a uniform solid sphere of radius R, initially heated to a uniformly distributed temperature $T_{\rm ini}$, and at $t = 0$ placed into a heat bath that keeps its surface at temperature T_0.

Problem 6.14. Suggest a reasonable definition of the entropy production rate (per unit volume), and calculate this rate for a stationary thermal conduction, assuming that it obeys the Fourier law, in a material with negligible thermal expansion. Give a physical interpretation of the result. Does the stationary temperature distribution in a sample correspond to the minimum of the total entropy production in it?

References

[1] Sze S 1981 *Physics of Semiconductor Devices* 2nd edn (Wiley)
[2] Pierret R 1996 *Semiconductor Device Fundamentals* 2nd edn (Addison Wesley)
[3] Pathria R and Beale P 2011 *Statistical Mechanics* 3rd edn (Elsevier)
[4] Chen K-T and Lee P 2009 *Phys. Rev.* B **79** 18
[5] Rowe D (ed) 2005 *Thermoelectrics Handbook: Macro to Nano* (CRC Press)
[6] Pitaevskii L and Lifshitz E 1981 *Physical Kinetics* (Butterworth-Heinemann)
[7] Harris S 2011 *An Introduction to the Theory of the Boltzmann equation* (Dover)
[8] Bergman T *et al* 2011 *Fundamentals of Heat and Mass Transfer* 7th edn (Wiley)

[77] I am sorry for using for the cross-section the same letter as for the electric Ohmic conductivity. (Both notations are very traditional.) Let me hope this would not lead to confusion, because the conductivity is not discussed in this problem.

Appendix A

Selected mathematical formulas

This appendix lists selected mathematical formulas that are used in this lecture course series, but not always remembered by students (and some instructors :-).

A.1 Constants

- Euclidean circle's *length-to-diameter ratio*:

$$\pi = 3.141\,592\,653\ldots; \qquad \pi^{1/2} \approx 1.77. \tag{A.1}$$

- *Natural logarithm base*:

$$e \equiv \lim_{n\to\infty}\left(1 + \frac{1}{n}\right)^n = 2.718\,281\,828\ldots; \tag{A.2a}$$

from that value, the logarithm base conversion factors are as follows ($\xi > 0$):

$$\frac{\ln \xi}{\log_{10}\xi} = \ln 10 \approx 2.303, \qquad \frac{\log_{10}\xi}{\ln \xi} = \frac{1}{\ln 10} \approx 0.434. \tag{A.2b}$$

- The *Euler* (or 'Euler–Mascheroni') *constant*:

$$\gamma \equiv \lim_{n\to\infty}\left(1 + \frac{1}{2} + \frac{1}{3} + \ldots \frac{1}{n} - \ln n\right) = 0.577\,156\,649\,0\ldots; \qquad e^\gamma \approx 1.781. \tag{A.3}$$

A.2 Combinatorics, sums, and series

(i) *Combinatorics*

- The number of different *permutations*, i.e. *ordered* sequences of k elements selected from a set of n distinct elements ($n \geqslant k$), is

$${}^nP_k \equiv n\cdot(n-1)\cdots(n-k+1) = \frac{n!}{(n-k)!}; \tag{A.4a}$$

in particular, the number of different permutations of *all* elements of the set ($n = k$) is

$$^{k}P_{k} = k \cdot (k-1) \cdots 2 \cdot 1 = k!. \tag{A.4b}$$

- The number of different *combinations*, i.e. *unordered* sequences of k elements from a set of $n \geqslant k$ distinct elements, is equal to the binomial coefficient

$$^{n}C_{k} \equiv \binom{n}{k} \equiv \frac{^{n}P_{k}}{^{k}P_{k}} = \frac{n!}{k!(n-k)!}. \tag{A.5}$$

In an alternative, very popular 'ball/box language', $^{n}C_{k}$ is the number of different ways to put in a box, in an arbitrary order, k balls selected from n distinct balls.

- A generalization of the binomial coefficient notion is the multinomial coefficient,

$$^{n}C_{k_1,k_2,\ldots k_l} \equiv \frac{n!}{k_1!k_2!\ldots k_l!}, \qquad \text{with } n = \sum_{j=1}^{l} k_j, \tag{A.6}$$

which, in the standard mathematical language, is a number of different permutations in a multiset of l distinct element types from an n-element set which contains k_j ($j = 1, 2, \ldots l$) elements of each type. In the 'ball/box language', the coefficient (A.6) is the number of different ways to distribute n distinct balls between l distinct boxes, each time keeping the number (k_j) of balls in the jth box fixed, but ignoring their order inside the box. The binomial coefficient $^{n}C_{k}$ (A.5) is a particular case of the multinomial coefficient (A.6) for $l = 2$ - counting the explicit box for the first one, and the remaining space for the second box, so that if $k_1 \equiv k$, then $k_2 = n - k$.

- One more important combinatorial quantity is the number $M_n^{(k)}$ of ways to place n *indistinguishable* balls into k distinct boxes. It may be readily calculated from Eq. (A.5) as the number of different ways to select ($k - 1$) partitions between the boxes in an imagined linear row of ($k - 1 + n$) 'objects' (balls in the boxes *and* partitions between them):

$$M_n^{(k)} = {}^{n-1+k}C_{k-1} \equiv \frac{(k-1+n)!}{(k-1)!n!}. \tag{A.7}$$

(ii) *Sums and series*

- *Arithmetic progression*:

$$r + 2r + \cdots + nr \equiv \sum_{k=1}^{n} kr = \frac{n(r+nr)}{2}; \tag{A.8a}$$

in particular, at $r = 1$ it is reduced to the sum of n first natural numbers:

$$1 + 2 + \cdots + n \equiv \sum_{k=1}^{n} k = \frac{n(n+1)}{2}. \tag{A.8b}$$

- Sums of squares and cubes of n first natural numbers:

$$1^2 + 2^2 + \cdots + n^2 \equiv \sum_{k=1}^{n} k^2 = \frac{n(n+1)(2n+1)}{6}; \tag{A.9a}$$

$$1^3 + 2^3 + \cdots + n^3 \equiv \sum_{k=1}^{n} k^3 = \frac{n^2(n+1)^2}{4}. \tag{A.9b}$$

- The *Riemann zeta function*:

$$\zeta(s) \equiv 1 + \frac{1}{2^s} + \frac{1}{3^s} + \cdots \equiv \sum_{k=1}^{\infty} \frac{1}{k^s}; \tag{A.10a}$$

the particular values frequently met in applications are

$$\begin{aligned} &\zeta\left(\frac{3}{2}\right) \approx 2.612, \quad \zeta(2) = \frac{\pi^2}{6}, \quad \zeta\left(\frac{5}{2}\right) \approx 1.341, \\ &\zeta(3) \approx 1.202, \quad \zeta(4) = \frac{\pi^4}{90}, \quad \zeta(5) \approx 1.037. \end{aligned} \tag{A.10b}$$

- Finite geometric progression (for real $\lambda \neq 1$):

$$1 + \lambda + \lambda^2 + \cdots + \lambda^{n-1} \equiv \sum_{k=0}^{n-1} \lambda^k = \frac{1-\lambda^n}{1-\lambda}; \tag{A.11a}$$

in particular, if $\lambda^2 < 1$, the progression has a finite limit at $n \to \infty$ (called the *geometric series*):

$$\lim_{n\to\infty} \sum_{k=0}^{n-1} \lambda^k = \sum_{k=0}^{\infty} \lambda^k = \frac{1}{1-\lambda}. \tag{A.11b}$$

- *Binomial sum* (or the 'binomial theorem'):

$$(1 + a)^n = \sum_{k=0}^{n} {}^nC_k a^k, \tag{A.12}$$

where ${}^nC_k$ are the binomial coefficients defined by Eq. (A.5).

- The *Stirling formula*:

$$\lim_{n\to\infty} \ln(n!) = n(\ln n - 1) + \frac{1}{2}\ln(2\pi n) + \frac{1}{12n} - \frac{1}{360n^3} + \ldots; \tag{A.13}$$

for most applications in physics, the first term[1] is sufficient.

- The *Taylor* (or 'Taylor–Maclaurin') *series*: for any infinitely differentiable function $f(\xi)$:

$$\lim_{\tilde{\xi}\to 0} f(\xi + \tilde{\xi}) = f(\xi) + \frac{df}{d\xi}(\xi)\ \tilde{\xi} + \frac{1}{2!}\frac{d^2f}{d\xi^2}(\xi)\ \tilde{\xi}^2 + \cdots$$
$$= \sum_{k=0}^{\infty} \frac{1}{k!}\frac{d^kf}{d\xi^k}(\xi)\ \tilde{\xi}^k; \tag{A.14a}$$

note that for many functions this series converges only within a limited, sometimes small range of deviations $\tilde{\xi}$. For a function of several arguments, $f(\xi_1, \xi_2, \ldots, \xi_N)$, the first terms of the Taylor series are

$$\lim_{\tilde{\xi}_k \to 0} f(\xi_1 + \tilde{\xi}_1,\ \xi_2 + \tilde{\xi}_2,\ \cdots) = f(\xi_1,\ \xi_2,\ \cdots)$$
$$+ \sum_{k=1}^{N} \frac{\partial f}{\partial \xi_k}(\xi_1,\ \xi_2,\ \cdots)\ \tilde{\xi}_k \tag{A.14b}$$
$$+ \frac{1}{2!}\sum_{k,k'=1}^{N} \frac{\partial^2 f}{\partial_k \xi\ \partial \xi_{k'}} \tilde{\xi}_k \tilde{\xi}_{k'} + \cdots$$

- The *Euler–Maclaurin formula*, valid for any infinitely differentiable function $f(\xi)$:

$$\sum_{k=1}^{n} f(k) = \int_0^n f(\xi)d\xi + \frac{1}{2}[f(n) - f(0)] + \frac{1}{6}\cdot\frac{1}{2!}\left[\frac{df}{d\xi}(n) - \frac{df}{d\xi}(0)\right]$$
$$- \frac{1}{30}\cdot\frac{1}{4!}\left[\frac{d^3f}{d\xi^3}(n) - \frac{d^3f}{d\xi^3}(0)\right] \tag{A.15a}$$
$$+ \frac{1}{42}\cdot\frac{1}{6!}\left[\frac{d^5f}{d\xi^5}(n) - \frac{d^5f}{d\xi^5}(0)\right] + \cdots;$$

the coefficients participating in this formula are the so-called *Bernoulli numbers*[2]:

$$B_1 = \frac{1}{2},\quad B_2 = \frac{1}{6},\quad B_3 = 0,\quad B_4 = \frac{1}{30},\quad B_5 = 0,$$
$$B_6 = \frac{1}{42},\quad B_7 = 0,\quad B_8 = \frac{1}{30},\quad \cdots \tag{A.15b}$$

[1] Actually, this leading term was derived by A de Moivre in 1733, before J Stirling's work.

[2] Note that definitions of B_k (or rather their signs and indices) vary even among the most popular handbooks.

A.3 Basic trigonometric functions

- Trigonometric functions of the sum and the difference of two arguments[3]:

$$\cos(a \pm b) = \cos a \cos b \mp \sin a \sin b, \tag{A.16a}$$

$$\sin(a \pm b) = \sin a \cos b \pm \cos a \sin b. \tag{A.16b}$$

- Sums of two functions of arbitrary arguments:

$$\cos a + \cos b = 2\cos\frac{a+b}{2}\cos\frac{b-a}{2}, \tag{A.17a}$$

$$\cos a - \cos b = 2\sin\frac{a+b}{2}\sin\frac{b-a}{2}, \tag{A.17b}$$

$$\sin a \pm \sin b = 2\sin\frac{a \pm b}{2}\cos\frac{\pm b - a}{2}. \tag{A.17c}$$

- Trigonometric function products:

$$2\cos a \cos b = \cos(a+b) + \cos(a-b), \tag{A.18a}$$

$$2\sin a \cos b = \sin(a+b) + \sin(a-b), \tag{A.18b}$$

$$2\sin a \sin b = \cos(a-b) - \cos(a+b); \tag{A.18c}$$

for the particular case of equal arguments, $b = a$, these three formulas yield the following expressions for the squares of trigonometric functions, and their product:

$$\cos^2 a = \frac{1}{2}(1 + \cos 2a), \quad \sin a \cos a = \frac{1}{2}\sin 2a, \quad \sin^2 a = \frac{1}{2}(1 - \cos 2a). \tag{A.18d}$$

- Cubes of trigonometric functions:

$$\cos^3 a = \frac{3}{4}\cos a + \frac{1}{4}\cos 3a, \quad \sin^3 a = \frac{3}{4}\sin a - \frac{1}{4}\sin 3a. \tag{A.19}$$

- Trigonometric functions of a complex argument:

$$\sin(a+ib) = \sin a \cosh b + i\cos a \sinh b, \quad \cos(a+ib) = \cos a \cosh b - i\sin a \sinh b. \tag{A.20}$$

[3] I am confident that the reader is quite capable of deriving the relations (A.16) by representing the exponent in the elementary relation $e^{i(a \pm b)} = e^{ia}e^{\pm ib}$ as a sum of its real and imaginary parts, Eqs. (A.18) directly from Eqs. (A.16), and Eqs. (A.17) from Eqs. (A.18) by variable replacement; however, I am still providing these formulas to save his or her time. (Quite a few formulas below are included because of the same reason.)

- Sums of trigonometric functions of n equidistant arguments:

$$\sum_{k=1}^{n}\begin{Bmatrix}\sin\\ \cos\end{Bmatrix}k\xi = \begin{Bmatrix}\sin\\ \cos\end{Bmatrix}\left(\frac{n+1}{2}\xi\right)\sin\left(\frac{n}{2}\xi\right)\Big/\sin\left(\frac{\xi}{2}\right). \qquad \text{(A.21)}$$

A.4 General differentiation

- Full differential of a product of two functions:

$$d(fg) = (df)g + f(dg). \qquad \text{(A.22)}$$

- Full differential of a function of several independent arguments, $f(\xi_1, \xi_2,\dots, \xi_n)$:

$$df = \sum_{k=1}^{n}\frac{\partial f}{\partial \xi_k}d\xi_k. \qquad \text{(A.23)}$$

- Curvature of the Cartesian plot of a 1D function $f(\xi)$:

$$\kappa \equiv \frac{1}{R} = \frac{|\, d^2f/d\xi^2\,|}{[1 + (df/d\xi)^2]^{3/2}}. \qquad \text{(A.24)}$$

A.5 General integration

- Integration *by parts* - immediately follows from Eq. (A.22):

$$\int_{g(A)}^{g(B)} f\ dg = fg\,\Big|_A^B - \int_{f(A)}^{f(B)} g\ df. \qquad \text{(A.25)}$$

- Numerical (approximate) integration of 1D functions: the simplest *trapezoidal rule*,

$$\begin{aligned}\int_a^b f(\xi)d\xi &\approx h\left[f\left(a+\frac{h}{2}\right) + f\left(a+\frac{3h}{2}\right) + \cdots + f\left(b-\frac{h}{2}\right)\right]\\ &= h\sum_{n=1}^{N} f\left(a - \frac{h}{2} + nh\right), \qquad h \equiv \frac{b-a}{N}.\end{aligned} \qquad \text{(A.26)}$$

has relatively low accuracy (error of the order of $(h^3/12)d^2f/d\xi^2$ per step), so that the following *Simpson formula*,

$$\begin{aligned}\int_a^b f(\xi)d\xi &\approx \frac{h}{3}[f(a) + 4f(a+h) + 2f(a+2h) + \cdots + 4f(b-h) + f(b)],\\ h &\equiv \frac{b-a}{2N},\end{aligned} \qquad \text{(A.27)}$$

whose error per step scales as $(h^5/180)d^4f/d\xi^4$, is used much more frequently[4].

A.6 A few 1D integrals[5]

(i) *Indefinite integrals*:

- Integrals with $(1 + \xi^2)^{1/2}$:

$$\int (1 + \xi^2)^{1/2} d\xi = \frac{\xi}{2}(1 + \xi^2)^{1/2} + \frac{1}{2} \ln|\xi + (1 + \xi^2)^{1/2}|, \tag{A.28}$$

$$\int \frac{d\xi}{(1 + \xi^2)^{1/2}} = \ln|\xi + (1 + \xi^2)^{1/2}|, \tag{A.29a}$$

$$\int \frac{d\xi}{(1 + \xi^2)^{3/2}} = \frac{\xi}{(1 + \xi^2)^{1/2}}. \tag{A.29b}$$

- Miscellaneous indefinite integrals:

$$\int \frac{d\xi}{\xi(\xi^2 + 2a\xi - 1)^{1/2}} = \arccos \frac{a\xi - 1}{|\xi|(a^2 + 1)^{1/2}}, \tag{A.30a}$$

$$\int \frac{(\sin\xi - \xi\cos\xi)^2}{\xi^5} d\xi = \frac{2\xi \sin 2\xi + \cos 2\xi - 2\xi^2 - 1}{8\xi^4}, \tag{A.30b}$$

$$\int \frac{d\xi}{a + b\cos\xi} = \frac{2}{(a^2 - b^2)^{1/2}} \tan^{-1}\left[\frac{(a - b)}{(a^2 - b^2)^{1/2}} \tan\frac{\xi}{2}\right], \quad \text{for } a^2 > b^2. \tag{A.30c}$$

$$\int \frac{d\xi}{1 + \xi^2} = \tan^{-1}\xi. \tag{A.30d}$$

(ii) *Semi-definite integrals*:

- Integrals with $1/(e^\xi \pm 1)$:

$$\int_a^\infty \frac{d\xi}{e^\xi + 1} = \ln(1 + e^{-a}), \tag{A.31a}$$

---

[4] Higher-order formulas (e.g. the *Bode rule*), and other guidance including ready-for-use codes for computer calculations may be found, for example, in the popular reference texts by W H Press *et al* [1]. In addition, some advanced codes are used as subroutines in the software packages listed in the same section. In some cases, the Euler–Maclaurin formula (A.15) may also be useful for numerical integration.

[5] A powerful (and free) interactive online tool for working out indefinite 1D integrals is available at http://integrals.wolfram.com/index.jsp.

$$\int_{a>0}^{\infty} \frac{d\xi}{e^{\xi} - 1} = \ln \frac{1}{1 - e^{-a}}. \qquad \text{(A.31}b\text{)}$$

(iii) *Definite integrals*:

- Integrals with $1/(1 + \xi^2)$:[6]

$$\int_0^{\infty} \frac{d\xi}{1 + \xi^2} = \frac{\pi}{2}, \qquad \text{(A.32}a\text{)}$$

$$\int_0^{\infty} \frac{d\xi}{(1 + \xi^2)^{3/2}} = 1; \qquad \text{(A.32}b\text{)}$$

more generally,

$$\int_0^{\infty} \frac{d\xi}{(1 + \xi^2)^n} = \frac{\pi}{2} \frac{(2n - 3)!!}{(2n - 2)!!} \equiv \frac{\pi}{2} \frac{1 \cdot 3 \cdot 5 \ldots (2n - 3)}{2 \cdot 4 \cdot 6 \ldots (2n - 2)}, \qquad \text{(A.32}c\text{)}$$
$$\text{for } n = 2,\ 3,\ \ldots$$

- Integrals with $(1 - \xi^{2n})^{1/2}$:

$$\int_0^{1} \frac{d\xi}{(1 - \xi^{2n})^{1/2}} = \frac{\pi^{1/2}}{2n} \Gamma\left(\frac{1}{2n}\right) \Big/ \Gamma\left(\frac{n + 1}{2n}\right), \qquad \text{(A.33}a\text{)}$$

$$\int_0^{1} (1 - \xi^{2n})^{1/2} d\xi = \frac{\pi^{1/2}}{4n} \Gamma\left(\frac{1}{2n}\right) \Big/ \Gamma\left(\frac{3n + 1}{2n}\right), \qquad \text{(A.33}b\text{)}$$

where $\Gamma(s)$ is the *gamma-function,* which is most often defined (for $\text{Re}\, s > 0$) by the following integral:

$$\int_0^{\infty} \xi^{s-1} e^{-\xi}\, d\xi = \Gamma(s). \qquad \text{(A.34}a\text{)}$$

The key property of this function is the recurrence relation, valid for any $s \neq 0, -1, -2, \ldots$:

$$\Gamma(s + 1) = s\Gamma(s). \qquad \text{(A.34}b\text{)}$$

Since, according to Eq. (A.34*a*), $\Gamma(1) = 1$, Eq. (A.34*b*) for non-negative integers takes the form

$$\Gamma(n + 1) = n!, \quad \text{for } n = 0,\ 1,\ 2,\ \cdots \qquad \text{(A.34}c\text{)}$$

[6] Eq. (A.32*a*) follows immediately from Eq. (A.30*d*), and Eq. (A.32*b*) from Eq. (A.29*b*)—a couple more examples of the (intentional) redundancy in this list.

(where $0! \equiv 1$). Because of this, for integer $s = n + 1 \geqslant 1$, Eq. (A.34*a*) is reduced to

$$\int_0^\infty \xi^n e^{-\xi} d\xi = n!. \qquad \text{(A.34}d\text{)}$$

Other frequently met values of the gamma-function are those for positive semi-integer arguments:

$$\Gamma\left(\frac{1}{2}\right) = \pi^{1/2}, \quad \Gamma\left(\frac{3}{2}\right) = \frac{1}{2}\pi^{1/2}, \quad \Gamma\left(\frac{5}{2}\right) = \frac{1}{2}\cdot\frac{3}{2}\pi^{1/2},$$
$$\Gamma\left(\frac{7}{2}\right) = \frac{1}{2}\cdot\frac{3}{2}\cdot\frac{5}{2}\pi^{1/2}, \quad \ldots. \qquad \text{(A.34}e\text{)}$$

- Integrals with $1/(e^\xi \pm 1)$:

$$\int_0^\infty \frac{\xi^{s-1}d\xi}{e^\xi + 1} = (1 - 2^{1-s})\,\Gamma(s)\zeta(s), \quad \text{for } s > 0, \qquad \text{(A.35}a\text{)}$$

$$\int_0^\infty \frac{\xi^{s-1}d\xi}{e^\xi - 1} = \Gamma(s)\zeta(s), \quad \text{for } s > 1, \qquad \text{(A.35}b\text{)}$$

where $\zeta(s)$ is the Riemann zeta-function—see Eq. (A.10). Particular cases: for $s = 2n$,

$$\int_0^\infty \frac{\xi^{2n-1}d\xi}{e^\xi + 1} = \frac{2^{2n-1} - 1}{2n}\pi^{2n}B_{2n}, \qquad \text{(A.35}c\text{)}$$

$$\int_0^\infty \frac{\xi^{2n-1}d\xi}{e^\xi - 1} = \frac{(2\pi)^{2n}}{4n}B_{2n}. \qquad \text{(A.35}d\text{)}$$

where B_n are the Bernoulli numbers—see Eq. (A.15). For the particular case $s = 1$ (when Eq. (A.35*a*) yields uncertainty),

$$\int_0^\infty \frac{d\xi}{e^\xi + 1} = \ln 2. \qquad \text{(A.35}e\text{)}$$

- Integrals with $\exp\{-\xi^2\}$:

$$\int_0^\infty \xi^s e^{-\xi^2} d\xi = \frac{1}{2}\Gamma\left(\frac{s+1}{2}\right), \quad \text{for } s > -1; \qquad \text{(A.36}a\text{)}$$

for applications the most important particular values of s are 0 and 2:

$$\int_0^\infty e^{-\xi^2} d\xi = \frac{1}{2}\Gamma\left(\frac{1}{2}\right) = \frac{\pi^{1/2}}{2}, \qquad \text{(A.36}b\text{)}$$

$$\int_0^\infty \xi^2 e^{-\xi^2} d\xi = \frac{1}{2}\Gamma\left(\frac{3}{2}\right) = \frac{\pi^{1/2}}{4}, \qquad \text{(A.36}c\text{)}$$

although we will also run into the cases $s = 4$ and $s = 6$:

$$\int_0^\infty \xi^4 e^{-\xi^2} d\xi = \frac{1}{2}\Gamma\left(\frac{5}{2}\right) = \frac{3\pi^{1/2}}{8},$$
$$\int_0^\infty \xi^6 e^{-\xi^2} d\xi = \frac{1}{2}\Gamma\left(\frac{7}{2}\right) = \frac{15\pi^{1/2}}{16}; \qquad \text{(A.36}d\text{)}$$

for odd integer values $s = 2n + 1$ (with $n = 0, 1, 2,\ldots$), Eq. (A.36a) takes a simpler form:

$$\int_0^\infty \xi^{2n+1} e^{-\xi^2} d\xi = \frac{1}{2}\Gamma(n + 1) = \frac{n!}{2}. \qquad \text{(A.36}e\text{)}$$

- Integrals with cosine and sine functions:

$$\int_0^\infty \cos(\xi^2)\, d\xi = \int_0^\infty \sin(\xi^2)\, d\xi = \left(\frac{\pi}{8}\right)^{1/2}. \qquad \text{(A.37)}$$

$$\int_0^\infty \frac{\cos\xi}{a^2 + \xi^2} d\xi = \frac{\pi}{2a} e^{-a}. \qquad \text{(A.38)}$$

$$\int_0^\infty \left(\frac{\sin\xi}{\xi}\right)^2 d\xi = \frac{\pi}{2}. \qquad \text{(A.39)}$$

- Integrals with logarithms:

$$\int_0^1 \ln\frac{a + (1 - \xi^2)^{1/2}}{a - (1 - \xi^2)^{1/2}} d\xi = \pi[a - (a^2 - 1)^{1/2}], \quad \text{for } a \geqslant 1. \qquad \text{(A.40)}$$

$$\int_0^1 \ln\frac{1 + (1 - \xi)^{1/2}}{\xi^{1/2}} d\xi = 1. \qquad \text{(A.41)}$$

- Integral representations of the Bessel functions of integer order:

$$J_n(\alpha) = \frac{1}{2\pi}\int_{-\pi}^{+\pi} e^{i(\alpha\sin\xi - n\xi)} d\xi,$$
$$\text{so that } e^{i\alpha\sin\xi} = \sum_{k=-\infty}^{\infty} J_k(\alpha) e^{ik\xi}; \qquad \text{(A.42}a\text{)}$$

$$I_n(\alpha) = \frac{1}{\pi}\int_0^\pi e^{\alpha\cos\xi}\cos n\xi \; d\xi. \qquad \text{(A.42}b\text{)}$$

A.7 3D vector products

(i) *Definitions*:

- *Scalar* ('dot-') *product*:

$$\mathbf{a} \cdot \mathbf{b} = \sum_{j=1}^{3} a_j b_j, \tag{A.43}$$

where a_j and b_j are vector components in any orthogonal coordinate system. In particular, the vector squared (the same as the norm squared):

$$a^2 \equiv \mathbf{a} \cdot \mathbf{a} = \sum_{j=1}^{3} a_j^2 \equiv \| \mathbf{a} \|^2. \tag{A.44}$$

- *Vector* ('cross-') *product*:

$$\begin{aligned} \mathbf{a} \times \mathbf{b} &\equiv \mathbf{n}_1(a_2b_3 - a_3b_2) + \mathbf{n}_2(a_3b_1 - a_1b_3) + \mathbf{n}_3(a_1b_2 - a_2b_1) \\ &= \begin{vmatrix} \mathbf{n}_1 & \mathbf{n}_2 & \mathbf{n}_3 \\ a_1 & a_2 & a_3 \\ b_1 & b_2 & b_3 \end{vmatrix}, \end{aligned} \tag{A.45}$$

where $\{\mathbf{n}_j\}$ is the set of mutually perpendicular unit vectors[7] along the corresponding coordinate system axes[8]. In particular, Eq. (A.45) yields

$$\mathbf{a} \times \mathbf{a} = 0. \tag{A.46}$$

(ii) *Corollaries* (readily verified by Cartesian components):

- Double vector product (the so-called *bac minus cab* rule):

$$\mathbf{a} \times (\mathbf{b} \times \mathbf{c}) = \mathbf{b}(\mathbf{a} \cdot \mathbf{c}) - \mathbf{c}(\mathbf{a} \cdot \mathbf{b}). \tag{A.47}$$

- Mixed scalar–vector product (the *operand rotation rule*):

$$\mathbf{a} \cdot (\mathbf{b} \times \mathbf{c}) = \mathbf{b} \cdot (\mathbf{c} \times \mathbf{a}) = \mathbf{c} \cdot (\mathbf{a} \times \mathbf{b}). \tag{A.48}$$

- Scalar product of vector products:

$$(\mathbf{a} \times \mathbf{b}) \cdot (\mathbf{c} \times \mathbf{d}) = (\mathbf{a} \cdot \mathbf{c})(\mathbf{b} \cdot \mathbf{d}) - (\mathbf{a} \cdot \mathbf{d})(\mathbf{b} \cdot \mathbf{c}); \tag{A.49a}$$

---

[7] Other popular notations for this vector set are $\{\mathbf{e}_j\}$ and $\{\hat{\mathbf{r}}_j\}$.

[8] It is easy to use Eq. (A.45) to check that the direction of the product vector corresponds to the well-known 'right-hand rule' and to the even more convenient *corkscrew rule*: if we rotate a corkscrew's handle from the first operand toward the second one, its axis moves in the direction of the product.

in the particular case of two similar operands (say, $\mathbf{a} = \mathbf{c}$ and $\mathbf{b} = \mathbf{d}$), the last formula is reduced to

$$(\mathbf{a} \times \mathbf{b})^2 = (ab)^2 - (\mathbf{a} \cdot \mathbf{b})^2. \qquad \text{(A.49}b\text{)}$$

A.8 Differentiation in 3D Cartesian coordinates

- Definition of the *del* (or 'nabla') vector-operator ∇:[9]

$$\nabla \equiv \sum_{j=1}^{3} \mathbf{n}_j \frac{\partial}{\partial r_j}, \qquad \text{(A.50)}$$

where r_j is a set of linear and orthogonal (*Cartesian*) coordinates along directions $\mathbf{n}_j$. In accordance with this definition, the operator ∇ acting on a *scalar* function of coordinates, $f(\mathbf{r})$,[10] gives its gradient, i.e. a new *vector*:

$$\nabla f \equiv \sum_{j=1}^{3} \mathbf{n}_j \frac{\partial f}{\partial r_j} \equiv \mathbf{grad}\, f. \qquad \text{(A.51)}$$

- The *scalar product* of del by a *vector* function of coordinates (a *vector field*),

$$\mathbf{f}(\mathbf{r}) \equiv \sum_{j=1}^{3} \mathbf{n}_j f_j(\mathbf{r}), \qquad \text{(A.52)}$$

compiled formally following Eq. (A.43), is a *scalar* function—the *divergence* of the initial function:

$$\nabla \cdot \mathbf{f} \equiv \sum_{j=1}^{3} \frac{\partial f_j}{\partial r_j} \equiv \operatorname{div} \mathbf{f}, \qquad \text{(A.53)}$$

while the *vector product* of ∇ and $\mathbf{f}$, formed in a formal accordance with Eq. (A.45), is a new vector - the *curl* (in European tradition, called rotor and denoted **rot**) of $\mathbf{f}$:

$$\nabla \times \mathbf{f} \equiv \begin{vmatrix} \mathbf{n}_1 & \mathbf{n}_2 & \mathbf{n}_3 \\ \dfrac{\partial}{\partial r_1} & \dfrac{\partial}{\partial r_2} & \dfrac{\partial}{\partial r_3} \\ f_1 & f_2 & f_3 \end{vmatrix} = \mathbf{n}_1\left(\frac{\partial f_3}{\partial r_2} - \frac{\partial f_2}{\partial r_3}\right) + \mathbf{n}_2\left(\frac{\partial f_1}{\partial r_3} - \frac{\partial f_3}{\partial r_1}\right) + \mathbf{n}_3\left(\frac{\partial f_2}{\partial r_1} - \frac{\partial f_1}{\partial r_2}\right) \equiv \mathbf{curl}\, \mathbf{f}. \qquad \text{(A.54)}$$

[9] One can run into the following notation: $\nabla \equiv \partial/\partial \mathbf{r}$, which is convenient is some cases, but may be misleading in quite a few others, so it will be not used in these notes.

[10] In this, and four next sections, all scalar and vector functions are assumed to be differentiable.

- One more frequently met 'product' is $(\mathbf{f}\cdot\nabla)\mathbf{g}$, where $\mathbf{f}$ and $\mathbf{g}$ are two arbitrary vector functions of $\mathbf{r}$. This product should be also understood in the sense implied by Eq. (A.43), i.e. as a vector whose jth Cartesian component is

$$[(\mathbf{f}\cdot\nabla)\,\mathbf{g}]_j = \sum_{j'=1}^{3} f_{j'}\frac{\partial g_j}{\partial r_{j'}}. \tag{A.55}$$

A.9 The Laplace operator $\nabla^2 \equiv \nabla\cdot\nabla$

- Expression in Cartesian coordinates—in the formal accordance with Eq. (A.44):

$$\nabla^2 = \sum_{j=1}^{3}\frac{\partial^2}{\partial r_j^2}. \tag{A.56}$$

- According to its definition, the Laplace operator acting on a *scalar* function of coordinates gives a new scalar function:

$$\nabla^2 f \equiv \nabla\cdot(\nabla f) = \mathrm{div}(\mathbf{grad}\, f) = \sum_{j=1}^{3}\frac{\partial^2 f}{\partial r_j^2}. \tag{A.57}$$

- On the other hand, acting on a *vector* function (A.52), the operator ∇^2 returns another *vector*:

$$\nabla^2\mathbf{f} = \sum_{j=1}^{3}\mathbf{n}_j\nabla^2 f_j. \tag{A.58}$$

Note that Eqs. (A.56)–(A.58) are only valid in Cartesian (i.e. orthogonal and linear) coordinates, but generally not in other (even orthogonal) coordinates—see, e.g. Eqs. (A.61), (A.64), (A.67) and (A.70) below.

A.10 Operators ∇ and ∇^2 in the most important systems of orthogonal coordinates[11]

(i) *Cylindrical*[12] *coordinates* $\{\rho, \varphi, z\}$ (see figure below) may be defined by their relations with the Cartesian coordinates:

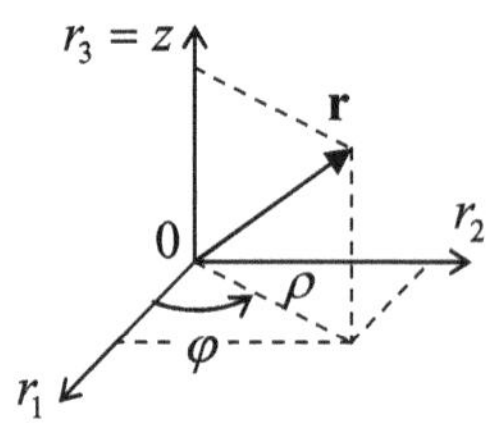

$$r_1 = \rho\cos\varphi,\quad r_2 = \rho\sin\varphi,\quad r_3 = z. \tag{A.59}$$

[11] Some other orthogonal curvilinear coordinate systems are discussed in *Part EM*, section 2.3.

[12] In the 2D geometry with fixed coordinate z, these coordinates are called *polar*.

- Gradient of a scalar function:

$$\nabla f = \mathbf{n}_\rho \frac{\partial f}{\partial \rho} + \mathbf{n}_\varphi \frac{1}{\rho}\frac{\partial f}{\partial \varphi} + \mathbf{n}_z \frac{\partial f}{\partial z}. \tag{A.60}$$

- The Laplace operator of a scalar function:

$$\nabla^2 f = \frac{1}{\rho}\frac{\partial}{\partial \rho}\left(\rho \frac{\partial f}{\partial \rho}\right) + \frac{1}{\rho^2}\frac{\partial^2 f}{\partial \varphi^2} + \frac{\partial^2 f}{\partial z^2}, \tag{A.61}$$

- Divergence of a vector function of coordinates ($\mathbf{f} = \mathbf{n}_\rho f_\rho + \mathbf{n}_\varphi f_\varphi + \mathbf{n}_z f_z$):

$$\nabla \cdot \mathbf{f} = \frac{1}{\rho}\frac{\partial(\rho f_\rho)}{\partial \rho} + \frac{1}{\rho}\frac{\partial f_\varphi}{\partial \varphi} + \frac{\partial f_z}{\partial z}. \tag{A.62}$$

- Curl of a vector function:

$$\nabla \times \mathbf{f} = \mathbf{n}_\rho \left(\frac{1}{\rho}\frac{\partial f_z}{\partial \varphi} - \frac{\partial f_\varphi}{\partial z}\right) + \mathbf{n}_\varphi \left(\frac{\partial f_\rho}{\partial z} - \frac{\partial f_z}{\partial \rho}\right) + \mathbf{n}_z \frac{1}{\rho}\left(\frac{\partial(\rho f_\varphi)}{\partial \rho} - \frac{\partial f_\rho}{\partial \varphi}\right). \tag{A.63}$$

- The Laplace operator of a vector function:

$$\nabla^2 \mathbf{f} = \mathbf{n}_\rho \left(\nabla^2 f_\rho - \frac{1}{\rho^2} f_\rho - \frac{2}{\rho^2}\frac{\partial f_\varphi}{\partial \varphi}\right) + \mathbf{n}_\varphi \left(\nabla^2 f_\varphi - \frac{1}{\rho^2} f_\varphi + \frac{2}{\rho^2}\frac{\partial f_\rho}{\partial \varphi}\right) + \mathbf{n}_z \nabla^2 f_z. \tag{A.64}$$

(ii) *Spherical coordinates* $\{r, \theta, \varphi\}$ (see figure below) may be defined as:

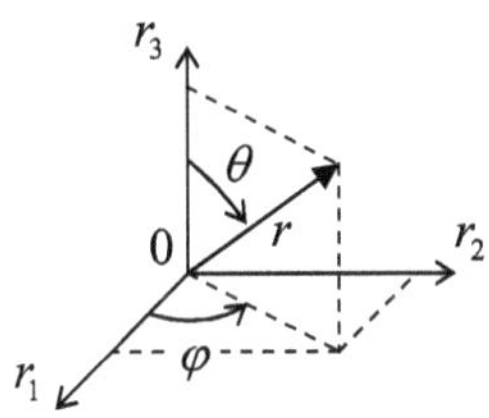

$$r_1 = r \sin\theta \cos\varphi,$$
$$r_2 = r \sin\theta \sin\varphi, \tag{A.65}$$
$$r_3 = r \cos\theta.$$

- Gradient of a scalar function:

$$\nabla f = \mathbf{n}_r \frac{\partial f}{\partial r} + \mathbf{n}_\theta \frac{1}{r}\frac{\partial f}{\partial \theta} + \mathbf{n}_\varphi \frac{1}{r \sin\theta}\frac{\partial f}{\partial \varphi}. \tag{A.66}$$

- The Laplace operator of a scalar function:

$$\nabla^2 f = \frac{1}{r^2}\frac{\partial}{\partial r}\left(r^2 \frac{\partial f}{\partial r}\right) + \frac{1}{r^2 \sin\theta}\frac{\partial}{\partial \theta}\left(\sin\theta \frac{\partial f}{\partial \theta}\right) + \frac{1}{(r \sin\theta)^2}\frac{\partial^2 f}{\partial \varphi^2}. \tag{A.67}$$

- Divergence of a vector function $\mathbf{f} = \mathbf{n}_r f_r + \mathbf{n}_\theta f_\theta + \mathbf{n}_\varphi f_\varphi$:

$$\nabla \cdot \mathbf{f} = \frac{1}{r^2}\frac{\partial(r^2 f_r)}{\partial r} + \frac{1}{r \sin\theta}\frac{\partial(f_\theta \sin\theta)}{\partial\theta} + \frac{1}{r\sin\theta}\frac{\partial f_\varphi}{\partial\varphi}. \tag{A.68}$$

- Curl of a similar vector function:

$$\begin{aligned}\nabla \times \mathbf{f} = \mathbf{n}_r \frac{1}{r\sin\theta}\left(\frac{\partial(f_\varphi \sin\theta)}{\partial\theta} - \frac{\partial f_\theta}{\partial\varphi}\right) + \mathbf{n}_\theta \frac{1}{r}\left(\frac{1}{\sin\theta}\frac{\partial f_r}{\partial\varphi} - \frac{\partial(r f_\varphi)}{\partial r}\right)\\ + \mathbf{n}_\varphi \frac{1}{r}\left(\frac{\partial(r f_\theta)}{\partial r} - \frac{\partial f_r}{\partial\theta}\right).\end{aligned} \tag{A.69}$$

- The Laplace operator of a vector function:

$$\begin{aligned}\nabla^2\mathbf{f} = \mathbf{n}_r\left(\nabla^2 f_r - \frac{2}{r^2}f_r - \frac{2}{r^2\sin\theta}\frac{\partial}{\partial\theta}(f_\theta\sin\theta) - \frac{2}{r^2\sin\theta}\frac{\partial f_\varphi}{\partial\varphi}\right)\\ + \mathbf{n}_\theta\left(\nabla^2 f_\theta - \frac{1}{r^2\sin^2\theta}f_\theta + \frac{2}{r^2}\frac{\partial f_r}{\partial\theta} - \frac{2\cos\theta}{r^2\sin^2\theta}\frac{\partial f_\varphi}{\partial\varphi}\right)\\ + \mathbf{n}_\varphi\left(\nabla^2 f_\varphi - \frac{1}{r^2\sin^2\theta}f_\varphi + \frac{2}{r^2\sin\theta}\frac{\partial f_r}{\partial\varphi} + \frac{2\cos\theta}{r^2\sin^2\theta}\frac{\partial f_\theta}{\partial\varphi}\right).\end{aligned} \tag{A.70}$$

A.11 Products involving ∇

(i) *Useful zeros*:

- For any scalar function $f(\mathbf{r})$,

$$\nabla \times (\nabla f) \equiv \mathbf{curl}(\mathbf{grad} f) = 0. \tag{A.71}$$

- For any vector function $\mathbf{f(r)}$,

$$\nabla \cdot (\nabla \times \mathbf{f}) \equiv \mathrm{div}(\mathbf{curl} f) = 0. \tag{A.72}$$

(ii) The *Laplace operator* expressed via the curl of a curl:

$$\nabla^2\mathbf{f} = \nabla(\nabla\cdot\mathbf{f}) - \nabla\times(\nabla\times\mathbf{f}). \tag{A.73}$$

(iii) Spatial differentiation of a product of a *scalar* function by a *vector* function:

- The scalar 3D generalization of Eq. (A.22) is

$$\nabla \cdot (f\,\mathbf{g}) = (\nabla f)\cdot\mathbf{g} + f(\nabla\cdot\mathbf{g}). \tag{A.74a}$$

- Its vector generalization is similar:

$$\nabla \times (f\, \mathbf{g}) = (\nabla f) \times \mathbf{g} + f(\nabla \times \mathbf{g}). \tag{A.74b}$$

(iv) Spatial differentiation of products of *two vector* functions:

$$\nabla \times (\mathbf{f} \times \mathbf{g}) = \mathbf{f}(\nabla \cdot \mathbf{g}) - (\mathbf{f} \cdot \nabla)\mathbf{g} - (\nabla \cdot \mathbf{f})\mathbf{g} + (\mathbf{g} \cdot \nabla)\,\mathbf{f}, \tag{A.75}$$

$$\nabla(\mathbf{f} \cdot \mathbf{g}) = (\mathbf{f} \cdot \nabla)\mathbf{g} + (\mathbf{g} \cdot \nabla)\,\mathbf{f} + \mathbf{f} \times (\nabla \times \mathbf{g}) + \mathbf{g} \times (\nabla \times \mathbf{f}), \tag{A.76}$$

$$\nabla \cdot (\mathbf{f} \times \mathbf{g}) = \mathbf{g} \cdot (\nabla \times \mathbf{f}) - \mathbf{f} \cdot (\nabla \times \mathbf{g}). \tag{A.77}$$

A.12 Integro-differential relations

(i) For an *arbitrary surface S* limited by closed contour *C*:

- The *Stokes theorem*, valid for any differentiable vector field $\mathbf{f}(\mathbf{r})$:

$$\int_S (\nabla \times \mathbf{f}) \cdot d^2\mathbf{r} \equiv \int_S (\nabla \times \mathbf{f})_n d^2 r = \oint_C \mathbf{f} \cdot d\mathbf{r} \equiv \oint_C f_\tau dr, \tag{A.78}$$

where $d^2\mathbf{r} \equiv \mathbf{n} d^2 r$ is the elementary area vector (normal to the surface), and $d\mathbf{r}$ is the elementary contour length vector (tangential to the contour line).

(ii) For an *arbitrary volume V* limited by closed surface *S*:

- *Divergence* (or 'Gauss') *theorem*, valid for any differentiable vector field $\mathbf{f}(\mathbf{r})$:

$$\int_V (\nabla \cdot \mathbf{f}\,)\, d^3 r = \oint_S \mathbf{f} \cdot d^2\mathbf{r} \equiv \oint_S f_n d^2 r. \tag{A.79}$$

- *Green's theorem*, valid for two differentiable scalar functions $f(\mathbf{r})$ and $g(\mathbf{r})$:

$$\int_V (f\, \nabla^2 g\; - g\nabla^2 f\,)\, d^3 r = \oint_S (f\, \nabla g\, - g\nabla f\,)_n d^2 r. \tag{A.80}$$

- An identity valid for any two scalar functions f and g, and a vector field $\mathbf{j}$ with $\nabla\cdot\mathbf{j} = 0$ (all differentiable):

$$\int_V [f(\mathbf{j} \cdot \nabla g) + g(\mathbf{j} \cdot \nabla f\,)]\; d^3 r = \oint_S f g j_n d^2 r. \tag{A.81}$$

A.13 The Kronecker delta and Levi-Civita permutation symbols

- The *Kronecker delta symbol* (defined for integer indices):

$$\delta_{jj'} \equiv \begin{cases} 1, & \text{if } j' = j, \\ 0, & \text{otherwise.} \end{cases} \tag{A.82}$$

- The *Levi-Civita permutation symbol* (most frequently used for 3 integer indices, each taking one of values 1, 2, or 3):

$$\varepsilon_{jj'j''} \equiv \begin{cases} +1, & \text{if the indices follow in the 'correct' ('even')} \\ & \text{order: } 1 \to 2 \to 3 \to 1 \to 2 \ldots, \\ -1, & \text{if the indices follow in the 'incorrect' ('odd')} \\ & \text{order: } 1 \to 3 \to 2 \to 1 \to 3 \ldots, \\ 0, & \text{if any two indices coincide.} \end{cases} \quad \text{(A.83)}$$

- Relation between the Levi-Civita and the Kronecker delta products:

$$\varepsilon_{jj'j''}\varepsilon_{kk'k''} = \sum_{l,l',l''=1}^{3} \begin{vmatrix} \delta_{jl} & \delta_{jl'} & \delta_{jl''} \\ \delta_{j'l} & \delta_{j'l'} & \delta_{j'l''} \\ \delta_{j''l} & \delta_{j''l'} & \delta_{j''l''} \end{vmatrix}; \quad \text{(A.84}a\text{)}$$

summation of this relation, written for 3 different values of $j = k$, over these values yields the so-called *contracted epsilon identity*:

$$\sum_{j=1}^{3} \varepsilon_{jj'j''}\varepsilon_{jk'k''} = \delta_{j'k'}\delta_{j''k''} - \delta_{j'k''}\delta_{j''k'}. \quad \text{(A.84}b\text{)}$$

A.14 Dirac's delta-function, sign function, and theta-function

- Definition of 1D *delta-function* (for real $a < b$):

$$\int_a^b f(\xi)\delta(\xi)d\xi = \begin{cases} f(0), & \text{if } a < 0 < b, \\ 0, & \text{otherwise,} \end{cases} \quad \text{(A.85)}$$

where $f(\xi)$ is any function continuous near $\xi = 0$. In particular (if $f(\xi) = 1$ near $\xi = 0$), the definition yields

$$\int_a^b \delta(\xi)d\xi = \begin{cases} 1, & \text{if } a < 0 < b, \\ 0, & \text{otherwise.} \end{cases} \quad \text{(A.86)}$$

- Relation to the *theta-function* $\theta(\xi)$ and *sign function* $\text{sgn}(\xi)$

$$\delta(\xi) = \frac{d}{d\xi}\theta(\zeta) = \frac{1}{2}\frac{d}{d\xi}\text{sgn}(\xi), \quad \text{(A.87}a\text{)}$$

where

$$\theta(\xi) \equiv \frac{\text{sgn}(\xi) + 1}{2} = \begin{cases} 0, & \text{if } \xi < 0, \\ 1, & \text{if } \xi > 1, \end{cases}$$
$$\text{sgn}(\xi) \equiv \frac{\xi}{|\xi|} = \begin{cases} -1, & \text{if } \xi < 0, \\ +1, & \text{if } \xi > 1. \end{cases} \quad \text{(A.87}b\text{)}$$

- An important integral[13]:

$$\int_{-\infty}^{+\infty} e^{is\xi}ds = 2\pi\delta(\xi). \quad \text{(A.88)}$$

- 3D generalization of the delta-function of the radius-vector (the 2D generalization is similar):

$$\int_V f(\mathbf{r})\delta(\mathbf{r})d^3r = \begin{cases} f(0), & \text{if } 0 \in V, \\ 0, & \text{otherwise}; \end{cases} \quad \text{(A.89)}$$

  it may be represented as a product of 1D delta-functions of Cartesian coordinates:

$$\delta(\mathbf{r}) = \delta(r_1)\delta(r_2)\delta(r_3). \quad \text{(A.90)}$$

A.15 The Cauchy theorem and integral

Let a complex function $f(z)$ be analytic within a part of the complex plane z, that is limited by a closed contour C and includes point z'. Then

$$\oint_C f(z)dz = 0, \quad \text{(A.91)}$$

$$\oint_C f(z)\frac{dz}{z - z'} = 2\pi i f(z') \quad \text{(A.92)}$$

The first of these relations is usually called the *Cauchy integral theorem* (or the 'Cauchy–Goursat theorem'), and the second one—the *Cauchy integral* (or the 'Cauchy integral formula').

A.16 Literature

(i) Properties of some *special functions* are briefly discussed at the relevant points of the lecture notes; in the alphabetical order:

- Airy functions: *Part QM* section 2.4;
- Bessel functions: *Part EM* section 2.7;
- Fresnel integrals: *Part EM* section 8.6;
- Hermite polynomials: *Part QM* section 2.9;
- Laguerre polynomials (both simple and associated): *Part QM* section 3.7;

[13] The coefficient in this relation may be readily recalled by considering its left-hand part as the Fourier-integral representation of function $f(s) \equiv 1$, and applying Eq. (A.85) to the reciprocal Fourier transform

$$f(s) \equiv 1 = \frac{1}{2\pi}\int_{-\infty}^{+\infty} e^{-is\xi}[2\pi\delta(\xi)]d\xi.$$

- Legendre polynomials, associated Legendre functions: *Part EM* section 2.8, and *Part QM* section 3.6;
- Spherical harmonics: *Part QM* section 3.6;
- Spherical Bessel functions: *Part QM* sections 3.6 and 3.8.

(ii) For *more formulas*, and their discussion, I can recommend the following handbooks[14]:

- *Handbook of Mathematical Formulas* [2];
- *Tables of Integrals, Series, and Products* [3];
- *Mathematical Handbook for Scientists and Engineers* [4];
- *Integrals and Series* volumes 1 and 2 [5];
- A popular textbook *Mathematical Methods for Physicists* [6] may be also used as a formula manual.

Many formulas are also available from the symbolic calculation modules of the commercially available software packages listed in section (iv) below.

(iii) Probably the most popular collection of *numerical calculation codes* are the twin manuals by W Press *et al* [1]:

- *Numerical Recipes in Fortran 77*;
- *Numerical Recipes* [in C++—KKL].

My lecture notes include very brief introductions to numerical methods of differential equation solution:

- ordinary differential equations: *Part CM*, section 5.7;
- partial differential equations: *Part CM* section 8.5 and *Part EM* section 2.11, which include references to literature for further reading.

(iv) The following are the most popular *software packages* for numerical and symbolic calculations, all with plotting capabilities (in the alphabetical order):

- Maple (www.maplesoft.com/products/maple/);
- MathCAD (www.ptc.com/engineering-math-software/mathcad/);
- Mathematica (www.wolfram.com/mathematica/);
- MATLAB (www.mathworks.com/products/matlab.html).

References

[1] Press W *et al* 1992 *Numerical Recipes in Fortran 77* 2nd edn (Cambridge: Cambridge University Press)
Press W *et al* 2007 *Numerical Recipes* 3rd edn (Cambridge: Cambridge University Press)

[2] Abramowitz M and Stegun I (eds) 1965 *Handbook of Mathematical Formulas* (New York: Dover), and numerous later printings. An updated version of this collection is now available online at http://dlmf.nist.gov/.

[14] On a personal note, perhaps 90% of all formula needs throughout my research career were satisfied by a tiny, wonderfully compiled old book [7], used copies of which, rather amazingly, are still available on the Web.

[3] Gradshteyn I and Ryzhik I 1980 *Tables of Integrals, Series, and Products* 5th edn (New York: Academic)
[4] Korn G and Korn T 2000 *Mathematical Handbook for Scientists and Engineers* 2nd edn (New York: Academic)
[5] Prudnikov A *et al* 1986 *Integrals and Series* vol 1 (Boca Raton, FL: CRC Press)
Prudnikov A *et al* 1986 *Integrals and Series* vol 2 (Boca Raton, FL: CRC Press)
[6] Arfken G *et al* 2012 *Mathematical Methods for Physicists* 7th edn (New York: Academic)
[7] Dwight H 1961 *Tables of Integrals and Other Mathematical Formulas* 4th edn (London: Macmillan)

Appendix B

Selected physical constants

The listed numerical values of the constants are from the most recent (2014) International CODATA recommendation (see, e.g. http://physics.nist.gov/cuu/Constants/index.html), besides a newer result for k_B—see [1]. *Please note the recently announced (but, by this volume's press time, not yet official) adjustment of the SI values - see, e.g. https://www.nist.gov/si-redefinition/meet-constants. In particular, the Planck constant will also get a definite value (within the interval specified in table B.1), enabling a new, fundamental standard of the kilogram.*

Table B.1.

Symbol	Quantity	SI value and unit	Gaussian value and unit	Relative rms uncertainty
c	speed of light in free space	$2.99\ 792\ 458 \times 10^{8}$ m s^{-1}	$2.99\ 792\ 458 \times 10^{10}$ cm s^{-1}	0 (defined value)
G	gravitation constant	6.6741×10^{-11} m^{3} kg^{-1} s^{-2}	6.6741×10^{-8} cm ^{3} g^{-1} s^{-2}	$\sim 5 \times 10^{-5}$
$\hbar$	Planck constant	$1.05\ 457\ 180 \times 10^{-34}$ J s	$1.05\ 457\ 180 \times 10^{-27}$ erg s	$\sim 2 \times 10^{-8}$
e	elementary electric charge	$1.6\ 021\ 762 \times 10^{-19}$ C	$4.803\ 203 \times 10^{-10}$ statcoulomb	$\sim 6 \times 10^{-9}$
m_e	electron's rest mass	$0.91\ 093\ 835 \times 10^{-30}$ kg	$0.91\ 093\ 835 \times 10^{-27}$ g	$\sim 1 \times 10^{-8}$
m_p	proton's rest mass	$1.67\ 262\ 190 \times 10^{-27}$ kg	$1.67\ 262\ 190 \times 10^{-24}$ g	$\sim 1 \times 10^{-8}$
μ_0	magnetic constant	$4\pi \times 10^{-7}$ N A^{-2}	–	0 (defined value)
ε_0	electric constant	$8.854\ 187\ 817 \times 10^{-12}$ F m^{-1}	–	0 (defined value)
k_B	Boltzmann constant	$1.380\ 649 \times 10^{-23}$ J K^{-1}	$1.3\ 806\ 490 \times 10^{-16}$ erg K^{-1}	$\sim 2 \times 10^{-6}$

Comments:

1. The fixed value of c was defined by an international convention in 1983, in order to extend the official definition of the second (as 'the duration of 9 192 631 770 periods of the radiation corresponding to the transition between the two hyperfine levels of the ground state of the cesium-133 atom') to that of the meter. The values are back-compatible with the legacy definitions of the meter (initially, as 1/40 000 000th of the Earth's meridian length) and the second (for a long time, as $1/(24 \times 60 \times 60) = 1/86\,400$th of the Earth's rotation period), within the experimental errors of those measures.
2. ε_0 and μ_0 are not really the fundamental constants; in the SI system of units one of them (say, μ_0) is selected arbitrarily[1], while the other one is defined via the relation $\varepsilon_0\mu_0 = 1/c^2$.
3. The Boltzmann constant k_B is also not quite fundamental, because its only role is to comply with the independent definition of the kelvin (K), as the temperature unit in which the triple point of water is exactly 273.16 K. If temperature is expressed in energy units $k_B T$ (as is done, for example, in *Part SM* of this series), this constant disappears altogether.
4. The dimensionless *fine structure* ('Sommerfeld's') *constant* α is numerically the same in any system of units:

$$\alpha \equiv \left\{ \begin{matrix} e^2/4\pi\varepsilon_0\hbar c & \text{in SI units} \\ e^2/\hbar c & \text{in Gaussian units} \end{matrix} \right\} \approx 7.297\,352\,566 \times 10^{-3}$$
$$\approx \frac{1}{137.035\,999\,14},$$

and is known with a much smaller relative rms uncertainty (currently, $\sim 3 \times 10^{-10}$) than those of the component constants.

References

[1] Gaiser C *et al* 2017 *Metrologia* **54** 280

[2] Newell D 2014 *Phys. Today* **67** 35–41

[1] Note that the selected value of μ_0 may be changed (a bit) in a few years—see, e.g., [2].

Bibliography

This section presents a partial list of textbooks and monographs used in the work on the EAP series[1,2].

Part CM: Classical Mechanics

Fetter A L and Walecka J D 2003 *Theoretical Mechanics of Particles and Continua* (New York: Dover)

Goldstein H, Poole C and Safko J 2002 *Classical Mechanics* 3rd edn (Reading, MA: Addison Wesley)

Granger R A 1995 *Fluid Mechanics* (New York: Dover)

José J V and Saletan E J 1998 *Classical Dynamics* (Cambridge: Cambridge University Press)

Landau L D and Lifshitz E M 1976 *Mechanics* 3rd edn (Oxford: Butterworth-Heinemann)

Landau L D and Lifshitz E M 1986 *Theory of Elasticity* (Oxford: Butterworth-Heinemann)

Landau L D and Lifshitz E M 1987 *Fluid Mechanics* 2nd edn (Oxford: Butterworth-Heinemann)

Schuster H G 1995 *Deterministic Chaos* 3rd edn (New York: Wiley)

Sommerfeld A 1964 *Mechanics* (New York: Academic)

Sommerfeld A 1964 *Mechanics of Deformable Bodies* (New York: Academic)

Part EM: Classical Electrodynamics

Batygin V V and Toptygin I N 1978 *Problems in Electrodynamics* 2nd edn (New York: Academic)

Griffiths D J 2007 *Introduction to Electrodynamics* 3rd edn (Englewood Cliffs, NJ: Prentice-Hall)

Jackson J D 1999 *Classical Electrodynamics* 3rd edn (New York: Wiley)

Landau L D and Lifshitz E M 1984 *Electrodynamics of Continuous Media* 2nd edn (Auckland: Reed)

Landau L D and Lifshitz E M 1975 *The Classical Theory of Fields* 4th edn (Oxford: Pergamon)

[1] The list does not include the sources (mostly, recent original publications) cited in the lecture notes and problem solutions, and the mathematics textbooks and handbooks listed in section A.16.

[2] Recently several high-quality teaching materials on advanced physics became available online, including R. Fitzpatrick's text on *Classical Electromagnetism* (farside.ph.utexas.edu/teaching/jk1/Electromagnetism.pdf), B Simons' 'lecture shrunks' on *Advanced Quantum Mechanics* (www.tcm.phy.cam.ac.uk/~bds10/aqp.html), and D Tong's lecture notes on several advanced topics (www.damtp.cam.ac.uk/user/tong/teaching.html).

Panofsky W K H and Phillips M 1990 *Classical Electricity and Magnetism* 2nd edn (New York: Dover)
Stratton J A 2007 *Electromagnetic Theory* (New York: Wiley)
Tamm I E 1979 *Fundamentals of the Theory of Electricity* (Paris: Mir)
Zangwill A 2013 *Modern Electrodynamics* (Cambridge: Cambridge University Press)

Part QM: Quantum Mechanics

Abers E S 2004 *Quantum Mechanics* (London: Pearson)
Auletta G, Fortunato M and Parisi G 2009 *Quantum Mechanics* (Cambridge: Cambridge University Press)
Capri A Z 2002 *Nonrelativistic Quantum Mechanics* 3rd edn (Singapore: World Scientific)
Cohen-Tannoudji C, Diu B and Laloë F 2005 *Quantum Mechanics* (New York: Wiley)
Constantinescu F, Magyari E and Spiers J A 1971 *Problems in Quantum Mechanics* (Amsterdam: Elsevier)
Galitski V *et al* 2013 *Exploring Quantum Mechanics* (Oxford: Oxford University Press)
Gottfried K and Yan T-M 2004 *Quantum Mechanics: Fundamentals* 2nd edn (Berlin: Springer)
Griffith D 2005 *Quantum Mechanics* 2nd edn (Englewood Cliffs, NJ: Prentice Hall)
Landau L D and Lifshitz E M 1977 *Quantum Mechanics (Nonrelativistic Theory)* 3rd edn (Oxford: Pergamon)
Messiah A 1999 *Quantum Mechanics* (New York: Dover)
Merzbacher E 1998 *Quantum Mechanics* 3rd edn (New York: Wiley)
Miller D A B 2008 *Quantum Mechanics for Scientists and Engineers* (Cambridge: Cambridge University Press)
Sakurai J J 1994 *Modern Quantum Mechanics* (Reading, MA: Addison-Wesley)
Schiff L I 1968 *Quantum Mechanics* 3rd edn (New York: McGraw-Hill)
Shankar R 1980 *Principles of Quantum Mechanics* 2nd edn (Berlin: Springer)
Schwabl F 2002 *Quantum Mechanics* 3rd edn (Berlin: Springer)

Part SM: Statistical Mechanics

Feynman R P 1998 *Statistical Mechanics* 2nd edn (Boulder, CO: Westview)
Huang K 1987 *Statistical Mechanics* 2nd edn (New York: Wiley)
Kubo R 1965 *Statistical Mechanics* (Amsterdam: Elsevier)
Landau L D and Lifshitz E M 1980 *Statistical Physics, Part 1* 3rd edn (Oxford: Pergamon)
Lifshitz E M and Pitaevskii L P 1981 *Physical Kinetics* (Oxford: Pergamon)
Pathria R K and Beale P D 2011 *Statistical Mechanics* 3rd edn (Amsterdam: Elsevier)
Pierce J R 1980 *An Introduction to Information Theory* 2nd edn (New York: Dover)
Plishke M and Bergersen B 2006 *Equilibrium Statistical Physics* 3rd edn (Singapore: World Scientific)
Schwabl F 2000 *Statistical Mechanics* (Berlin: Springer)
Yeomans J M 1992 *Statistical Mechanics of Phase Transitions* (Oxford: Oxford University Press)

Multidisciplinary/specialty

Ashcroft W N and Mermin N D 1976 *Solid State Physics* (Philadelphia, PA: Saunders)
Blum K 1981 *Density Matrix and Applications* (New York: Plenum)

Breuer H-P and Petruccione E 2002 *The Theory of Open Quantum Systems* (Oxford: Oxford University Press)
Cahn S B and Nadgorny B E 1994 *A Guide to Physics Problems, Part 1* (New York: Plenum)
Cahn S B, Mahan G D and Nadgorny B E 1997 *A Guide to Physics Problems, Part 2* (New York: Plenum)
Cronin J A, Greenberg D F and Telegdi V L 1967 *University of Chicago Graduate Problems in Physics* (Reading, MA: Addison Wesley)
Hook J R and Hall H E 1991 *Solid State Physics* 2nd edn (New York: Wiley)
Joos G 1986 *Theoretical Physics* (New York: Dover)
Kaye G W C and Laby T H 1986 *Tables of Physical and Chemical Constants* 15th edn (London: Longmans Green)
Kompaneyets A S 2012 *Theoretical Physics* 2nd edn (New York: Dover)
Lax M 1968 *Fluctuations and Coherent Phenomena* (London: Gordon and Breach)
Lifshitz E M and Pitaevskii L P 1980 *Statistical Physics, Part 2* (Oxford: Pergamon)
Newbury N *et al* 1991 *Princeton Problems in Physics with Solutions* (Princeton, NJ: Princeton University Press)
Pauling L 1988 *General Chemistry* 3rd edn (New York: Dover)
Tinkham M 1996 *Introduction to Superconductivity* 2nd edn (New York: McGraw-Hill)
Walecka J D 2008 *Introduction to Modern Physics* (Singapore: World Scientific)
Ziman J M 1979 *Principles of the Theory of Solids* 2nd edn (Cambridge: Cambridge University Press)

www.ingramcontent.com/pod-product-compliance
Ingram Content Group UK Ltd.
Pitfield, Milton Keynes, MK11 3LW, UK
UKHW051900100926
8556IPUK00005B/705